## The Disappearance of the Social in American Social Psychology

*The Disappearance of the Social in American Social Psychology* is a critical conceptual history of American social psychology. In this challenging work, John Greenwood demarcates the original conception of the social dimensions of cognition, emotion, and behavior, and of the discipline of social psychology itself, that was embraced by early twentieth-century American social psychologists. He documents how this fertile conception of social psychological phenomena came to be progressively neglected as the century developed, to the point that scarcely any trace of the original conception of the social remains in contemporary American social psychology. In a penetrating analysis, Greenwood suggests a number of subtle historical reasons why the original conception of the social came to be abandoned, stressing that none of these were particularly good reasons for the neglect of the original conception of the social. By demonstrating the historical contingency of this neglect, Greenwood indicates that what has been lost may once again be regained. This engaging work will appeal to social psychologists, sociologists, anthropologists, and other social scientists, and historians and philosophers of social and psychological science.

John D. Greenwood is Professor of Philosophy and Psychology at the City College and Graduate Center of the City University of New York. He is the author of *Explanation and Experiment in Social Psychological Science* (1989), *Relations and Representations* (1991), and *Realism, Identity and Emotion* (1994).

# The Disappearance of the Social in American Social Psychology

JOHN D. GREENWOOD
*City University of New York*

CAMBRIDGE UNIVERSITY PRESS
Cambridge, New York, Melbourne, Madrid, Cape Town, Singapore, São Paulo, Delhi

Cambridge University Press
The Edinburgh Building, Cambridge CB2 8RU, UK

Published in the United States of America by Cambridge University Press, New York

www.cambridge.org
Information on this title: www.cambridge.org/9780521830140



First published 2004
This digitally printed version 2008

*A catalogue record for this publication is available from the British Library*

*Library of Congress Cataloguing in Publication data*
Greenwood, John D.
The disappearance of the social in American social psychology / John D. Greenwood.
p. cm.
ISBN 0-521-83014-1
1. Social psychology – United States – History. 2. Social psychology. I. Title.
HM1027.U6G74 2003
302′.0973 – dc21 2003048567

ISBN 978-0-521-83014-0 hardback
ISBN 978-0-521-09954-7 paperback

*For my good friends at Kent Ridge*

To understand the intimacy and separateness between individual and group we must grasp the unusual process that gives rise to groups at the human level. It is a process in which individuals play an extraordinary role, confronting us with a type of part-whole relation unprecedented in nature. It is the only part-whole relation that depends on the recapitulation of the structure of the whole in the part. Only because individuals are capable of encompassing group relations and possibilities can they create a society that eventually faces them as an independent, or even hostile, set of conditions.

Solomon E. Asch, *Social Psychology*

# Contents

*Preface* *page* ix

Introduction: What Happened to the "Social" in Social Psychology? 1

1 The Lost World 18

2 Wundt and Völkerpsychologie 43

3 Durkheim and Social Facts 68

4 The Social and the Psychological 87

5 Social Psychology and the "Social Mind" 109

6 Individualism and the Social 136

7 Crowds, Publics, and Experimental Social Psychology 160

8 Crossroads 185

9 Crisis 214

10 The Rediscovery of the Social? 245

*References* 267

*Index* 303

# Preface

This work is about a peculiar historical anomaly – the neglect and eventual abandonment of the rich and theoretically fertile conception of the social embraced by early American social psychologists – that I stumbled upon almost by accident.

Rom Harré and Paul Secord originally stimulated my interest in the social dimensions of human psychology and behavior and the special problems they generate for a scientific and experimental social psychology. Since my graduate days in Oxford, much of my professional career has been devoted to the exploration of these issues, developed in a number of books and journal articles. My more recent interest in the history of psychology came about as a result of having to substitute for a teaching colleague overtaken by motherhood. Although I immediately fell in love with the subject, which I have taught for the past fifteen years, for a long time the overlap with my metatheoretical work in social psychology was minimal.

However, some years ago I was asked to review Margaret Gilbert's book *On Social Facts* (Princeton University Press, 1991). In consequence, I was forced to recognize that I had been cheerfully talking about the social dimensions of behavior, emotion, groups, identity, and the like for many years without reflecting critically on my own conception of the social. As I explored this issue, I was pleased to discover that something very close to my own conception had been embraced by early American social psychologists. At the same time I realized that that this conception had been almost completely abandoned by contemporary social psychologists. Why had this rich and promising conception of the social been abandoned? The present work is the outcome of my attempt to answer this puzzling question.

I first tried out some of the historical ideas that form the basis of this work in a paper that I gave at the 30th Meeting of *Cheiron* at the University of San Diego in June 1998. My thanks to David Leary for encouraging me to develop these ideas and to Kurt Danziger, Ian Lubek, Franz Samelson, Paul Secord, and Andrew Winston for critical feedback on earlier drafts of the work. My thanks also to audiences at the National University of Singapore and the University of North Carolina at Greensboro for their critical responses to early versions of my historical thesis. Thanks also to Mitchell C. Ash and Bill Woodward, General Editors of the Cambridge Series in the History of Psychology, and to Mary Childs and Frank Smith, at Cambridge University Press, New York, for their encouragement and support.

My research was greatly aided by a Rifkind Fellowship from the City College of New York, City University of New York, and a Senior Visiting Fellowship from the National University of Singapore. I am deeply indebted to both institutions.

Thanks to Taylor and Francis Publishing Company for permission to employ material from my paper "From *Völkerpsychologie* to cultural psychology: The once and future discipline?" *Philosophical Psychology*, 12 (1999), pp. 503–514; to John Wiley & Co. for permission to employ material from my paper "Individualism and the social in early American social psychology," *Journal for the History of the Behavioral Sciences*, 36 (2000), pp. 443–456; and to the American Psychological Association to employ material from my paper "Wundt, Völkerpsychologie, and experimental social psychology," *History of Psychology*, 6 (2003), pp. 70–88.

The production of this work turned out to be a voyage of discovery and rediscovery. From a new historical perspective, I found myself returning to many of the themes of the "crisis" in social psychology that had engaged me as a graduate student at Oxford in the 1970s. I also had the pleasure of drafting the first version of this work at the National University of Singapore, where I had drafted my first book (*Explanation and Experiment in Social Psychological Science*, Springer-Verlag, 1989) some fifteen years earlier. I hope the reader finds the work as rewarding as my own experience in writing it.

# Introduction

## *What Happened to the "Social" in Social Psychology?*

In this work I document the historical abandonment of the distinctive conception of the social dimensions of cognition, emotion, and behavior, and of the discipline of social psychology itself,[1] that was recognized in the early decades of twentieth century American social psychology.[2] This conception was progressively neglected from the 1930s onward, to the extent that scarcely a trace of the original conception of the social remains in contemporary American "social" psychology. I also suggest some explanations, albeit partial and tentative, of this historical neglect and eventual abandonment.

On the face of it, this is a remarkable and surprising claim to make. American social psychology is a well-established discipline with an almost hundred-year history and a present professional membership in the thousands. However, the fact that a discipline calls itself social psychology does not guarantee the social nature of whatever is considered to be its subject matter. In this work, I argue that contemporary American social psychology has virtually abandoned the study of the social dimensions of psychological states and behavior.

Of course, whether one is inclined to accept this claim will largely depend upon one's conception of the social. Those who embrace a different conception of the social from the one advocated in this work might very

[1] By a distinctive conception of the social dimensions of cognition, emotion, and behavior, I mean a conception that distinguishes between socially and individually engaged psychological states and behavior and that treats their distinction as the justification for recognizing social psychology as a discipline distinct from individual psychology. The distinction is explicated in the following chapters (especially Chapter 1).

[2] By early decades of the twentieth century, I mean the first three decades.

well hold that American social psychology has never been more social than it is today. For better or worse, most contemporary American social psychologists do in fact embrace a different conception of the social. It is to the historical explanation of this peculiar fact that the present work is directed.

## I

The founding fathers of scientific psychology in Germany and the United States and the early American pioneers of social psychology held a distinctive conception of the social dimensions of cognition, emotion, and behavior and of the discipline of social psychology itself. They recognized psychological states and behavior grounded in the membership of social groups, or social "collectivities" or "communities." Social psychology, or "group" or "collective" psychology, as it was sometimes called, was identified as that branch of psychological science concerned with the study of psychological states and behavior oriented to the represented psychology and behavior of members of social groups. Individual psychology, by contrast, was held to be concerned with the study of psychological states engaged independently of the represented psychology and behavior of members of social groups, e.g., those grounded in genetic endowment or nonsocial forms of learning.

Wilhelm Wundt is generally acknowledged as the institutional founding father of academic scientific psychology. Wundt founded the discipline of scientific psychology at the University of Leipzig in Germany in the 1880s by appropriating the experimental methods of the newly developed discipline of physiology and applying them to the study of conscious experience. However, Wundt also thought that the experimental study of conscious experience ought to be supplemented by the comparative-historical study of socially embedded psychological states and behavior, and he spent his later years developing this form of psychology in the ten-volume *Völkerpsychologie* (1900–1920), variously translated as "social psychology," "folk psychology," or "cultural psychology."[3]

That is, Wundt clearly acknowledged forms of cognition, emotion, and behavior grounded in the membership of social groups: "All such mental products of a general character presuppose as a condition the existence of a mental *community* composed of many individuals" (Wundt,

[3] There is some dispute about how the term "*Völkerpsychologie*" is best translated. The issue is discussed in Chapter 2.

1897/1902, p. 23). Wundt also distinguished "social" from "individual" or "experimental" psychology on the grounds that the objects of "social" as opposed to "individual" or "experimental" psychology are grounded in the membership of social groups:

Because of this dependence on the community, in particular the social community, this whole department of psychological investigation is designated as *social psychology*, and distinguished from individual, or as it may be called because of its predominating method, *experimental* psychology. (Wundt, 1897/1902, p. 23)

Similarly, Wundt's student Oswald Külpe, despite his later disagreements with his former teacher over the experimental analysis of thought processes, maintained that "social psychology treats of the mental phenomena dependent upon a community of individuals; it is already a special department of study, if not a fully developed science" (Külpe, 1895, p. 7).

Although Wundt had many American doctoral students who returned to found the first psychology departments and laboratories in the United States and Canada, few returned to enthusiastically promote the study of *Völkerpsychologie*. Nonetheless, many early American scientific psychologists, including both so-called structuralist psychologists such as Edward B. Titchener and functionalist psychologists such as James R. Angell, followed Wundt in recognizing the distinct identity as well as the value of social psychology conceived as a discipline concerned with those psychological states and behavior that are grounded in the membership of social groups:[4]

Just as the scope of psychology extends beyond man to the animals, so does it extend from the individual man to groups of men, to societies. The subject-matter of psychology is human experience considered as dependent upon the individual. But since the individuals of the same race and epoch are organized in much the same way, and since they live together in a society where their conduct affects and is affected by the conduct of others, their view of experience under its dependent aspect naturally becomes, in certain main features, a common or general view;

[4] The same conception of social psychological phenomena is also to be found in some early European psychologists, such as Jean Piaget (1932) and Frederic K. Bartlett (1932). For example, Bartlett (1932) maintained that cognitive processes such as memory are frequently grounded in socially engaged beliefs and attitudes:

Several of the factors influencing the individual observer are social in origin and character ... many of the transformations which took place as a result of the repeated reproductions of prose passages were directly due to the influence of social conventions and beliefs current in the group to which the individual subject belonged. (p. 118)

Discussion of the development of social psychology in Europe is, however, beyond the scope of the present work.

and this common view is embodied in those social institutions to which we have referred above, – in language, religion, law and custom. (Titchener, 1910, p. 28)[5]

*Social psychology*, in its broadest sense, has to do mainly with the psychological principles involved in those expressions of mental life which take form in social relations, organizations, and practices. (Angell, 1908, p. 4)

This conception of social psychological phenomena and of the province of social psychology is clearly evident in the early textbooks on social psychology, such as Edward Ross's *Social Psychology* (1908):

Social psychology, as the writer conceives it, studies the psychic planes and currents that come into existence among men in consequence of their association.... The aligning power of association triumphs over diversity of temperament and experience.... The individuality that each has received from the hand of nature is largely effaced, and we find people gathered into great planes of uniformity. (p. 1)[6]

Analogously, William McDougall (1920) maintained that "social" or "group" mentality is the proper subject matter of "social" or "group" psychology, the aim of which is to "display the general principles of collective mental life which are incapable of being deduced from the laws of the mental life of isolated individuals" (pp. 7–8).

Yet by the late 1920s and 1930s, this distinctive conception of the social dimensions of psychological states and behavior and of the discipline of social psychology was beginning to be abandoned by American social psychologists. Floyd Allport (1924a) was vigorous in his rejection of "social" or "group" forms of cognition, emotion, and behavior as the subject matter of a distinctive social psychology, and indeed he famously denied that social psychology forms a separate discipline distinct from individual psychology:

There is no psychology of groups which is not essentially and entirely a psychology of individuals. Social psychology must not be placed in contradistinction to the

[5] Titchener is often portrayed by historians as a dismissive critic of Wundt's *Völkerpsychologie*, largely on the basis of negative comments about its role in Wundt's system that he made in his obituary on Wundt (Titchener, 1921). Yet Titchener retained an active and critical interest in the project of a *Völkerpsychologie* and was an astute commentator on the methodological problems of any form of comparative-historical psychology that dealt with different social and cultural communities. See, for example, his critical commentary on the psychological findings of the Torres Straits expedition (Titchener, 1916), whose intellectual goals he nonetheless clearly supported.

[6] Although Ross himself claimed (1908, p. 2) that *Social Psychology* omitted the "psychology of groups" (which he held to be closely tied to the "morphology" of groups, the subject matter of "psychological sociology"), his detailed discussions of fashion, conventionality, and custom generally relate these phenomena to specific social groups.

> psychology of the individual; *it is part of the psychology of the individual*, whose behavior it studies in relation to that sector of his environment comprised by his fellows. (p. 4)

From the 1930s onward, the social dimensions of psychological states and behavior came to be increasingly neglected by American social psychologists.

There were lots of exceptions, such as Asch (1951, 1952), Asch, Block, and Hertzman (1938), Cantril (1941), Charters and Newcomb (1952), Converse and Campbell (1953), Festinger (1947), Festinger, Riecken, and Schachter (1956), Festinger, Schachter, and Back (1950), French (1944), Kelley (1955), Kelley and Volkart (1952), Kelley and Woodruff (1956), Lewin (1947a), Lewin, Lippitt, and White (1939), Newcomb (1943), Sherif (1935, 1936, 1948), Sherif and Cantril (1947), Siegel and Siegel (1957), Stouffer, Lunsdame, et al. (1949), and Stouffer, Suchman, De Vinney, Star, and Williams (1949). The original conception of the subject matter of social psychology can still be identified in some works published in the 1950s and 1960s, and some of the clearest theoretical statements of this conception were in fact advanced during the 1950s (e.g., Asch, 1952). However, these works appear to have represented the vestiges of the earlier social tradition, not the increasingly asocial tradition that developed from the 1930s onward.

Trying to establish the exact date of the abandonment of the original conception of the subject matter of social psychology is of course a fruitless and arbitrary exercise – and one that I don't attempt in this work. What I suggest is that, although the original conception was developed and sustained in the first four decades of the twentieth century, by the late 1920s and 1930s it was being abandoned by many social psychologists in favor of Floyd Allport's alternative asocial vision. While the original conception continued to be represented in articles and books in the 1950s and 1960s and arguably reached a high-water mark in the 1950s, it was rapidly displaced by the narrow experimental paradigm that came to dominate American social psychology in the 1950s and 1960s.

Whenever exactly the original conception was abandoned, it is very clear that it is no longer maintained by contemporary American social psychology. In early American studies of social beliefs and attitudes, for example, beliefs and attitudes were held to be social by virtue of their orientation to the represented beliefs and attitudes of members of social groups, irrespective of the types of objects to which they were directed (i.e., the adjective "social" was employed to qualify beliefs and attitudes themselves). In contrast, in contemporary American social psychology, cognition is characterized as social merely by virtue of the objects to which it is directed, namely, other persons or social groups, not by virtue of its orientation to the represented cognition of members of social groups (i.e., the adjective "social" is employed to qualify only the objects of cognition, not cognition itself): "The study of social cognition concerns how people make sense of other people and themselves" (Fiske & Taylor, 1991, p. 17).[7]

Early American social psychologists maintained that the causal dynamics of social cognition (and emotion and behavior) are different from the causal dynamics of individual cognition (and emotion and behavior). As McDougall (1920) put it, "the thinking and acting of each man, insofar as he thinks and acts as a member of a society, are very different from his thinking or acting as an isolated individual" (pp. 9–10).[8] However, it is a general presumption of contemporary studies of social cognition that the basic cognitive processes engaged in the perception and cognition of nonsocial objects, such as tables, trees, and tarantulas, are also engaged in the perception and cognition of social objects, such as other persons and social groups. In consequence, the contemporary study of social cognition is essentially the application of the principles of individual cognitive psychology to the domain of "social objects," namely, other persons and social groups:

> As one reviews research on social cognition, the analogy between the perception of things and the perception of people becomes increasingly clear. The argument is made repeatedly: the principles that describe how people think in general also describe how people think about people. (Fiske & Taylor, 1991, p. 18)[9]

7 Compare the various definitions of social cognition offered in Devine, Hamilton, and Ostrom (1994), Higgins, Ruble, and Hartup (1983), Ross and Nisbett (1991), and Wegner and Vallacher (1977).

8 This passage was quoted by McDougall from his earlier work *Psychology: The Science of Behavior* (1912).

9 Although it is often recognized that cognitive processes relating to persons are likely to differ from cognitive processes relating to things, these differences are generally conceived in terms of modifications of individual cognitive processing to fit distinctive features of the human objects of cognition, not in terms of any fundamental distinction between individual as opposed to social *forms of cognition*:

> Social cognition, of course, differs from the general principles of cognition in some ways. Compared to objects, people are more likely to be causal agents, to perceive as well as being perceived, and intimately to involve the observer's self. They are difficult targets of cognition; because they adjust themselves upon being perceived, many of their important attributes (e.g., traits) must be inferred, and the accuracy of observations is hard to determine. People frequently change, and are unavoidably complex as targets of cognition. Hence those who study social cognition must adapt the ideas of cognitive psychology to suit the special features of cognitions about people. (Fiske & Taylor, 1991, p. 20)

Similar sorts of points can be made about contemporary American social psychological research on social behavior and social groups. Social behavior, for example, was originally conceived as behavior oriented to the represented behavior of members of social groups, irrespective of the objects to which it is directed, which might include trees, rivers, rubbish bins, domestic animals, or fellow humans. However, from the 1930s onward social behavior came to be characterized as behavior directed toward other persons or groups, independently of whether such behavior is oriented to the represented behavior of members of social groups (F. H. Allport, 1924a, 1933; G. W. Allport, 1954; Aronson, 1972; Krech & Crutchfield, 1948; Murphy & Murphy, 1931; Murphy, Murphy, & Newcomb, 1937; Smith, 1945; Znaniecki, 1925, 1936). Most social psychologists came to adopt Floyd Allport's (1924a) *interpersonal*[10] definition of social behavior:

> Behavior in general may be regarded as the interplay of stimulation and reaction between the individual and his environment. Social behavior comprises the stimulations and reactions arising between an individual and the *social* portion of his environment; that is, between the individual and his fellows. Examples of such behavior would be the reactions to language, gestures and other movements of our fellow men, in contrast with our reactions towards non-social objects, such as plants, minerals, tools, and inclement weather. (pp. 3–4)

In general, it may be said that the domain of contemporary social psychology remains the same restricted and fundamentally asocial domain defined (or, strictly speaking, redefined) by Floyd Allport in the 1920s and reaffirmed by Gordon Allport's oft-quoted definition from the 1950s:

> Social psychology is the science which studies the behavior of the individual in so far as his behavior stimulates other individuals, or is itself a reaction to their behavior; and which describes the consciousness of the individuals insofar as it is a consciousness of social objects and social reactions. (F. H. Allport, 1924a, p. 12)

> With few exceptions, social psychologists regard their discipline as *an attempt to understand and explain how the thought, feeling, and behavior of individuals are influenced by the actual, imagined, or implied presence of other human beings.* (G. W. Allport, 1954, p. 5)

Why was the original conception of social psychological phenomena and of the discipline of social psychology abandoned by later generations

[10] Many social behaviors are of course also interpersonal, but the two categories are not equivalent. The distinction between social and interpersonal behavior is discussed in detail in Chapter 1.

of American social psychologists? In this work I suggest a number of explanations. In part the abandonment appears to have been a product of the unfortunate association of theories of the social dimensions of psychological states and behavior with theories about the emergent properties of supraindividual "group minds," which were anathema to those social psychologists who were committed empiricists and experimentalists. In part it appears to have been a product of the apparent threat posed by the social dimensions of psychological states and behavior to cherished principles of autonomy and rationality, which were integral to the special form of moral and political individualism embraced by many American social psychologists. And in part it appears to have been a product of the impoverished concept of the social that some American social psychologists inherited from European "crowd" theorists such as Gabriel Tarde (1890/1903) and Gustav Le Bon (1895/1896), which provided the asocial paradigm for the experimental analysis of "social groups" developed by Floyd Allport, Dashiell (1930, 1935), and Murphy and Murphy (1931; Murphy, Murphy, & Newcomb 1937). While the original conception of social psychological phenomena was retained until the 1960s, it was beginning to be replaced by the asocial experimental paradigm in the late 1920s and 1930s. It was displaced almost completely by the increasingly narrow conception of experimentation in social psychology that developed in the 1950s and 1960s, which was itself a development of the asocial experimental tradition initiated by Floyd Allport in the 1920s.

A number of historians of the social sciences have recently argued that the formative years for American social science were the decades between 1870 and 1930 (Manicas, 1987; Ross, 1991). In this book, I suggest that much the same is true of American social psychology, in a number of significant ways. It was during this period that social psychology came to be recognized as a distinct discipline, and it was during this period that the original conception of the social dimensions of psychological states and behavior was formulated. It was also during this period that the alternative asocial theoretical and experimental paradigm in social psychology was developed by Floyd Allport.

While the two positions retained their advocates during the 1930s and 1940s, and while the original conception of the social enjoyed a brief postwar renaissance in the 1950s, the asocial theoretical and experimental paradigm quickly displaced the original conception of the social in the postwar years. Although American social psychology expanded dramatically as a scientific discipline after World War II (Cartwright, 1979; Farr, 1996), and in an institutional sense only came to full maturity after the war (with the development of independent departments of social psychology, graduate programs in social psychology, and so forth), this amounted to the expansion of an essentially asocial theoretical and methodological paradigm that was already securely in place by the late 1930s. Or so I argue in this work.

## II

It is perhaps worth stressing at the outset that this work does not aim to provide a comprehensive history of twentieth century American social psychology. Franz Samelson (1974) has claimed that an adequate history of social psychology still remains to be written. I agree that it does, and this work makes no pretense of offering such a general history. The aim is much more narrowly focused: to chart the historical neglect of the original conception of social psychological phenomena[11] to be found in early American social psychology and suggest some explanations of this neglect.

It is perhaps also worth stressing that this work does not attempt to develop a detailed critique of the theoretical and empirical achievements of twentieth century American social psychology. It is not hard to discern an (at least implicit) condemnation of the theoretical and empirical achievements of late twentieth century social psychology in the work of some recent historians and social constructionist critics who complain about the asocial nature of contemporary social psychology. No such condemnation is intended by the present work, the aim of which is simply to argue that, whatever the merits of the post-1930 tradition of theoretical and empirical work that came to dominate American social psychology (which I believe to be have been considerable),[12] this tradition no longer constitutes a tradition of distinctively *social* psychology. That said, this work

[11] Throughout the rest of this work I use the term "social psychological phenomena" as shorthand for social (i.e., socially engaged) forms of cognition, emotion, and behavior, and "individual psychological phenomena" as shorthand for individual (i.e., individually engaged) forms of cognition, emotion, and behavior. The use of the term "phenomena" is not intended to suggest that there is anything esoteric (or especially phenomenal) about social and individual psychological states and behavior. The term is just preferred over more theoretically loaded cognates such as "factors," "components," "elements," and the like.

[12] Although I believe these achievements to have been considerable, I also recognize the special epistemological and methodological problems of the discipline, especially the special problems of laboratory experimentation in social psychology. I have discussed these issues in detail elsewhere (Greenwood, 1989).

makes no pretence at theoretical neutrality. I believe the original conception of social psychological states and behavior shared by early American social psychologists had much to recommend it and consequently believe that something important was lost when the original conception of the social was abandoned.

The focus of this work is restricted to American "psychological" social psychology, defined as the form of social psychology practiced within departments of psychology at academic institutions in North America (the United States and Canada). This is because, although academic social psychology did develop in a somewhat different fashion in other countries, the American paradigm has come to dominate social psychology worldwide.[13] The question of whether the original conception of social psychological phenomena was retained within American "sociological" psychology, defined as social psychology practiced within departments of sociology at academic institutions in North America, is left largely open.[14] For whatever vestiges of the original conception of social psychological phenomena can be discovered in American departments of sociology, it is certainly the case that academic psychologists have come to dominate the journal, handbook, and textbook markets in social psychology, and significantly outnumber sociological social psychologists at both the faculty and student levels (Burgess, 1977; Collier, Minton, & Reynolds, 1991; E. E. Jones, 1985, 1998; Liska, 1977).[15]

I don't pretend to be the first person to complain about the neglect of the social in American social psychology or the first to offer putative explanations of it. A number of other critics have complained about the neglect of the social in American social psychology (Farr, 1996; Graumann, 1986; Moscovici, 1972; Pepitone, 1976, 1981; Post, 1980; Stroebe, 1979) and have offered historical accounts of the "individualization" of

[13] Even the so-called European alternative looks increasingly American, and the new "third-force" Asian vision of social psychology ("Editor's, Preface," *Asian Journal of Social Psychology*, 1998) appears to simply appropriate the North American paradigm to the study of Asian peoples.

[14] With the exception of the "symbolic interactionist" tradition, which is discussed at length in Chapter 4.

[15] Nonetheless, it is worth acknowledging that many of the early American social psychologists who recognized the social dimensions of psychological states and behavior were institutionally located in departments of sociology rather than departments of psychology. These include Luther Bernard (1926a, 1931), Emory Bogardus (1918, 1924a, 1924b), Charles Ellwood (1917, 1924, 1925), Franklin Giddings (1896, 1924), Robert Park (1902; Park & Burgess, 1921), Edward Ross (1906, 1908), William I. Thomas (1904; Thomas & Znaniecki, 1918), and Kimball Young (1925, 1930, 1931).

American social psychology (Farr, 1996; Graumann, 1986). However, my own account differs from these others in two fundamental respects.

In the first place, most of these critics fail to specify what exactly is supposed to have been neglected or "individualized" in American social psychology. They provide rather vague and amorphous characterizations of the social in terms of "trans- or supra-individual structures" (Graumann, 1986, p. 97), "relationalism" (Pepitone, 1981, p. 972), or "the relationship between the individual and the community (or society)" (Farr, 1996, p. 117), and they do not provide illustrative examples of what exactly they take to have been neglected or individualized. This makes it very hard to assess their historical claims and to conceive of their implied alternative to contemporary social psychology.[16] In contrast, I try to spell out in some detail the specific conception of the social dimensions of psychological states and behavior held by early American social psychologists but neglected from the 1930s onward.

The common complaint about the individualization of the social is especially misleading, because it tends to suggest that social psychology ought to concern itself with the emergent properties of supraindividual social groups as opposed to the psychological properties of individuals who constitute social groups. Graumann (1986, p. 97), for example, complains that social psychology "is not a social science" because it deals with intra- as opposed to interpersonal psychological states and fails to deal with "trans- or supra-individual structures." However, as will be argued in some detail in the following chapters, the fundamental distinction between social and individual psychological states and behavior (and thus the fundamental distinction between social and individual psychology) is grounded in a postulated difference *in the manner in which the psychological states and behavioral dispositions of individual persons are engaged.* It is not a distinction grounded in any postulated difference in the objects – social groups as opposed to individuals – to which psychological properties are ascribed.

Any account of the distinctive social nature of the subject matter of social psychology has to recognize that social psychological states and behavioral dispositions, as much as individual psychological states and behavioral dispositions, are the psychological states and behavioral

[16] Many of these critics also neglect the substantive conception of social psychological states and behavior that can be identified in early American social psychology, as do most of the "social constructionist" critics who complain of the continuing "crisis" in social psychology (Gergen, 1973, 1982, 1985, 1989; Parker, 1989; Parker & Shotter, 1990).

dispositions of *individual persons* (and possibly some animals). Many critical analyses of the asocial nature of contemporary social psychology appear to neglect this fundamental feature of the social psychological and present the quite misleading impression that the only alternative to contemporary social psychology is an appeal to (metaphysically dubious) emergent entities and processes, such as "trans- or supra-individual structures."

In the second place, many of these critics locate the source of the neglect of the social in American social psychology in the commitment by its practitioners to experimental science. This commitment is itself often represented as a historical function of the perceived need by practitioners of the fledgling science to present social psychology as an objective, experimental science to university administrators, government agencies, grant-awarding bodies, and the public at large. Many critics also appeal to the role played by distinctively American commitments to "pragmatism" and "individualism."

While I do not deny that these factors played a major role in shaping the development of American social psychology, I do not think that they account for the specific neglect of the social in American social psychology. The neglect of the social in American social psychology does not appear to have been a direct product of the undoubted commitment by many of its practitioners to experimental science. This is important to stress, because it seems to be assumed by many historians and recent "social constructionist" critics that such a commitment *precludes* the study of the social dimensions of cognition, emotion, and behavior. Yet this cannot be the case, since (as will be noted in later chapters) there are exemplary experimental studies of the social dimensions of psychological states and behavior to be found in the social psychological literature. What needs to be explained is the relative *paucity* of such studies: how the legitimate commitment to experimental science came to be distorted by other conceptual constraints to generate an asocial theoretical and experimental social psychology.

There is little doubt that characteristically American commitments to pragmatism (White, 1973) and individualism (Arieli, 1964; Bellah, Madsen, Sullivan, Swidler, & Tipton, 1985) played a significant role in shaping the development of American social psychology. However, such commitments cannot adequately explain the neglect of the social, as many early social psychologists who explored the social dimensions of psychological states and behavior, such as Daniel Katz, Richard Schanck, Muzafer Sherif, William I. Thomas, and Junius F. Brown, were also committed pragmatists. Their commitment to the social utility of social psychology was at least as strong as (if not stronger than) that of later generations of social psychologists. Similarly, both early advocates and critics of a distinctively social conception of cognition, emotion, and behavior, and of social psychology, such as William McDougall and Floyd Allport (to take a famous advocate and famous critic), were committed individualists, both philosophically and morally.

## III

In charting the historical neglect of the social in American social psychology, and advancing some tentative explanations of this neglect, I offer a *critical conceptual history* of American social psychology: a new animal, perhaps, for many historians of the social and behavioral sciences. As I hope to illustrate in the following chapters, there are no intrinsic conceptual impediments to the objective and experimental study of the social dimensions of cognition, emotion, and behavior – in other words, to the development of social psychology as a genuinely scientific and experimental discipline. Yet, as I also hope to illustrate in the following chapters, the promotion of such a discipline in America was thwarted by historically local meta-theoretical positions and associations (Amundson, 1985), which shaped the peculiarly asocial development of American social psychology from the 1920s and 1930s onward. The point of offering such an account of the essential *contingency* of the asocial development of American social psychology is in the hope that the recognition of the historically local nature of these conceptual commitments and associations may enable some contemporary practioners to surmount them.

The present work is thus fundamentally "internalist" in orientation, insofar as it advances an account of the neglect of the social in American social psychology primarily in terms of the conceptual commitments and associations of twentieth century American social psychologists. It is not, however, an internalist account in the sense that it is written by an insider, and indeed much of the conventional internal history of the discipline offered by social psychologists such as Gordon Allport (1954, 1968a, 1985), Dorwin Cartwright (1979), and Edward E. Jones (1985, 1998) is disputed in the following chapters. My own professional background is someone peculiar. As a professional philosopher of social science who recently developed an interest in the history of the social and psychological sciences, I count as neither a conventional insider nor a conventional outsider (being neither a professional social psychologist nor historian).

Whether this constitutes an advantage or disadvantage I leave to the reader to judge.

The present work also aims to provide a generally "contextualist" account of the neglect of the social in American social psychology insofar as it tries to render the neglect of the social intelligible from the point of view of American social psychologists working in the 1930s and later decades.[17] While ultimately unjustified, the abandonment of the social by many later generations of American social psychologists is not hard to understand given their historically developed (and culturally sedimented) conceptual commitments and associations.

It has become common in recent years to lay greater emphasis on the role of "external" social and political factors in the historical development of the sciences, including the social sciences and psychology (Altman, 1987; Buss, 1975; Furomoto, 1989), and such factors have indeed been emphasized in historical accounts of the development of American social psychology (Cartwright, 1973; Collier, Minton, & Reynolds, 1991; Lubek, 1986; Morawski, 1979). There is also little doubt that external social and political factors did play an important role in the development of social psychology as an academic discipline and an experimental science, and in the development of particular types of theories and areas of research.

The development of American social psychology was undoubtedly shaped by the roles played by grant-funding agencies (e.g., the Carnegie, Rockefeller, Ford, and Russell Sage Foundations and the Social Science

[17] The present work also aims to provide a generally contextualist account in the following respect. The historical account offered is not approached from a so-called Whig (Butterfield, 1951) or presentist (Stocking, 1965) perspective, which would treat social psychology as gradually approximating the idealized perspective of the present moment: the sort of "house history" (Woodward, 1987) developed by writers such as Gordon Allport (1954, 1968a, 1985), Dorwin Cartwright (1979), and Edward E. Jones (1985, 1998). On the contrary, it is maintained that in one important respect the development of social psychology from the 1930s onward has been *regressive*: it has come to neglect the genuinely social conception of human psychology and behavior that it originally recognized. Thus, the present historical account, although restricted in scope, also hopefully illustrates that the development of social scientific disciplines is not always a linear progression to a richer and more sophisticated theoretical conception of their subject matters (contra Wetterstein, 1975). Nonetheless, as noted earlier, the present account, unlike most contextualist accounts, makes no pretense of neutrality. It suggests that, in the case of American social psychology, an originally rich and sophisticated conception of social cognition, emotion, and behavior was lost. The point of a critical conceptual history is to insist that there is nothing inevitable or final about this: what once was lost can also be regained.

Research Council), by the academic competition among the fledgling social sciences, notably between the newly developed disciplines of sociology and psychology (Haskell, 1977; Samelson, 1985), and, as noted earlier, by the distinctively American ideological commitments to pragmatism and individualism. Many of the topics studied by American social psychologists in the twentieth century, such as conflict, prejudice, aggression, group decision-making, and productivity, for example, do appear to have been a product of distinctively American interests and concerns (Apfelbaum & Lubek, 1976; Lubek, 1979; Moscovici, 1972), and the specific research focus on small groups in the period during and immediately following World War II appears to have been significantly influenced by the policies and interests of major funding agencies, such as the Office of Naval Research (Cina, 1981; Steiner, 1974).

However, there are a number of reasons why I have mainly focused on internal factors in the present historical account. In the first place, the account is both partial and critical. It represents a limited conceptual history of the neglect of the social in American social psychology, not a general history of twentieth century American social psychology. It is offered as a critical challenge to traditional practitioners and historians of social psychology, to "social constructionist" critics, and to historians who have complained about the neglect of the social in social psychology. Although it is undoubtedly narrow and partial and tentative, it explicates the conception of the social dimensions of cognition, emotion, and behavior embraced by early American social psychologists and suggests specific historical reasons for the neglect and eventual abandonment of this original conception. My hope is that it can be defended, modified, and extended in the light of further critical and historical responses. While recognizing its partiality, I hope that something can be learned about the nature and history of American social psychology by pressing this internal conceptual history to its limits.

In the second place, although external social and political factors undoubtedly played a major role in the general development of twentieth century American social psychology, these factors seem insufficient to explain the specific neglect of the social in American social psychology (although they very likely exacerbated changes produced by largely internal conceptual factors).[18] Social and political factors may explain why American social psychologists focused on certain topics at the expense of

[18] And of course such internal conceptual factors are, in the last analysis, socially constructed and historically sedimented conceptual factors.

others but do not explain the neglect of the social dimensions of the topics studied. For example, social and political factors may explain why American social psychologists focused on conflict and group decision-making but do not explain their neglect of the social dimensions of conflict and group decision-making (Plon, 1974). While there are no doubt distinctive external (and distinctly American) reasons why aggression became a focal research concern in American social psychology, there are no obvious external reasons to explain why American social psychology has systematically neglected the social dimensions of aggression, such as the grounding of at least some forms of aggression in the represented behavior of members of social groups (e.g., other gang members). Yet the social dimensions of aggression have been neglected by post-1930 American social psychological theoretical approaches to aggression, such as the "frustration-aggression" theory of Dollard, Doob, Miller, Mowrer, and Sears (1939) and the "social-learning" theories of Bandura (1973) and Berkowitz (1962).[19]

## IV

One final comment before embarking on the details of the history. Throughout much of the twentieth century, practicing social psychologists in the United States have often gone out of their way to eschew what have been conceived as pointless and sterile philosophical discussions of what is "social" about social forms of cognition, emotion, and behavior, or about social psychology itself. Thus, contemporary social psychologists might object that reflexively focusing on the definition of the social or on the distinction between social and individual psychological phenomena is to abandon scientific psychology for philosophy, and that the task of the scientific social psychologist is to focus on the *phenomena* referenced by our concepts of the social, not the *content* of our concepts of the social.[20]

This is a peculiar attitude, for which there is little justification beyond caricatures of the distinction between philosophy and science. Many of the great advances in the much-admired "hard" physical sciences were

[19] These dimensions, however, are manifest in the work of European social psychologists such as Marsh and Campbell (1982), Marsh, Rosser, and Harré (1978), Siann (1985), and Wolfgang and Ferracuti (1967). Compare Pepitone's (1999, p. 175) complaint that American social psychologists have neglected socially orientated motives in the study of aggression: "In aggression, for example, there was no room for honor or shame in the provocation of aggression."

[20] For this complaint, see for example Zajonc (1966, p. 8) and McGuire (1986, p. 102).

achieved in part through critical reflection on and development of concepts such as "inertia," "acceleration," and "simultaneity," and significant changes in the content of such concepts produced significant changes in their theoretical and empirical referents. We deceive ourselves if we imagine that we can communicate effectively in social psychological science without some shared grasp of the content of our fundamental concepts or that real progress in social psychology (or any other scientific discipline) can be achieved without critical reflection on and development of these concepts.

We deceive ourselves doubly if we imagine that American social psychology developed as a discipline *independently* of changes in practitioners' concepts of the social. Later generations of American social psychologists referenced different sets of cognition, emotion, and behavior from those referenced by early generations of American social psychologists because they changed their concept of the social: the original subject matter as well as the original concept of the social was lost in the process. In refusing to confront the concept of the social, contemporary practitioners and critics of social psychological science blind themselves to the original vision of the social dimensions of human psychology and behavior and of social psychological science. This critical conceptual history hopes to shed a little light where presently there is much darkness.

# I

# The Lost World

The aim of this work is to document and suggest some explanations of the historical neglect and eventual abandonment of the distinctive conception of the social dimensions of cognition, emotion, and behavior, and of the discipline of social psychology itself, held by early American social psychologists. In this chapter I try to explicate and critically develop this distinctive conception of the social psychological in order to provide the reader with a clearer sense of what exactly came to be neglected and eventually abandoned by American social psychology.

According to this conception, social (or "collective" or "group") cognition, emotion, and behavior are forms of cognition, emotion, and behavior engaged by individual persons (and possibly some animals)[1] *because and on condition that they represent other members of a social group as engaging these (or other)*[2] *forms of cognition, emotion, and behavior in*

[1] In this book, I leave it as an entirely open question whether animals have the capacity for social forms of cognition, emotion, and behavior: that is, whether animals as a matter of fact satisfy the conditions for social cognition, emotion, and behavior discussed in this chapter. There is some reason to doubt this (see, e.g., the discussion of the work of Tomasello, Kruger, and Ratner, 1993, at the end of this chapter). However, social psychological states and behaviors ought not to be denied of animals solely on the grounds that they lack language or consciousness, since neither language nor consciousness appears to be necessary for sociality (although they undoubtedly enrich it in myriad ways).

[2] The reference to "other" forms of cognition, emotion, and behavior is designed to cover instances of cooperative, competitive, and combative forms of human psychology and behavior: where I push (only) when you pull, when I return (only) when you serve, when I fight you (only) when you insult me, and so forth. Compare, for example, Bernard (1931), who treats "collective behavior" as a synonym for "social behavior":

*similar circumstances.*[3] As Katz and Schanck (1938) put it, they are the attitudes and practices "prescribed" by group membership. According to this conception, social groups themselves are populations of individuals who share[4] socially engaged forms of cognition, emotion, and behavior.[5]

Collective behavior is only individual behavior in its collective aspects. It may consist of the multiplication of identical or similar acts, or it may represent the cooperative adjustment of unlike, but complementary, behaviors. It does not represent the behavior of a new and independent organism, self-functioning as a unit. The behavior of the collectivity centers on the several individual units of the collectivity, although the behavior of each unit may be conditioned or determined by the similar or dissimilar behavior of the other units. (pp. 62–63)

However, not every interpersonal sequence of action and reaction counts as social interaction, as, for example, in the case of two persons embroiled in an escalating dispute over the true boundary between their yards and who each respond to the other's movement of the boundary fence by including more of the other's land. For an interactive sequence to constitute a social interaction, the actions and reactions of the participants must be oriented to the represented actions and reactions of members of a social group, even if this is only the dyad constituted by the two participants. In this example, it is unlikely that the participants to the dispute represent their behaviors as oriented to the behavior of members of the dyad constituted by the pair of them. Contrast this with the case of two friends traveling together on a train who pursue an escalating competition to pay for the food and drink.

3 The terms "engage," "engaged," and "engaging" are used in a quasi-technical sense throughout this work, to refer to the actualization or instantiation of forms of cognition, emotion, and behavior and thus to describe the different ways in which they may be actualized or instantiated (socially as opposed to individually). The phrases "social dimensions of cognition, emotion, and behavior," "socially engaged cognition, emotion, and behavior," "social psychological states and behavior," "social dimensions of human psychology and behavior," "social psychological phenomena," and "psychological states and behavior oriented to the represented psychology and behavior of members of social groups" are used interchangeably to refer to *forms of cognition, emotion, and behavior engaged by individuals because and on condition that they represent members of a social group as engaging these (or other) forms of cognition, emotion, and behavior in appropriate circumstances.* Also, the phrase "social psychological states and behavior" should be read as including psychological processes and behavioral dispositions.

4 *Shared* cognition, emotion, and behavior is here understood to include cognition, emotion, or behavior engaged jointly with members of a social group and represented by members of a social group as engaged jointly with other members of a social group.

5 This characterization of a social group may appear circular, since social forms of cognition, emotion, and behavior are themselves characterized as social by reference to their orientation to the represented psychology and behavior of members of social groups. However, the circularity involved is natural and not vicious: It is merely a reflection of the fact that the social engagement of psychological states and behavior and the constitution of social groups are generally two moments of the same psychological process. As Simmel (1908/1959) aptly described the constitution of social groups, "The consciousness of constituting with the others a unity is actually all there is to that unity" (p. 7). Moreover, the circularity involved (while entirely natural) is not strictly necessary and could be eliminated

On this account of social psychological states and behavior, a belief is a social belief, for example, if and only if an individual holds that belief because and on condition that other members of a social group are represented as holding that (or another) belief. The belief held by a member of a millennium sect that "The Guardians" will descend from space to save the sect on a particular day is a social belief if and only if it is held because and on condition that other members of the sect are represented as holding that belief.[6] On this account, a behavior is a social behavior if and only if an individual behaves in a particular way because and on condition that other members of a social group are represented as behaving in that (or another) way in similar circumstances. Wearing blue jeans is a social behavior if and only if an individual wears blue jeans because and on condition that other members of a social group are represented as wearing blue jeans.[7]

Social psychological states and behaviors are social by virtue of *the manner in which they are engaged by individual persons* (with their unique personalities, spatiotemporal locations, and life histories). They are not social by virtue of their contents or objects or their being engaged by social groups (or "social collectives" or "social communities")[8] as opposed to individuals. A social belief or attitude, for example, is a belief or attitude that is held by an individual (or individuals) *socially*: that is, because and on condition that other members of a social group are represented as holding that belief or attitude.[9] An individual belief or attitude is a belief or

(see Greenwood, 2003). I have included strictly unnecessary references to social forms of cognition, emotion, and behavior and social groups in their definitions because I do want to emphasize the generally joint nature of their constitution.

[6] For this example, see Festinger, Riecken, and Schachter (1956).

[7] For this reason, fashion perhaps represents the purest if also the least noble form of socially engaged cognition, emotion, and behavior. Certain fashion items (e.g., rings through the nose) are worn for no reason other than the fact that other members of a social group are represented as wearing them.

[8] Or societies, for that matter. In this book, I reserve the term "society" for referencing the intersecting aggregations of smaller social groupings (such as occupational, religious, and political groupings) that compose the populations of nations: Thus, one talks about British as opposed to French or European society but not (or not usually) about Catholic or professional psychologist society. However, the term "society" is sometimes also used, and was frequently used by early American social psychologists, to reference smaller social groupings, such as the "societies" of Catholics, bankers, and Republicans.

[9] Or another belief or attitude. This qualification, noted in the original definition of social forms of cognition, emotion, and behavior, is left out for the sake of convenience, but it should be understood as holding in all consequent discussion of social forms of cognition, emotion, and behavior.

attitude that is held by an individual (or individuals) *individually*: that is, independently of whether any member of any social group is represented as holding that belief or attitude. For example, an individual Catholic's belief that abortion is wrong is a social belief if and only if it is held socially – if and only if it is held because and on condition that other Catholics are represented as holding this belief. An individual Catholic's belief that abortion is wrong is an individual belief if and only if it is held individually, for reasons or causes independent of whether any other Catholic (or any member of any social group) is represented as holding this belief – if, for example, it is held because the person has accepted rational arguments or evidence in favor of this belief.

Since the difference between social and individual beliefs or attitudes is a difference with respect to how beliefs or attitudes are held, an individual may hold one belief or attitude socially and another belief or attitude individually, or may hold one and the same belief or attitude both socially and individually. An individual may hold one belief or attitude socially, qua member of a social group, and another or different belief or attitude individually, without reference to any social group. For example, some college professors may approve of affirmative action socially, because and on condition that other college professors are represented as approving of it, but disapprove of it individually, because they believe it to be one injustice replacing another.[10] Or an individual may hold one and the same belief or attitude socially, qua member of a social group, and individually, independently of any social group. Some Catholics may disapprove of abortion (at least in part) because and on condition that other Catholics are represented as disapproving of abortion and (at least in part) because they have been convinced by rational arguments and evidence.[11]

[10] Compare William James (1890) on contrary social and individual attitudes:

> A judge, a statesman, are in like manner debarred by the honor of their cloth from entering into pecuniary relations perfectly honorable to persons in private life. Nothing is commoner than to hear people discriminate between their different selves of this sort: "As a man I pity you, but as an official I must show you no mercy; as a politician I regard him as an ally, but as a moralist I loathe him"; etc., etc. (p. 295)

[11] This might explain the conservative nature of many beliefs and attitudes, which may be resistant to change because they are held in part socially. In many cases, we might not recognize the social component of our belief or attitude, or find it easy to deny, since we can often cite some individually held reason(s) for maintaining the relevant belief or attitude. Rationalizations may be especially effective when genuine reasons can be offered (even when these reasons are insufficient to warrant a particular belief or attitude).

## I

It may be useful to illustrate some of these points by reference to a concrete example drawn from the early American social psychological literature. The distinction between social and individual beliefs and attitudes can be illustrated by reference to a study conducted by Schanck (1932) concerning the preferences for forms of baptism among Methodists and Baptists. Among the Methodists, for example, 90 percent expressed a preference for sprinkling (as opposed to immersion) when asked for a statement of their attitude *as church members*, whereas 16 percent expressed a preference for sprinkling when asked for a statement of their own *private* feelings. Thus, we may say that while most Methodists held this social preference (held this preference socially), only a few held this individual preference (held this preference individually) – a good many held a different individual preference. Given the figures, we may also say that some Methodists held this preference both as a social and an individual preference: that is, both socially and individually.

Two qualifications concerning this example are perhaps in order. In the first place, it has been assumed that *individual* attitudes (attitudes held individually) can be equated with *private* attitudes. In the context of the Schanck questionnaire, this is probably legitimate. It is likely that being asked for one's private attitude would have been interpreted by interviewees as being asked for one's individual attitude, since by answering the question one is in fact making one's attitude public. However, the public/private distinction cannot be generally equated with the social/individual distinction. Private beliefs or attitudes (beliefs or attitudes that one keeps to oneself) that might be unpopular or "politically incorrect," such as the belief that African-Americans are intellectually or morally inferior, might very well be held socially (and much evidence suggests that they are). Conversely, persons may go out of their way to publicly express their individually held attitudes about the injustice of income taxes or the immorality of eating meat.

In the second place, it has also been assumed that the Methodists' verbal reports of their social and individual attitudes were honest and accurate. However, La Piere (1934) noted that verbal reports of beliefs and attitudes may be dishonest and may not be accurate even when they are honest: they may be belied by the behavior of the individuals who sincerely avow them. La Piere demonstrated the gulf between verbally avowed attitudes and behavior in his study of the differences between the verbally avowed attitudes and behaviors of hoteliers and restaurateurs with respect to their service of Chinese customers. Schanck (1932) himself noted that his Methodists often avowed one attitude but behaved in accord with another. They condemned smoking and card playing as good Methodists but did both in Schanck's company behind closed curtains and doors.

The methodological prescription that La Piere derived from his study remains sound. Whenever possible, beliefs and attitudes should be measured via their behavioral expression instead of, or at least in addition to, measuring them via verbal responses to questionnaires.[12] However, neither the La Piere study nor the Schanck study demonstrated that individuals never act in accord with their social attitudes, and Schanck also stressed that the Methodists did often act in accord with their avowed social attitudes: when and where they did appeared to be a function of the perceived relevance of the particular situations to their social group memberships.

These qualifications aside, the Schanck example can be adapted to illustrate some critical points about social beliefs and attitudes. It is not sufficient for a belief or attitude to be social that it is held by the majority of members of a social group, far less a mere plurality of individuals. Most Methodists, for instance, will maintain a preference for sprinkling qua Methodists if they hold this attitude socially, and the fact that a preference is held socially by Methodists explains its generality among Methodists. However, the members of a congregation of Methodists coming out of church on a Sunday morning may all believe that it is raining by virtue of the liquid evidence falling from the skies, or they may all believe that New York is east of Los Angeles because this is how their positions are represented on all available maps. Yet these beliefs are not social beliefs, for they are (presumably) not held by Methodists because and on condition that they represent other Methodists as holding them.

That is, many common beliefs and attitudes are held individually, even among members of social groups. Conversely, a social belief or attitude need not be restricted to members of a particular social group but may be held socially by members of other social groups. For example, many Baptists as well as many Catholics might hold a negative social attitude toward abortion. As Durkheim (1895/1982a) succinctly put it, social beliefs and attitudes are general because they are social (held socially); they are not social because they are general. For Durkheim, a social fact, including

[12] Except, according to La Piere (1934), in the case of purely "symbolic" attitudes such as religious attitudes, where "an honest answer to the question 'Do you believe in God?' reveals all there is to be measured" (p. 235).

any social form of cognition, emotion, or behavior, is general "because it is collective (i.e., more or less obligatory); but it is very far from being collective because it is general. It is a condition of the group repeated in individuals because it imposes itself upon them" (p. 56).[13]

Edward Ross (1908), the author of one of the first social psychology texts in America, was also very clear on this point:

> Social psychology pays no attention to the non-psychic parallelisms among human beings (an epidemic of disease or the prevalence of chills and fever among the early settlers of river-bottom lands), or to the psychic parallelisms that result therefrom (melancholia or belief in eternal punishment). It neglects the uniformities among people that are produced by the direct action of a common physical environment (superstitiousness of sailors, gayety of open-air peoples, suggestibility of dwellers on monotonous plains, independent spirit of mountaineers), or by subjection to similar conditions of life (dissipatedness of tramp printers, recklessness of cowboys, preciseness of elderly school teachers, suspiciousness of farmers). (p. 2)
>
> Social psychology deals only with uniformities due to *social* causes, *i.e*, to *mental contacts* or *mental interactions*. In each case we must ask, "Are these human beings aligned by their common instincts and temperament, their common geographical situation, their identical conditions of life, or by *their interpsychology, i.e.*, the influences they have received from one another or from a common human source?" The fact that a mental agreement extends through society bringing into a common plane great numbers of men does not make it *social*. It is *social* only in so far as it arises out of the interplay of minds. (p. 3)[14]

Or, as Ellsworth Faris (1925) put it, "social" or "group" attitudes refer to "collective phenomena that are not mere summations" (p. 406).

Early American social psychologists maintained that social (as opposed to merely common) beliefs and attitudes are held *conditionally* in relation to the represented beliefs and attitudes of members of a social group, and

[13] This is important to stress, because many contemporary "social representation" theorists (who frequently avow a Durkheimian ancestry) appear to treat the widespread nature of a representation as a sufficient condition of its sociality (see, e.g., Moscovici, 1998b).

[14] Ross (1908) also noted that common but individually engaged uniformities that are a product of biological inheritance are not social:

> Social psychology ignores uniformities arising directly or indirectly out of race endowment – negro volubility, gypsy nomadism, Malay vindictiveness, Singhalese treachery, Magyar passion for music, Slavic mysticism, Teutonic venturesomeness, and American restlessness. (p. 3)

He also doubted whether many of these uniformities are in fact genetically determined:

> How far such common characters are really racial in origin and how far merely social is a matter yet to be settled. Probably they are much less congenital than we love to imagine. "Race" is the cheap explanation tyros offer for any collective trait that they are too stupid or lazy to trace to its origin in the physical environment, the social environment, or historical conditions. (p. 3)

they regularly stressed the "reciprocity," "interstimulation," and "interconditioning" of social beliefs and attitudes. They maintained that beliefs and attitudes are social by virtue of their orientation to the represented beliefs and attitudes of members of particular social groups. Thus Bogardus (1924a), for example, identified "occupational attitudes" as those social attitudes associated with particular occupations or professions:

> Each occupation has its characteristic attitudes, which, taken in the large, may be referred to here as the occupational attitude. . . . each occupation is characterized by social attitudes and values peculiar to itself.
>
> It would seem that two persons might start with about the same inherited predispositions, the same mental equipment, and by choosing different occupations, for example, one, a money-making occupation, and the other, a service occupation, such as missionary work, at the end of twenty years have become "successful," but have drifted so far apart in occupational and social attitudes as to have almost nothing in common. (pp. 172–173)

Bogardus (1924b, p. 3) explicitly employed the term "social attitudes" to mean "socialized attitudes": that is, attitudes held socially, because and on condition that other members of an occupational group are represented as holding these attitudes. Analogously, W. S. Watson and Hartmann (1939) talked of religious attitudes as social attitudes oriented to the represented attitudes of members of religious groups, and Edwards (1941) talked of political attitudes as social attitudes oriented to the represented attitudes of members of political groups.

The types of social groups toward which social beliefs and attitudes might be oriented were recognized as many and various, ranging from simple friendship dyads to whole societies and including families, clubs, professions, religious groups, political parties, and the like. Ellwood (1925), for example, listed "the family, the neighborhood group, kinship groups, cities, states and nations" and "political parties, religious sects, trade unions, industrial corporations, and the like" (p. 117). McDougall (1920, chap. 5) counted churches, trade unions, occupational groups, colleges, castes, kinship groups, and nations.

Franklin Giddings (1896, 1924) noted that the higher discriminative capacities of humans, including their developed linguistic skills, not only facilitated the development of human social groups but also promoted the differentiation of social groups based on different forms of associated psychological states and behavior:

> With discriminations talked about came sortings, the beginnings of classification, of distinctions of kind; and among these the most important by far was a talked about discrimination of "own kind" from "other kind," of "my kind" and "our kind" from "your kind," "his kind" and "their kind." (Giddings, 1924, p. 454)

This led Giddings to characterize "consciousness of kind" as the primary basis of socially engaged psychological states and behavior, transforming merely homogeneous physical associations into genuine social groups:

When men attained it [consciousness of kind] they began to be *social* as already they had been *gregarious*. Now they not only *consorted* by kind, but also they began to *associate*, picking and choosing companions and confirming their likes and dislikes by talking about them. It was, in short, "the consciousness of kind," or at any rate, the "talked about" distinctions of kind *that converted the animal herd into human society*, a reconditioning of all behavior second in its tremendous importance only to the effects of speech itself. (1924, p. 454, original emphasis)

Analogously, Knight Dunlap (1925) characterized "social consciousness" as "consciousness (in the individual, of course) of *others* in *the group*, and consciousness of them, as *related, in the group*, to oneself; in other words, consciousness of *being a member of the group*" (p. 19). Dunlap, like many other early American social psychologists, maintained that this characteristic feature of humans constitutes the primary rationale for a distinctively social psychology:

One of the outstanding characteristics of the human individual is his associating in groups of various kinds. These groups are not mere collections of people, but possess psychological characteristics binding the individuals together or organizing them in complicated ways. The family, the tribe, the nation, and the religious group are the most important of these organizations, but many other types are found. Industrial groups and secret societies have their important and fundamental psychological characteristics, and the various groups dependent upon local contiguity are also psychologically organized. The numerous special groups, such as athletic teams, festal parties, and welfare agencies are possible only through mental organization....

Human groups are the manifestation of the social nature of man, that is to say, of his tendency to form societies. Or rather, that "tendency" is merely the abstract fact that he does organize himself in groups. The psychological study of man is therefore not complete until we have investigated his groupings, and analyzed the mental factors involved therein. This study is *social psychology*, or *group psychology*. (1925, p. 11)

W. I. Thomas and Florin Znaniecki (1918) similarly maintained that

psychology is not exclusively individual psychology. We find numerous monographs listed as psychological, but studying conscious phenomena which are not supposed to have their source in "human nature" in general, but in special social conditions, which can vary with the variation of these conditions and still be common to all individuals.... To this sphere of psychology belong all investigations that concern conscious phenomena peculiar to races, nationalities, religious, political, professional groups, corresponding to special occupations and interests, provoked by special influences of a social milieu, developed by educational activities and legal measures, etc. The term "social psychology" has become current for this type of investigation. (p. 27)

Most early American social psychologists were also very clear that social psychological states and behavior are not social just because they are directed to social objects (i.e., other persons and social groups). Social beliefs and attitudes, for example, were not held to be restricted in any way by their contents and objects. Thus, one might have social beliefs and attitudes about nonsocial objects, such as the weather, snakes, the Eiffel Tower, and the orbits of the planets, as well as social beliefs and attitudes about social objects, such as one's fiancée, one's father, Muslims, or the federal government. Thomas and Znaniecki (1918) were particularly clear on this point:

And thus social psychology, when it undertakes to study the conscious phenomena found in a given social group, has no reasons a priori which force it to limit itself to a certain class of such phenomena to the exclusion of others; any manifestation of the conscious life of any member of the group is an attitude when taken in connection with the values which constitute the sphere of experience of this group, and this sphere includes data of the natural environment as well as artistic works or religious beliefs, technical products and economic relations as well as scientific theories. (p. 28)

Of course, as Thomas and Znaniecki also noted, it is entirely legitimate for social psychologists to focus on socially held beliefs and attitudes about social objects (such as other persons or political, racial or religious groups), since these are of special social interest to psychologists and laypersons. Nonetheless, they insisted that social beliefs and attitudes can be directed toward any type of object and that social beliefs and attitudes directed towards nonsocial objects (e.g., colors) are entirely legitimate and appropriate objects of social psychological investigation:

The field of social psychology practically comprises first of all the attitudes which are more or less generally found among the members of a social group, have a real importance in the life-organization of the individuals who have developed them, and manifest themselves in social activities of these individuals.... the field of social psychology may be extended to such attitudes as manifest themselves with regard, not to the social, but to the physical, environment of the individual, as soon as they show themselves affected by social culture; for example, the perception of colors would become a socio-psychological problem if it proved to have evolved during the cultural evolution under the influence of the decorative arts. (pp. 30–31)

Furthermore, in the case of those social beliefs and attitudes that are directed toward other persons or social groups, early American social psychologists insisted that such beliefs and attitudes are social because they are oriented to the represented beliefs and attitudes of members of social groups and not merely because they are directed toward persons or social groups.[15] This is particularly clear in early American social psychological treatments of social prejudice and stereotyping, for example. Ellwood Faris (1925) explained how the "learning" of racial prejudice is conditional upon the social acceptance of group attitudes:

> The individual manifestation of race prejudice cannot be understood apart from a consideration of group attitudes. In collecting data it often happens that the investigator finds cases of the acquisition of a prejudice with astonishing suddenness and as the result of a single experience. But this could only happen in a *milieu* where there was a pre-existing group attitude. One who has no negro prejudice can acquire it from a single unpleasant encounter but it is the group attitude that makes it possible for him to acquire it. An exactly similar experience with a red-headed person would not result in the same sort of red-head-prejudice in the absence of any defining group attitude. (p. 406)

Analogously, E. L. Horowitz (1936/1947a) reported that

> young children were found to be not devoid of prejudice; contact with a "nice" negro is not a universal panacea; living as neighbors, going to a common school, were found to be insufficient; Northern children were found to differ very, very slightly from Southern children. It seems that attitudes toward Negroes are now chiefly determined not by contact with negroes, but by contact with the prevalent attitude toward negroes. (p. 507)[16]

Many early American social psychologists linked social beliefs and attitudes with personality, effectively equating socially held beliefs and attitudes with the social dimensions of personality: "Defined in this way, social attitudes may be spoken of as the elements of personality. Personality consists of attitudes organized with reference to a group in a system more or less complete" (Faris, 1925, p. 408). Many also recognized that individual persons are normally members of a variety of different groups (such as family, occupational, religious, and political groups) and that the social orientation of much of our psychology and behavior to the represented psychology and behavior of members of a *variety* of different social groups creates a distinctive management problem in our everyday

[15] That is, they held that orientation to the represented beliefs and attitudes of members of social groups is necessary and sufficient for social beliefs and attitudes and that direction toward social objects (other persons or social groups) is neither necessary nor sufficient.

[16] Or, as Asch (1952) later put it, "The racial sentiment of Southerners is only in part directed towards Negroes; it is also a function of their most significant ties to family, neighborhood and group" (p. 575).

personal and social lives (Cooley, 1902; Dewey, 1927; Faris, 1925; James, 1890; La Piere, 1938):

It is a commonplace observation that the socialized individual is not one "person" but many "people"; which of these latter he will be depends upon time and circumstance. In his professional relationships, the doctor may be a calm, austere person, capable of operating upon his patients with the impersonality of a mechanic who works on an automobile. At home, putting his youngest child to bed, he may, however, play the nursery game of tweaking toes with abandon and evident relish and may break into a cold sweat while removing an infected toenail from one of these toes. On a fishing trip with male associates, he may be unwashed and unshaved for days; back in his office again, he may be the spotless, reserved physician.

The fact that the person is in part the function of external circumstances may be technically explained as a consequence of the fact that the human personality is a reactive mechanism and that there is no necessary relationship between the various reactions of a given individual. So viewed, the human personality may be described as consisting of a multitude of facets (reaction patterns) which, although never operating independently of the total personality, may have little in common one with the others. (La Piere, 1938, p. 15)

Famously, these sorts of considerations led William James (1890) to talk of the multiplicity of different "social selves" associated with any individual person:

We may practically say that he has as many social selves as there are distinct *groups* of persons about whose opinion he cares. He generally shows a different side of himself to each of these different groups. Many a youth who is demure enough before his parents and teachers, swears and swaggers like a pirate among his "tough" young friends. We do not show ourselves to our children as to our club companions, to our customers as to the laborers we employ, to our own masters and employers as to our intimate friends. From this there results what practically is a division of the man into several selves; and this may be a discordant splitting, as when one is afraid to let one set of his acquaintances know him as he is elsewhere; or it may be a perfectly harmonious division of labor, as where one tender to his children is stern to the soldiers or prisoners under his command. (p. 294)[17]

## II

While the discussion so far has focused on social beliefs and attitudes, similar points can be made about social forms of emotion and behavior.

[17] Compare Dewey (1927):

An individual as a member of different groups may be divided within himself, and in a true sense have conflicting selves, or be a relatively disintegrated individual. A man may be one thing as a church member and another thing as a member of the business community. The division may be carried in water tight compartments, or it may become such a division as to entail internal conflict. (p. 129)

Social behavior is intentional behavior engaged socially: that is, behavior engaged because and on condition that other members of a social group are represented as behaving in this (or another) way in appropriate circumstances. Individual behavior is intentional behavior engaged individually: that is, behavior engaged for reasons or causes independent of social group membership. For example, an aggressive behavior is a social behavior if and only if an individual behaves aggressively because and on condition that other members of a social group (e.g., a gang) are represented as behaving aggressively in similar circumstances (if, e.g., it is prescribed by "gang law"). An altruistic behavior is an individual behavior if and only if an individual behaves altruistically for reasons and causes independent of whether any member of any social group is represented as behaving altruistically in similar circumstances (e.g., in the hope of personal reward or out of instinctual feelings of sympathy for a victim).

As in the case of social beliefs and attitudes, one type of behavior (or disposition to behave) can be engaged socially (e.g., competitive behavior), and a quite different type of behavior (or disposition to behave) can be engaged individually (e.g., cooperative behavior), and a single behavior (or disposition to behave) can be engaged both socially and individually (e.g., joining a trade union). Both social and individual behaviors are the behaviors of individual persons. Social and individual behaviors are differentiated by reference to how they are engaged (socially as opposed to individually), not by virtue of the fact that one type of behavior is attributable to emergent or supraindividual entities such as social groups and the other attributable to the individual persons that compose them.

J. R. Kantor (1922) stressed this point by characterizing social behaviors as responses to institutional stimuli, that is, as responses to stimuli discriminated according to social group definitions of the situation[18] and associated behavioral prescriptions:

> If we are dealing exclusively with concrete responses to stimuli what else can we observe but the responses of individuals? Notice, however, that when we say that all psychological reactions are the responses of persons we are not blinding ourselves to the distinction between individual and social responses, for there is indeed all the difference in the world between the total behavior situation of an individual reaction and that of a social response.

[18] It was W. I. Thomas (1904) who introduced the quasi-theoretical term "definition of the situation." Thomas argued that social controls (in the form of conventions and mores) are organized mainly through social group definitions of situations.

> Incumbent upon us it is therefore to specify precisely wherein lies the difference between social and individual action. The distinguishing mark we assert lies not in the response factors but in the character of the stimulating situation. What exactly this difference is we may bring out in the following statement, namely that whereas an individual reaction is a response to some natural object or condition, the social or group reaction is a response to an institutional object or situation. Social psychology, therefore, is essentially institutional. (pp. 66–67)

As James (1890) noted, one and the same objective (or "natural") situation calls forth one type of response from members of one social group and a quite different type of response from members of a different social group. What counts as the appropriate social behavior for an individual person in any particular situation depends on the represented social group to which the behavior is oriented: "Thus a layman may abandon a city infected with cholera; but a priest or a doctor would think such an act incompatible with his honor. A soldier's honor requires him to fight or die under circumstances where another man can apologize or run away with no stain on his social self" (p. 295).

Like social beliefs and attitudes, social behaviors are not restricted to any type of purpose or object. So long as they are engaged socially, their purpose may be constructive or destructive, benign or malevolent, generous or miserly, and so forth, and social behaviors may be directed toward social objects, such as other persons and social groups, or toward nonsocial objects, such as animals, rivers, and the sun, moon, and stars.

A behavior is not social just because it is directed toward another person or social group or displayed by a plurality of members of a social group, either at the same time and place or at different times and places. Some interpersonal behaviors – that is, behaviors directed toward another person or persons – are not social behaviors even when they are displayed by a plurality of members of a social group. Acts of aggression and rape are interpersonal behaviors because they are directed toward other persons (the victims), but they are not social behaviors if individuals do not behave in these ways because and on condition that other members of a social group are represented as behaving in these ways in similar circumstances – if, for example, they are products of spontaneous aggression or lust (perhaps grounded in prolonged frustration). Many or most of the trade unionists assembled to elect their local president may rush off to the nearest hardware store to buy candles and salt when they hear word of the impending winter storm, but their action is not social if they do not behave in this way because and on condition that other members of the trade union are represented as doing so (as would seem unlikely).

In contrast, some social behaviors are not directed toward other persons but performed by single individuals in physical isolation from other members of a social group. The practice of solitary genuflection in front of a cross may be performed because and on condition that other members of a religious group are represented as behaving in this fashion in the presence of this religious symbol. A person may accept paper money as a means of exchange from an automatic teller machine in a deserted outlet at dead of night.[19] Solitary golfers may take as much pride in their fairway achievements as those who prefer the proximity of other golfers, and they may adhere to the conventions of the game as closely as the more gregarious types.[20]

Of course, many social behaviors are *also* interpersonal behaviors and are often displayed by a plurality of members of a social group, sometimes at the same time and place and sometimes at different times and places. Thus interpersonal acts of rape or aggression are also social behaviors when they are instances of "gang rape" or "gang warfare": when members of a gang behave in these ways because and on condition that other members of the gang are represented as behaving in these ways. Trade union members also often assemble together outside a workplace to form a picket line because and on condition that other trade unionists are represented as doing so (although some may do so independently of whether others are represented as doing so, in which case their behavior is not socially engaged). Many social behaviors are also displayed by a plurality of members of a social group genuflecting, withdrawing money from banks, attending funerals, and so forth, at the same time and place or at different times and places. However, such social behaviors are not social just because they are displayed by a plurality of members of a social group. Rather, as Durkheim would have said, they tend to be displayed by a plurality of members of a social group because they are social: because these forms of behavior are "imposed" upon members of social groups by virtue of their membership of these groups.

A behavior is not social just because it contributes to a goal that cannot be achieved unless others also behave in the same way (or some other way), as in the case of some cooperative collective enterprise. It may be the case that the citizens of some municipality can only avoid water rationing during a drought if they restrict their own individual use

[19] Accepting paper money as a means of exchange is cited by Weber (1922/1978, p. 22) as a paradigm example of a social action.

[20] The solitary golfer example comes from La Piere (1938, p. 8).

of water from their taps and hoses. However, citizens who restrict their individual use of water because they want to avoid rationing – that is, on rational grounds irrespective of whether others are represented as doing so – are not behaving socially.

The fact that a person recognizes that a behavior serves a collective goal, wants that collective goal to be achieved, and behaves rationally in attempting to achieve it is not sufficient for the behavior to constitute social behavior. Although the goal of the behavior relates to the collective or social good, the behavior may be performed individually. It may be the case (1) that a road cannot get laid unless all the members of a village play their prearranged parts and (2) that the villagers do play their parts. Yet someone who plays his or her part just because he or she thinks the road ought to be built because everyone will benefit – that is, on rational grounds irrespective of whether he or she represents other villagers as playing their parts – is not behaving socially. It is of course true that persons who individually behave in this way might not do so if they believed that others would not behave in the same way. Persons who want to avoid water rationing (or have the road built) might not restrict their own use of water (or play their own part) if they come to believe that others would not. However, if the only reason that they would not engage in the relevant behavior in such circumstances is because it would no longer be rational to do so (in relation to the collective goal), then the behavior – or behavioral restraint[21] – is not socially engaged.

This means that many so-called rational-choice and sociobiological theories of aggregate human behavior and its consequences are not theories of social behavior at all. If a person avoids living in certain neighborhoods because of an individual preference to reside among neighbors at least 40 percent of whom are of his or her own race (Schelling, 1978) irrespective of how the preferences of members of social groups are represented, or if a person avoids incest or helps others because of an innate disposition (Dawkins, 1976; Wilson, 1975) irrespective of how the behavior of other members of social groups is represented, such behavior is not social behavior.

This also has the consequence that many altruistic actions might not be social (if they are rationally motivated or instinctual in nature) and that some selfish actions might be social (if persons sometimes act in their

[21] I follow Weber in treating intentionally refraining from behavior (or action) as a form of behavior (or action). See Weber's (1922/1978) description of forms of behavior as "overt or covert, omission or acquiescence" (p. 4).

selfish interest because and on condition that other members of a social group are represented as doing so). Some social behaviors may involve conflict or competition: instances of conflict or competitive behavior will be social if individuals behave in these ways because and on condition that other members of a group are represented as behaving in these (or different) ways in similar circumstances, as in the case of an escalating conflict between committee members or competition for the leadership of a teenage gang. Conversely, some forms of cooperative behavior may be engaged individually and thus not constitute instances of social behavior, as when I vote in favor of your proposal at the department meeting (as everyone has asked me to) only because I reason that I will personally benefit from it.

Analogous points may be made about social emotions. An emotion is a social emotion if and only if it is experienced socially: that is, if and only if it is experienced because and on condition that other members of a social group are represented as experiencing this (or another) emotion in similar circumstances. An emotion is an individual emotion if and only if it is experienced individually: that is, if and only if it is experienced for reasons or causes independent of social group membership. Some emotions may be experienced socially, other contrary emotions may be experienced individually, and one and the same emotion may be experienced both socially and individually.

Both social and individual emotions are the emotions of individual persons. Social and individual emotions are differentiated by reference to how they are experienced (socially as opposed to individually), not by virtue of the fact that one type of emotion is attributable to emergent or supraindividual entities such as social groups and another attributable to the individual persons that compose them. Emotions are not social just because they are experienced by a plurality of members of a social group or because they are directed toward other persons or groups. Most of the trade unionists at an outdoor rally may become individually afraid of the spectacular lighting storm, and the meeting might break up when they individually flee to the safety of their own homes. A woman about to be raped by the gang members may be genuinely afraid of these other persons, but there is nothing social about her fear. Social emotions are not restricted in their contents and may be directed toward nonsocial as well as social objects. Thus persons can be socially ashamed of and angry at their epilepsy, the failure of their crops, and the size of their houses as well as their colleagues, their children, and the fellow members of their congregation. Their shame or anger is socially experienced if it is oriented to the represented reactions of other members of a social group – if it is experienced because and on condition that other members of the social group are represented as having these reactions in similar situations (if these are appropriate reactions for the members) irrespective of whether any other members are actually present.

The question of whether emotions are experienced socially or individually is independent of the question of whether there are universal as opposed to culturally and historically specific forms of emotional response. Many early American social psychologists maintained that the basic forms of many emotions are more or less universal:

> People are alike in their tonal responses because they have had about the same fundamental experiences of gain or loss. In the history of the human species, certain ways of doing have proved favorable to race development; and others unfavorable. Advantage is accompanied by agreeable tones or feelings, and disadvantage by a disagreeable tonal quality, ranging from a sense of complete loss (sorrow) to one of complete energization (angry determination). Upholders of race prejudice and race pride should observe that all races irrespective of color are characterized under similar circumstances by the same psychic tones or feelings. Social traditions have developed variations, but after all, the white, yellow and black races, alike experience joy, sorrow, and anger when responding to the respective types of stimuli. (Bogardus, 1924b, p. 12)

However, this did not incline them to suppose that all emotions are experienced individually. On the contrary, they recognized that many emotions are socially experienced because the appropriate objects of fear, shame, and pride for different social groups are matters of convention within different social groups. Professional psychologists take pride in their publications and cattle slaughterers in their clean kills; Catholics are ashamed of their lustful feelings and Dionysian revelers ashamed of their lack of them. As Katz and Schanck (1938) put it, social factors

> *determine the specific content of the habits of individuals*. All human beings, for example, learn to fear and avoid certain objects because of over strong stimulation in childhood, but the Baganda children in Africa and the Kwakiutl children in northwest America fear different things. (pp. 523–524)

## III

A number of qualifications and clarifications are perhaps in order at this point, to answer some natural questions and concerns. The various examples of social cognition, emotion, and behavior discussed in this chapter (and others) are employed for illustrative purposes only. Whether any particular psychological state or behavior is socially or individually engaged is a matter that can only be determined by empirical research; the examples have been selected merely for their prima facie plausibility or illustrative value. Nonetheless, there seems little doubt that there are genuinely social forms of belief, attitude, emotion, and behavior (and desire, as any parent with a child or teenager who wants the latest toy craze or expensive fashion accessory knows). Their extent is really the only serious empirical question.

Social forms of cognition, emotion and behavior are *psychological properties of individual persons*. Although social groups do exist and generally ground social psychological states and behavior (causally and developmentally), they are not strictly necessary to the dynamics of social psychological phenomena. This is because it is sufficient for an attitude, for example, be held socially that the individual's attitude is oriented to the represented attitude of members of a social group. Thus an attitude may be held socially even if the individual *misrepresents* the attitude shared by members of the social group: for example, new recruits assigned to combat units may socially adopt a gung-ho attitude toward war because they *mistakenly* represent this attitude as shared by members of combat groups (Stouffer, Lumsdane, et al., 1949).[22] In the limiting case, an attitude may be held socially even if the individual misrepresents a population of individuals *as* a social group: for example, an individual who is disabled may misrepresent the population of disabled persons as a social group (as bound by shared social forms of cognition, emotion, and behavior).[23]

Furthermore, social forms of cognition, emotion, and behavior need not be restricted to psychological states and behavior oriented to social groups of which the individual is an actual member (or believes him- or herself to be an actual member). Thus students who aspire to become members of some elite social group may adopt the distinctive attitudes of members of that social group before they become accepted members, and they may retain these attitudes even if they never attain membership of the elite social group (Newcomb, 1943). Indeed, an individual may orient his or her psychology and behavior to the represented psychology and behavior of members of a social group of which he or she has not the remotest (present) chance of becoming a member: for example,

[22] It is generally recognized that members of social groups vary in the degree to which they accurately represent the social forms of cognition, emotion, and behavior shared by members of a social group (Chowdhry & Chowdhry, 1952).

[23] For further discussion of these types of cases, see Chapter 3.

preadolescents and females may orient their psychology and behavior to the psychology and behavior of members of exclusively male teenage gangs. The term "reference group" (Hyman, 1942) was introduced in social psychology to designate those represented social groups to which individuals orient their psychology and behavior, independently of whether they are actual or potential members. Of course, in practice most social psychological states and behavior are oriented toward represented social groups of which individuals are actual members. A person's reference groups are thus usually but not invariably equivalent to his or her membership groups.

Social forms of cognition, emotion, and behavior need not be held *consciously* or *reflectively* (although some undoubtedly are). All that is required is that the psychology and behavior of an individual be oriented to the represented psychology and behavior of members of a social group. This requires certain discriminatory and representational capacities (possibly including metarepresentational capacities)[24] but does not appear to require consciousness per se. Certainly this seems to have been the view of early American social psychologists, who explicitly acknowledged unconscious forms of social cognition, emotion, and behavior.[25] Analogously, while language may enormously enrich and extend the basic discriminatory and representational capacities required for social forms of cognition, emotion, and behavior, it does not appear to be a prerequisite for social psychological states and behavior.

Social forms of cognition, emotion, and behavior are not social just because they are interpersonally learned or interpersonally transmitted or imitated. Social learning is not equivalent to interpersonal learning.[26] I may come to believe, for example, that the train station is two blocks away, or that homosexuality is unnatural, because of what someone tells me, or I may come to adopt some behavior, for example, a new way of casting a fishing line, or a method of attracting attention, because I

[24] See Tomasello, Kruger, and Ratner (1993), whose work is discussed at the end of this chapter.

[25] See, for example, the discussion in Dunlap (1925, pp. 16–17) and in Burrow (1924) and Finney (1926). Freud (1921/1955) famously developed his own distinctive theory of how attitudes, emotions, and behavioral dispositions could be adopted by an individual unconsciously and socially, as "a member of a race, of a nation, of a caste, of a profession, of an institution, or a component part of a crowd of people who have been organized into a group" (p. 70).

[26] The fact that social learning is not equivalent to interpersonal learning casts doubt upon the avowedly social nature of much of the "social interactionist" tradition in social psychology. See Chapter 4.

observe another displaying that form of behavior. However, the belief or behavior is not a social belief or behavior unless I adopt and maintain that belief or behavior (at least in part) because and on condition that other members of a social group are represented as doing so. Thus it appears likely that persons come to accept many beliefs and behave in many ways individually via instruction by or in imitation of others, including other members of a social group. They accept these beliefs and behave in these ways individually if they accept these beliefs and behave in these ways for reasons or causes independent of whether they represent members of any social group as accepting these beliefs and behaving in these ways: if, for example, they believe homosexuality is unnatural on evolutionary grounds, or that a new way of casting a fishing line is better than the old because it is more efficient.

Thus, although some instances of interpersonal learning and imitation are undoubtedly social, many are not: they are merely interpersonal. Which are which is of course an open and empirical question, although it is far too regularly assumed that interpersonally acquired beliefs and behaviors, particularly beliefs and behaviors acquired from family members, school peers, and the media, are also social beliefs and behaviors. Yet beliefs and behaviors may be interpersonally acquired from family members, school peers, and the media but be individually accepted, for reasons or causes independent of whether any members of any social group are represented as maintaining them. Children may, for example, accept the reasons advanced by their parents why smoking is unhealthy (or be deterred by the hacking cough of their parents) even when their parents continue to smoke. Fishing novices may recognize that the old salt's method of gutting fish is just the most efficient method of gutting fish, and so forth. Conversely, social beliefs, attitudes, desires, emotions, and behavior can be socially adopted via access to mass media or literature without any interpersonal contact or communication with any of the members of the social group represented as having these beliefs, attitudes, desires, emotions, and behavior (Maccoby & Wilson, 1957).

In a recent paper, Tomasello, Kruger, and Ratner (1993) distinguish three types of what they call "cultural learning": imitative learning, instructed learning, and collaborative learning. These forms of learning are held to be uniquely human (with the possible exception of some domesticated or "encultured" primates), since they are held to involve different degrees of "perspective-taking," which requires the development of a "theory of mind" (Astington, Olson, & Harris, 1988; Gopnick, 1993; Whiten, 1991) or the ability to "simulate" the psychological states of others (Gordon, 1986; P. L. Harris, 1991). *Imitative learning* is held to involve simple perspective-taking and thus to require the concept of an intentional agent. *Instructed learning* is held to involve alternating or coordinated perspective-taking (intersubjectivity) and thus to require the additional concept of a mental agent. *Collaborative learning* is held to involve integrated perspective-taking (reflective intersubjectivity) and thus to require the additional concept of a reflective agent.

Although Tomasello, Kruger, and Ratner (1993) characterize these forms of learning as "social-cognitive," they do not constitute genuine forms of social learning: they merely constitute cognitively sophisticated and reflective forms of coordinated interpersonal learning (requiring a developed theory of mind or the ability to simulate the psychological states of others). They are not social forms of learning because they are not oriented to the represented cognition, emotion, and behavior of members of social groups.

Tomasello, Kruger, and Ratner effectively acknowledge this when they admit the following:

> We have left out of account here the institutionalization of many human practices. It is often the institutionalized structure that the developing child encounters and the adult relies on. Unfortunately this is a dimension of the problem that would take us far beyond our current aims. (p. 511, n. 11)

Forms of "institutional structure" would of course be oriented to the psychology and behavior of members of social groups. This suggests that genuine social learning presupposes the metacognitive and reflective capacities of other forms of "cultural learning" but involves the additional capacity to engage the represented perspectives of members of social groups (families, religious and occupational groups, and so forth).[27]

## IV

In sum, early American social psychologists shared a distinctive conception of the social dimensions of human psychology and behavior and of the subject matter of social psychology. Social psychology was conceived as the study of psychological states and behavior oriented to the represented psychology and behavior of members of social groups. This social conception was shared by Bogardus (1918, 1924a, 1924b), Dunlap (1925),

[27] Consequently, if Tomasello, Kruger, and Ratner (1993) are correct in their cognitive-developmental claims, only humans are capable of genuine social learning (with the possible exception of some domesticated and "encultured" animals).

Ellwood (1917, 1924, 1925), Faris (1925), Judd (1925, 1926), Kantor (1922), D. Katz and Schanck (1938), McDougall (1908, 1920), E. A. Ross (1906, 1908), Thomas (1904; Thomas & Znaniecki, 1918), Wallis (1925), and Young (1925, 1930, 1931), for example.[28] For such theorists, social psychological states and behavior constitute both the primary form and the object of explanation in social psychology: Which psychological states and behaviors are best explained socially; and why do individuals socially engage psychological states and behavior?[29] Or as William McDougall (1908) succinctly put it, "Social psychology has to show how, given the native propensities and capacities of the individual human mind, all the complex mental life of societies is shaped by them and in turn reacts upon the course of their development and operation in the individual" (p. 18).

It seems clear enough that early American social psychologists held this distinctive conception of the social dimensions of human psychology and behavior. The remaining question is why it came to be neglected and eventually abandoned. In Chapters 5–9 of this book, I try to provide at least a partial explanation.

Some readers may remain dissatisfied with the characterization of the social presented in this chapter, even if they do not dispute its attribution to early American social psychologists. They might object that this account of the social is already too individualistic: Social psychological states and dispositions turn out to be just the psychological states and dispositions of individual persons, and there is no appeal to emergent structural properties of social groups or collectives. It might also be objected that little has been said about the relation between individuals and social groups – the so-called master problem of social psychology (F. H. Allport, 1962) – or about the much discussed distinction between "psychological" and "sociological" forms of social psychology (Collier, Minton, & Reynolds, 1991; Stephen & Stephen, 1991; Stryker, 1983). In effect, only psychological forms of social psychology have been recognized.

[28] William McDougall is included here as an (honorary) early American social psychologist because of the influence of his work on the early development of American social psychology, despite the fact that he did not emigrate to the United States until 1920 and never became a U.S. citizen.

[29] It is important to recognize that social forms of cognition, emotion, and behavior are also objects of explanation in social psychology. Thus, although many forms of interpersonal behavior and cognitive processes are not themselves social in nature, they will figure prominently in the explanation of how many social forms of cognition, emotion, and behavior are developed.

In the following three chapters I try to address some of these issues via a critical discussion of the metatheoretical positions of Wilhelm Wundt and Émile Durkheim and of the relation between the social and the psychological. I argue that so-called holists such as Wundt and Durkheim, like supposed individualists such as Max Weber and Georg Simmel, treated social psychological states and dispositions as the psychological states and dispositions of individual persons.

I also suggest that to identify the central (or "master") problem of social psychology as the problem of how individuals relate to social groups is already to misconstrue the nature of the social psychological. It is to treat the relation between the social and the psychological as an *extrinsic* (or external) relation rather than an *intrinsic* (or internal) relation of mutual constitution – as if the social and the psychological were literally independent variables. In consequence, I argue that there cannot be any principled distinction (in terms of mode of explanation or subject matter) between psychological and sociological social psychology. There is only one form of social psychology, which can of course be pursued in departments of psychology or sociology (or economics, political science, or anthropology, for that matter).

Many critics of the asocial nature of American social psychology treat Wilhelm Wundt and Émile Durkheim (often represented as the founders of academic psychology and sociology respectively) as early advocates of a genuinely social form of social psychology. Wundt's *Völkerpsychologie*, sometimes characterized as his "second psychology" (Cahan & White, 1992), is regularly cited by such critics for its prescient recognition of the social dimensions of human psychology and behavior. Wundt is also often praised as one of the earliest theorists to have attributed the inadequacy of individualistic approaches to social psychology to the misapplication of experimental methods to the study of social psychological phenomena (Danziger, 1983; Farr, 1996; Forgas, 1981; Graumann, 1986; Leary, 1979; Meuller, 1979).[30] Durkheim is generally acclaimed as one of the champions of the autonomy of the social with respect to the psychology of individuals. He is often also characterized as the intellectual

30 Others cite Wundt as the historical founder of the contemporary discipline known as "cultural psychology," which Cole (1996) characterizes as "the once and future discipline." For Cole, the discipline that once was is a historically reconstructed variant of Wundt's *Völkerpsychologie*. Of course, whether one treats the *Völkerpsychologie* as an early form of "social" or "cultural" psychology partly depends on whether one is inclined to translate the term *Völkerpsychologie* as "social psychology" or "cultural psychology." The question of how to best translate it is discussed in Chapter 2.

forefather of the avowedly more social French tradition of research on "social representations" developed by Moscovici (1961, 1984) and his followers.

Some theorists naturally link the social psychological positions of Wundt and Durkheim:

> There are links of a theoretical and historical nature between the collective representations of Durkheim and the objects of study in the ten volumes of Wundt's *Völkerpsychologie* (1900–20), i.e., language, religion, customs, myth, magic and cognate phenomena. (Farr, 1998, p. xiii)[31]

> Durkheim's distinction between *représentations individuelles* and *représentations collectives* parallels Wundt's individual and collective psychology. (Forgas, 1981, p. 7)

There is a modicum of truth and some justice in this ancestral celebration. However, in the following chapters I argue that although both Wundt and Durkheim recognized the social dimensions of human psychology and behavior, Wundt did not demonstrate the inappropriateness or impossibility of an experimental social psychology, and Durkheim did not advocate any supraindividual structural social psychology independent of the psychology of individual persons.

This is important to stress, because I want to insist throughout this work that an experimental social psychology directed to the exploration of socially engaged psychological states and behavior is both possible and legitimate, and that the commitment to the scientific study of the social dimensions of human psychology and behavior does not entail any commitment to the supraindividuality of social groups or structures.

[31] This historical claim is certainly warranted. Durkheim visited Wundt in Leipzig in the mid-1880s and was particularly excited by his *Völkerpsychologie* project. Other notable visitors were George Herbert Mead and W. I. Thomas, both prominent figures in the history of American social psychology.

# 2

# Wundt and Völkerpsychologie

Wilhelm Wundt is generally recognized as the pioneer and institutional founder of the academic study of scientific psychology in Europe. He effectively created the first laboratory, textbook, journal, and Ph.D. program in experimental psychology. What he introduced in Leipzig in the 1880s was a discipline concerned with the experimental analysis of *immediate* experience, in supposed contrast to the natural sciences, which he held to be concerned with *mediate* experience: "with the *objects* of experience, thought of as independent of the subject" (1897/1902, p. 3). Wundt's experimental work, conducted in the Leipzig laboratory and reported in *Philosophische Studien* (later *Psychologische Studien*), largely consisted of studies of psychophysics, reaction time, perception, and attention. This form of "individual" or "general" experimental psychology Wundt called *physiologische Psychologie* (physiological psychology), not because it was grounded in or directed toward physiological objects but because it appropriated the experimental methods of the newly and successfully developed science of physiology.

As is well known, Wundt also thought that this form of experimental individual psychology should be supplemented by a *Völkerpsychologie*: a "social" or "group" or "folk" or "cultural" psychology (depending on the favored translation) concerned with the complex "mental products" of "social communities," such as language, myth, and custom. The idea of a form of psychology grounded in social community had been suggested by Johann Friedrich Herbart (1816), who characterized it as a form of "political ethology," and was developed by Wilhelm von Humboldt (1836), who explicitly related differences in forms of cognition, emotion, and behavior to different social communities and associated linguistic modes of expression.[1] Humboldt played a major role in the development of the modern German university and personally ensured the creation of professorships in newly developing disciplines such as linguistics and eventually psychology.

The idea of a new discipline specifically called *Völkerpsychologie* and devoted to the comparative and historical study of the mental products of social communities was first articulated by Moritz Lazarus (1824–1903) in an 1851 paper entitled "On the Concept and Possibility of a *Völkerpsychologie*."[2] In 1860, with Hajm Steinthal (1823–1899), Lazarus founded the journal *Zeitschrift für Völkerpsychologie und Sprachwissenschaft*. However, the general notion of a comparative psychology focused on cultural and historical variations in social community had a much longer history, and Wundt's project (despite his differences with Lazarus and Steinthal and his own idiosyncratic interests) represented the continuation of an older tradition that can be traced back to Giambattista Vico (1725/1984) and Johann Gottfried Herder (1784/1969) and that found its partial expression in the "Anthropologie" of Immanuel Kant (1798/1974) and the "ethology" of John Stuart Mill (1843).[3]

This tradition was largely concerned with the historical development of the mentality or "spirit" (*Völksgeist*) of different peoples and how their disparate forms of mentality evolved under varied historical, geographical, climatical, economic, social, and political conditions. This type of inquiry evolved into the comparative and developmental analysis of the different forms of "group mentality" or "group mind" associated with different social communities, ranging from families and religious communities to whole nations.

Wundt's *Völkerpsychologie* followed this tradition in maintaining that distinctively social forms of human psychology and behavior are grounded in social groups or "communities": "All such mental products of a general character presuppose as a condition the existence of a mental *community* composed of many individuals" (Wundt, 1897/1902, p. 23). Moreover,

[1] Humboldt held this view because he believed that social-developmental transformations of individual psychological phenomena are mediated by language, a position later developed by Lev Vygotsky (1934/1986, 1978) and Alexander Luria (1931, 1976).

[2] The term "*Völkerpsychologie*" itself was apparently coined by von Humboldt (Robinson, 1997).

[3] For a useful discussion of Wundt's place in this tradition, see the papers in Bringmann and Tweney (1980).

Wundt based his distinction between "social" psychology and "individual" psychology on the distinction between socially as opposed to individually engaged psychological states and behavior:[4]

> Because of this dependence on the community, in particular the social community, this whole department of psychological investigation is designated as *social psychology*, and distinguished from individual, or as it may be called because of its predominating method, *experimental* psychology. (p. 23)

Most of Wundt's American students who returned to create laboratories and psychology programs in the United States and Canada rejected the theoretical details of Wundt's own "voluntaristic" individual psychology of creative "apperceptive" processes, grounded in the alien philosophical tradition of Leibniz and Kant, and almost completely ignored *Völkerpsychologie*.[5] Nonetheless, as noted in the Introduction, many of the early pioneers of scientific psychology in North America, such as William James, Edward B. Titchener, and James R. Angell, followed Wundt in maintaining the need for a social psychology distinct from individual psychology (even if most did little to practically promote such a discipline), and many early American social psychologists identified the study of socially engaged forms of human psychology and behavior as the distinctive subject matter of social psychology.

[4] The distinction was not original to Wundt and was made earlier by Steinthal (1855):

> For contemporary psychology is *individual* psychology, that is, it takes as its object the mental individual, as it manifests itself in all creatures with mind, not only in humans but, up to a certain point, in animals as well. It is however an essential attribute of the human mind that it does not stand by itself as an individual but that it belongs to society, and first of all, in body and soul, to a folk. Therefore individual psychology calls quite essentially for its extension through folk psychology. A man belongs by birth to a folk, and this determines his mental development in many respects. Therefore the individual cannot even as such be fully understood without taking into account the mental totality within which he arises and lives. (pp. 387–388, in Diamond, 1974, pp. 715–716)

[5] A notable exception was Charles Judd, who published a few works promoting the idea of a "psychology of social institutions" (1925, 1926), inspired by Wundt's *Völkerpsychologie*. These turned out to be no match for works such as Floyd Allport's *Social Psychology* (1924a), which represented social psychology as a form of individual psychology. Judd abandoned the project and devoted the rest of his academic career to educational psychology. Another exception of sorts was Granville Stanley Hall. Hall was dismissive of Wundt's *Völkerpsychologie*, along with much of the rest of the German psychological tradition. Nevertheless, he encouraged Franz Boas to visit Wundt and to offer a course on *Völkerpsychologie* at Clark University. Boas transferred to Columbia University and founded the school of "cultural anthropology," which later included prominent anthropologists such as Margaret Mead and Ruth Benedict.

## I

As noted earlier, a good many critics of the asocial nature of contemporary social psychology credit Wundt with prescient recognition of the inadequacy of individualistic approaches that later came to dominate American social psychology. Wundt is credited with the recognition that the inadequacy of individualistic approaches derives from the misapplication of experimental methods to social psychological phenomena, which Wundt is generally supposed to have held can only be explored via comparative-historical methods.

Wundt is often represented as having maintained this position because he held that complex social psychological states are emergent supraindividual phenomena that cannot be reduced to the intrapsychological states of individuals. Graumann (1986), for example, claims the following:

> According to Wundt, no individual cognition, social or non-social (thought, judgement, evaluation), may methodologically be isolated from its sociocultural basis, as is bound to happen in purely experimental analyses. That is why for Wundt the experimental approach, basic as it is, had to remain restricted to the most basic, that is, simple, phenomena of individual mental life. (p. 98)

In other words, the experimental approach is narrowly restricted because "*cognition presupposes* (supraindividual) *language* and culture into which the individual is 'socialized' and, from childhood on, firmly embedded" (p. 98).

Similar sorts of claims are made by Danziger (1983), Farr (1996), Forgas (1981), Leary (1979), and Meuller (1979). Perhaps the most explicit and extreme statement of this position is to be found in Shook (1995):

> Wundt believed that due to the extreme complexity involved in the higher socio-historical nature of humanity, which includes morality, language, and in general any portion of human life which essentially requires participation in the larger social sphere, *experimentation is simply impossible.* (pp. 351–352, emphasis added)

Those who hold such a view naturally tend to explain the neglect of the social by American social psychologists as a product (at least in part) of their developing commitment to an experimental science based on the model of natural science.[6]

There is little doubt that Wundt recognized the social dimensions of human psychology and behavior, although his own substantive contribution

[6] See Chapter 9 for a critical discussion of this explanation.

to the proto-discipline of social psychology did not live up to its original promise. There is also little doubt that Wundt did express the view that social psychological phenomena cannot be investigated experimentally, although exactly what he meant by this is open to interpretation. For example, one of the clearest statements of this view is the oft-quoted passage from the *Principles of Physiological Psychology* (Wundt, 1904):

> We may add that, fortunately for the science, there are other sources of objective psychological knowledge, which become accessible at the very point which the experimental method fails us. These are certain products of the common mental life, in which we may trace the operation of determinate psychical motives: chief among them are language, myth and custom. In part determined by historical conditions, they are also, in part dependent upon universal psychological laws; and the phenomena that are referable to these laws form the subject-matter of a special psychological discipline, *ethnic* psychology. The results of ethnic psychology constitute, at the same time, our chief source of information regarding the general psychology of the complex mental processes. (pp. 5–6)

Yet Wundt went on to claim that "*ethnic* psychology" has to come to the aid of "individual psychology" only under special circumstances, namely, when the "developmental forms of the complex mental processes are in question" (p. 6). This does not obviously preclude the experimental study of the social dimensions of *developed* forms of "complex mental processes."

In this chapter I argue that it is doubtful if many of the passages that are normally cited as evidence that Wundt held the view that social psychological phenomena cannot be investigated experimentally actually support such a view. I also argue that if Wundt did hold such a view, he does not appear to have held it for the reasons usually avowed, namely, the presumed complexity and supraindividuality of social psychological phenomena, and I argue further that it would have been inconsistent with his own general theoretical position and methodological practice.

One thing seems reasonably clear at least. Wundt did not turn to *Völkerpsychologie* and comparative-historical methods because in later years he came to recognize the inadequacies and limitations of the experimental method as applied to "higher" psychological states grounded in social community. Wundt became interested in *Völkerpsychologie* very early in his career. His first academic position was at the University of Heidelberg as privatdozent in the Department of Physiology, where he offered his first course (in 1857) on experimental physiology (to four students in his mother's apartment). In 1858, he became research assistant to Hermann von Helmholtz, recently appointed as head of the new

Institute of Physiology at the University of Heidelberg. Although the position was something of a disappointment to Wundt, since he spent most of his time teaching medical students the elements of physiology and laboratory techniques, in 1859 he was able to offer a course on "social" or "folk" psychology [entitled "Anthropologie (Natural History of Man)" (Leary, 1979)], explicitly concerned with the relation between the individual and the social. In his first book, *Beiträge zur Theorie der Sinneswahrnehmung* (*Contributions toward the Theory of Perception*), written during his years at Heidelberg and published in 1862, Wundt documented the essential program of both experimental psychology and *Völkerpsychologie*.[7] In the two-volume *Vorlesungen über die Menschen- und Thierseele* (*Lectures on the Human and Animal Mind*), published in 1863, sixteen years before Wundt founded the experimental psychology program at Leipzig, he devoted many pages to the explication and development of the project of *Völkerpsychologie* (Blumenthal, 1979).

There is some dispute about whether the term *Völkerpsychologie* is best translated as "social psychology." Blumenthal (1975) notes that "*Völkerpsychologie* is a unique German term, one that has generally been mistranslated as 'folk psychology.' However, the prefix, *Völker*, carries the meaning of 'ethnic' or 'cultural'" (p. 1081). In Charles H. Judd's edition of Wundt's *Elements of Psychology* (1897/1902), *Völkerpsychologie* is translated as "social psychology." Edward Leroy Schaub used the term "folk psychology"[8] in his translation of Wundt's *Elements of Folk Psychology: Outlines of a Psychological History of the Development of Mankind* (1916), following Wundt's own complaint that "social psychology" (*sozialpsychologie*) was not a suitable translation for *Völkerpsychologie*.[9] However, Wundt's reason for rejecting this translation must be taken in context. He objected to it because in his own day the term *sozialpsychologie* was restrictively associated with contemporary cultural phenomena: "'Sozialpsychologie' (social psychology) at once

[7] In this work he also advanced the project of a scientific metaphysics, conceived as a third branch of psychology, which would unite the empirical findings of psychology and the natural and social sciences, including the cultural and historical sciences (Blumenthal, 1985).

[8] It should be noted that the term "folk psychology" has a quite different meaning in most contemporary psychological and philosophical contexts, where it is generally employed to reference laypersons' theories about psychological states and processes (Greenwood, 1991).

[9] The *Völkerpsychologie* itself has never been translated into English, except for a few selections (e.g., in Blumenthal, 1970, and Wundt, 1900/1973).

reminds us of modern sociology, which, even in its psychological phases, usually deals exclusively with questions of modern cultural life" (1916, p. 4). In contrast, according to Wundt, *Völkerpsychologie* embraces "families, classes, clans and groups" as well as "more comprehensive, social groups" such as "peoples" (p. 4).[10]

Although Wundt's *Völkerpsychologie* represented an early form of social psychology insofar as it acknowledged the social dimensions of human psychology and behavior, the term *Völkerpsychologie* is probably not best translated as "social psychology." For Wundt was not much concerned with the social *dynamics* of human psychology and behavior in his *Völkerpsychologie*. His concerns were primarily historical and, at the end of the day, disappointingly individualistic.

## II

Wundt, like Durkheim, did talk about "mental products" such as language, myth, and custom as emergent supraindividual phenomena. However, he also agreed with Durkheim that they are dependent on (because they can only be instantiated via) the psychological (albeit social psychological) states and behavioral practices of individual persons:[11]

> All such mental products of a general character presuppose as a condition the existence of a mental *community* composed of many individuals, though, of course, their deepest sources are the psychical attributes of the individual. (Wundt, 1897/1902, p. 23)

> It must not be forgotten that... there can be no folk community apart from individuals who enter into reciprocal relations within it. (1916, p. 3)

In consequence, the emergent supraindividual nature of mental products such as language, myth, and custom does not in principle appear to constitute any impediment to the experimental investigation of their (social) psychological causes.

In any case, language, myth, and custom are only the *immediate* objects of *Völkerpsychologie*. For Wundt, the whole point of the comparative-historical study of these mental products was to provide an objective

[10] Compare McDougall (1920) and Dunlap (1925), for example, who maintained that social psychology embraces families and occupational groups as well as peoples and nations.

[11] Although Durkheim (1893/1947) maintained that the "conscience collective" is "an entirely different thing from particular consciences," he nevertheless insisted that "it can be realized only through them" (p. 78).

basis for making inferences about the psychological processes involved in their production. Wundt was not interested in studying these mental products purely for their own sake, as might a linguist, anthropologist, or ethnologist:

> The origin and development of these products depend in every case on general psychical conditions which may be inferred from their objective attributes. Psychological analysis can, consequently, explain the psychical processes operative in their formation and development. (1897/1902, p. 23)

Wundt quite naturally thought that experimental methods are inappropriate for the study of whole (established) languages, myths, and customs, since these cannot be manipulated or controlled or investigated via introspection. Yet although the supraindividual nature of whole languages, myths, and customs precludes their experimental investigation, at least via manipulation and introspection, it does not obviously preclude the experimental investigation of the social psychological processes of individuals that can be inferred from them (via the comparative-historical study of languages, myths, and customs).

Wundt certainly treated the comparative-historical methods of *Völkerpsychologie* as *supplements* to experimental methods:

> Psychological analysis of the most general mental products, such as language, mythological ideas, and laws of custom, is to be regarded as an aid to the understanding of all the more complicated psychical processes. (1897/1902, p. 10)

> Thus, then, in the analysis of the higher mental processes, folk psychology is an indispensable supplement to the psychology of individual consciousness. Indeed, in the case of some questions the latter already finds itself obliged to fall back on the principles of folk psychology. . . . The former [folk psychology], however, is an important supplement to the latter [individual psychology], providing principles for the interpretation of more complicated processes of individual consciousness. (1916, p. 3)

However, it would be misleading to suggest that Wundt advocated the comparative-historical methods of *Völkerpsychologie* for the *negative* reason that social psychological products such as language, myth, and custom (or their social psychological causes) cannot be investigated experimentally and thus have to be studied via comparative-historical methods. Wundt had independent *positive* reasons for advocating the comparative-historical study of languages, myths, and customs and for stressing the supraindividual or "thing-like" nature of such mental products.

Wundt thought that the comparative-historical methods of *Völkerpsychologie* are at least as objective and scientific as the methods of experimental psychology and are as closely analogous to the methods of natural science as are the methods of experimental psychology.[12] Wundt does appear to have believed that mental products such as language, myth, and custom are expressions of "higher" psychological processes that are grounded in social community. However, for Wundt the primary evidential value of mental products such as language, myth, and custom lay not so much in their social nature as in their *objective* or thing-like nature. According to Wundt, these mental products are more closely analogous to "objects of nature" than the psychological processes that are responsible for them (and can be inferred from them). In this respect, they provide a more secure evidential base for psychological theory than the fleeting evidential base of experimental or introspective psychology.

According to Wundt, the psychological objects of scientific psychology, unlike the physical objects of natural science, can only be directly studied via "experimental observation" because they are exclusively *processes*. In contrast, some of the objects of the natural sciences – objects of nature – are objects of "pure observation":

> The case is different with *objects* of nature. They are relatively constant; they do not have to be produced at a particular moment, but are always at the observer's disposal and ready for examination.... When... the only question is the actual nature of these objects, without reference to their origin or modification, mere observation is generally enough. Thus, mineralogy, botany, zoology, anatomy, and geography, are pure sciences of observation so long as they are kept free from the... problems... that... have to do with processes of nature, not with the objects in themselves. (1897/1902, p. 20)

Wundt always maintained that psychological phenomena, whether individually or socially engaged, are processes, not objects: exact observation of them is generally only possible under conditions of experimental production and control. The comparative-historical methods of *Völkerpsychologie* do not allow for the pure observation of psychological phenomena (social psychological or individual psychological) any more than the experimental method does. Nonetheless, Wundt held that the mental products of certain psychological states grounded in social

[12] It may be argued that the comparative-historical methods of *Völkerpsychologie* are more closely analogous to the methods of natural science than experimental psychology. A good case can be made on Wundtian grounds.

community have properties analogous to the properties of objects of nature and thus can be treated *as if* they were "psychical objects" of pure observation:

> *Pure observation*, such as is possible in many departments of natural science, is, from the very character of psychical phenomena, impossible in *individual* psychology. Such a possibility would be conceivable only under the condition that there existed permanent psychical objects, independent of our attention, similar to the relatively permanent objects of nature, which remain unchanged by our observation of them. There are, indeed, certain facts at the disposal of psychology, which, though they are not real objects, still have the character of psychical objects inasmuch as they possess these attributes of relative permanence, and independence of the observer. Connected with these characteristics is the further fact that they are unapproachable by means of experiment in the common acceptance of the term. These facts are the *mental products* that have been developed in the course of history, such as language, mythological ideas, and customs....
>
> So-called social psychology corresponds to the method of pure observation, the objects of observation in this case being the mental products. (1897/1902, pp. 22–23)[13]

Wundt's point was that mental products such as language, myth, and custom, unlike the psychological processes responsible for their production, have properties analogous to objects of nature and thus may be treated as objects of pure observation. And it is these products, not the psychological processes responsible for them, that Wundt claimed are unapproachable "by means of experiment in the common acceptance of the term."

Wundt recognized the social dimensions of such mental products and the psychological states responsible for them. However, his theoretical interest in the social dimensions of language, myth, and custom was primarily related to their avowed role in providing a *stable* evidential base for psychological inference by virtue of their supraindividual or thing-like nature. Wundt rejected the study of language, myth, and custom via their particular psychological and behavioral expressions in particular individuals, not because their particular psychological and behavioral expressions cannot be subject to experimental manipulation and control (which they fairly obviously can), but because they are *too variable* to serve as objects of pure observation:

[13] Wundt seems to have thought that such phenomena are approachable by experiment in an extended sense of the term. See the passage from the *Völkerpsychologie* quoted at the end of this section.

The necessary connection of these products with social communities, which has given social psychology its name, is due to the fact that the mental products of the individual are of too variable a character to be the subjects of objective observation. The phenomena gain the necessary degree of constancy only when they become collective. (1897/1902, p. 23)[14]

Nonetheless, as noted earlier, for Wundt the whole point of studying such mental products was to make causal explanatory inferences to the psychological processes responsible for them:

The origin and development of these products depend in every case on general psychical conditions which may be inferred from their objective attributes. Psychical analysis can, consequently, explain the psychical processes operative in their formation and development. (1897/1902, p. 23)

Wundt seems to have believed that he could make significant inferences about the different social psychological processes that causally ground the production of socially diverse languages, myths, and customs. He suggested, for example, that significant differences in psychological processes grounded in social community could be inferred from documented differences in their linguistic, mythical, and customary products. Thus, for example, Wundt suggested that differences in social psychological motives could be inferred from the different types of word orderings of sentences in different languages:

Two classical languages, Greek and Latin, offer better examples of the dependence of word order on psychological motives, for in these languages the force of tradition on the positioning of words is less. Word position is much freer. Thus it can more easily follow momentary prevailing psychological themes. It is possible, then, to test in an experimental way the psychological significance of various word orderings by virtue of this capacity for free variation. (Wundt, 1901, passage translated in Blumenthal, 1970, p. 28)[15]

[14] One may of course doubt whether social psychological phenomena such as established languages, myths, and customs are quite as "constant" as Wundt supposed, for if they really were, their individual expressions would not be as variable as Wundt supposed. Wundt may have merely meant that individual expressions are too variable to form an objective basis for inference about psychological states via comparative-historical methods, but that leaves entirely open the question of whether one can study social (like individual) psychological phenomena experimentally via their individual expressions.

[15] One should not read too much into Wundt's use of the term "experimental" here, since it appeals to a very broad notion of experiment, according to which differences in causal processes can be legitimately inferred from discriminated differences in effects. Compare, for example, Wundt's description of Copernicus's theoretical conjectures about the basis of planetary motions as "experimental" (1894, p. 10). Wundt never denied that *Völkerpsychologie* is experimental in this broad sense.

## III

It is commonly supposed that Wundt expressed his fundamental objections to the experimental analysis of social psychological phenomena in his 1907 critique of the experimental procedures developed by Oswald Külpe and his colleagues (such as Narzib Ach, Karle Bühler, Karle Marbe, August Mayer, August Messer, Johannes Orth, and Henry J. Watt) at the University of Würzburg. This is because most commentators follow Wundt in equating "higher" thought processes and social psychological processes.[16]

However, Wundt's (1907) objections to the Würzburg experiments devoted to the "new psychology of thought" were objections to the particular methodological practices of the Würzburg experimentalists, notably Ach and Bühler, and did not constitute in principle objections to the experimental analysis of higher thought processes per se (however Wundt may have conceived of them). Wundt was not opposed to the goals of the experimental program pursued by the Würzburg school. With respect to the Würzburg studies of the influence of the "task" (*Aufgabe*) or "set" (*Einstellunung*) on cognitive processing (Ach, 1905), for example, Wundt maintained that they deserved "further application and development in the same direction" (1911a, p. 449, quoted in Woodward, 1982, p. 187).

Wundt laid down four conditions of experimental adequacy: the observer should be in a position to observe the phenomena investigated; the observer should be in a state of anticipatory attention; the experiment should be repeated; and the conditions under which the observed phenomena occur should be determined via variation of the experimental conditions. Wundt claimed that all four conditions were in fact violated by the Würzburg experiments. In consequence, he claimed that the Würzburg experiments were "sham experiments which have the appearance of being systematic only because they take place in a psychological laboratory" (1907, p. 329).

[16] This equation was also common enough among Wundt's contemporaries. See Hobhouse (1913), for example, who maintained that "as soon as we begin to follow the track of the higher developments of mind in man the nature of the enquiry changes. The forces to be considered are now social rather than psychological, or, more accurately, are matters of social rather than individual psychology" (p. 12). In contrast, a good many American social psychologists cited reasonable grounds for equating the social dimensions of human psychology and behavior with "lower" and irrational forms of human psychology and behavior (F. H. Allport, 1924a, 1933, 1934; Janis, 1968), a position developed to great effect by European "crowd" theorists such as Gabriel Tarde (1890/1903) and Gustav Le Bon (1895/1896).

Yet Wundt's complaints about the Würzburg experiments were somewhat disingenuous. He admitted that his four conditions were idealized and only approximately satisfied by natural scientific experiments. He complained that these conditions were not in fact satisfied by the Würzburg experiments but never demonstrated that they *could not* be satisfied by experiments directed to the investigation of higher thought processes (or social psychological processes in general). Bühler, in his response to Wundt (Bühler, 1908), made a reasonable case that all four conditions were routinely satisfied by the Würzburg experiments, at least to the degree that they were satisfied by experimental studies conducted within Wundt's own program, a position later endorsed by independent commentators such as Woodworth (1938) and Humphrey (1951).

Wundt himself was not averse to appealing to "self-observation" in support of his own particular theories of higher cognitive processing. In taking theoretical issue with the Würzburgers' interpretation of some of their experimental results, Wundt offered his own theoretical account of the transformation of mental configurations into sequential linguistic representations, citing evidential support from his own "incidental" self-observations of the process:

> In such self-observations it became entirely clear to me that one does not form the thought at the moment when one starts to express it in a sentence. The thought is present as a whole in one's consciousness before the first word is uttered. At the outset the focus of consciousness does not contain a single one of the verbal and other images which make their appearance in running through the thought and giving it linguistic expression, and it is only in the process of unfolding the thought that its parts rise successively to distinct awareness. (Wundt, 1907, quoted in Woodworth, 1938, pp. 784–785)

In the beginning of Bühler's (1908) paper responding to Wundt's critique, he remarked how surprised he was by it, since the Würzburgers were basically following Wundt's own practice. Or as Woodworth (1938) less charitably put it, Wundt "first demolishes the method of the thought experiment by his critique, and then proceeds to employ the same method to reach the same results (as to 'imageless thought') which had been reached by the Külpe school and which had seemed so objectionable" (p. 785).[17]

[17] Wundt's critique may have been an overreaction to the work of original disciples who appeared to have strayed too far from the rigid path ordained by the master. Or it may have reflected Wundt's growing neglect of his own experimental program while working on the multi-volumed *Völkerpsychologie*. Significantly, at the end of his critique, Wundt offered his recommendation on how the Würzburgers ought to develop (not abandon) the psychology of thought: experimental self-observation ought to be combined with (not replaced by) comparative-historical studies.

Indeed, it may be reasonably argued that Wundt clearly recognized that higher cognitive processes, such as memory and thought, can be experimentally manipulated. These were, after all, precisely the phenomena studied experimentally and introspectively by Wundt's American and German students in the Leipzig laboratory (e.g., by Harry Wolfe in his dissertation, "Studies on the Memory of Tones," and by Edward Scripture in his study of "Thinking and Feeling"). It seems fairly clear that Wundt thought that at least some higher cognitive processes could be studied via the methods of both experimental psychology and *Völkerpsychologie*. Indeed, Wundt claimed that experimental psychology and *Völkerpsychologie* differ essentially in their methods, not their psychological objects:

In the present stage of the science these two branches of psychology are generally taken up in different treatises; still, they are not so much different departments as different *methods*. (1897/1902, p. 23)

Perhaps the clearest example of a higher cognitive process that was treated by Wundt as a legitimate object of both experimental psychology and *Völkerpsychologie* is the process of "apperception," or the active (selective and constructive) process of attention. This "central control" process was a major focus of *Völkerpsychologie*. It was treated by Wundt as the evolutionary advance in mental development that distinguishes mankind from the rest of the animal kingdom and that made possible the development of complex cultural forms such as language, myth, and custom (Blumenthal, 1975). However, this central control process also formed a major focus of Wundt's individual experimental introspective psychology. The first Ph.D. awarded in Wundt's Leipzig laboratory under his supervision was for Max Freidrich's experimental analysis of apperception, entitled "On the Duration of Apperception for Simple and Complex Visual Stimuli" and published in the first volume of Wundt's journal *Philosophische Studien*. As recent commentators have noted, Wundt's studies of this central attentional process are generally acknowledged as precursors of some of the experimental achievements of contemporary cognitive psychology (Blumenthal, 1975; Leahey, 1979).

Now it might be objected that such experimental studies are still only studies of fairly elemental psychological processes and that the complex apperceptive products of such processes, such as social forms of cognition and memory, cannot be studied experimentally.[18] However, this objection is based on the regularly presumed equation of higher or complex

[18] This objection, which gets to the heart of the matter, was raised by Kurt Danziger.

cognitive processes and social psychological processes, which may be questioned. Although it is certainly appropriate to treat mental products such as language, myth, and custom, along with the social forms of cognition, emotion, and behavior responsible for them, as higher or more complex than elementary psychological processes in evolutionary and developmental terms, there is no good reason for treating social forms of cognition, emotion, and behavior as higher or more complex than individual forms of cognition, emotion, and behavior *qua forms of cognition, emotion, and behavior.*

Any belief, attitude, or behavior may be engaged socially or individually (i.e., with or without reference to the represented beliefs, attitudes, or behavior of members of social groups), so there is no good reason to deny that elemental psychological phenomena may be engaged socially.[19] Wundt claimed that "the conditions of mental reciprocity produce new and specific expressions of mental forces" (1908, p. 227, cited in Kusch, 1999, p. 169) but gave no reason to suppose that these new and specific expressions are restricted to higher or more complex mental forces.[20] Certainly Wundt gave no grounds for denying that elemental psychological processes, such as foci of attention or tone memories, could be socially engaged, or for denying that such socially engaged psychological processes could be explored experimentally, as later experiments by Sherif (1935) demonstrated that they could.

In any case, Wundt's critique of the Würzburg experiments was restricted to the *introspective* analysis of psychological phenomena, a technique Wundt only rarely employed in his own psychological experiments. Many of the experiments conducted in his Leipzig laboratory were designed to license inferences about psychological processes on the basis of reaction times, or other verbal or behavioral measures, in response to manipulated stimulus conditions. As Wundt (1894) put it,

> In psychology we find that only those mental phenomena which are directly accessible to physical influences can be made the subject matter of experiment. We cannot experiment upon the mind itself, but only upon its outworks, the organs

[19] Conversely, many complex higher cognitive processes, such as logical reasoning and military planning, may be individually engaged (i.e., without reference to the represented cognitive processes of members of any social group).

[20] Moreover, Wundt maintained (with Baldwin, 1897 and Cooley, 1902) that the notion of an isolated individual is "only an arbitrary abstraction" (1908, p. 293, cited in Danziger, 2001) and that individual minds are themselves "products" of collective minds (1911b, p. 10, cited in Kusch, 1999), which would appear to license socially engaged complements of all forms of human psychology that are individually engaged (simple or complex).

of sense and movement which are functionally related to mental processes. So that every psychological experiment is at the same time physiological, just as there are physical processes corresponding to the mental processes of sensation, idea and will. (pp. 10–11)

Wundt maintained that higher cognitive processes, including social psychological processes, are physically incarnated in the neurophysiological systems of individuals. In consequence, inferences about them can be made on the basis of laboratory experiments or comparative-historical analyses of their mental products.[21] Since, for Wundt, any psychological process that can be manipulated through stimulus presentation can in principle become the object of experimental investigation, there appear to be no grounds (or at least none offered by Wundt) for denying that this can be the case with respect to higher cognitive and social psychological phenomena,[22] since there is no reason to suppose that they cannot be manipulated through stimulus presentation (whether or not they can be isolated or explored introspectively in the fashion of the Würzburg experiments).[23] The fact that certain psychological states and processes presuppose the existence of social communities, or "language and culture" (Graumann, 1986), presents no special problem so long as these states and processes are incarnated in the neurophysiological systems of individuals (which Wundt, like most contemporary social psychologists, never doubted). It presents no more of a problem than the fact that human physiology presupposes a supportive biosphere, which hardly precludes the experimental investigation of physiological processes.

[21] The fact that such higher cognitive (and social psychological) processes are neurophysiologically incarnated is itself no reason to deny their psychological – or social psychological – nature and thus no reason to deny that they can be explored via *psychological* experimentation. As Wundt (1894) put it, "This is, of course, no reason for denying to experiment the character of a psychological method. It is simply due to the general condition of our mental life, one aspect of which is its constant connection to the body" (p. 11).

[22] Wundt may have been right to insist that social psychological influences are not strictly external, since they depend on a "shared psychological community" (Danziger, 2001, p. 88) or at least the representation of such a community. However, this hardly precludes their experimental exploration via stimulus presentation, as illustrated by the classic experimental studies of Asch (Asch, 1951; Asch, Block, & Hertzman, 1938) and Sherif (1935).

[23] This is not to deny that experiments in social psychology may require the special adaptation of the experimental method to the distinctive features of the phenomena investigated (Greenwood, 1989). As Wundt (1894) himself noted, "We must remember that in every department of investigation the experimental method takes on an especial form, according to the nature of the facts investigated" (p. 10).

It is possible that Wundt only thought that the social dimensions of cognition, emotion, and behavior could not be studied experimentally via *experimental introspection*: that experimentation directed to the study of social forms of cognition, emotion, and behavior would lose "the peculiar significance which it possesses as an instrument of introspection," as Wundt argued with respect to studies in child and animal psychology (1904, p. 6). Wundt may have only thought that experimental introspection is not an appropriate means of exploring the social dimensions of human psychology and behavior, because experimental subjects are unlikely to introspectively recognize the social dimensions of their prejudices, for example. If this was all that Wundt thought, it does not represent much of a challenge to experimental social psychology, since most contemporary defenders of experimental social psychology share Wundt's skepticism about the utility of introspective reports of psychological states and processes (Nisbett & Ross, 1980; Nisbett & Wilson, 1977).

## IV

Despite his recognition of the social dimensions of psychological processes and their products, Wundt does not seem to have been much concerned with the *synchronic* social psychological dynamics of cognition, emotion, and behavior in his *Völkerpsychologie*. In consequence, he had very little to say about the possibility of and potential for their experimental investigation. Wundt was much more interested in the question of the *diachronic* historical development of the social psychological processes that ground the development of language, myth, and custom than he was in the synchronic social psychological dynamics of particular languages, myths, or customs in any place and time.

Moreover, it was largely because Wundt was concerned with these diachronic developmental questions that he rejected experimental introspection as an investigative method. Introspection appears to have been rejected not because *Völkerpsychologie* is concerned with the supraindividual social dimensions of psychological phenomena but because it is concerned with the historical evolution of (individual and social) psychological processes, which is of course not accessible to introspection:

> It is true that the attempt has frequently been made to investigate the complex functions of thought on the basis of mere introspection. These attempts, however, have always been unsuccessful. Individual consciousness is wholly incapable of giving us *a history of the development of human thought*, for it is conditioned

by an earlier history concerning which it cannot of itself give us any knowledge. (Wundt, 1916, p. 3, emphasis added)

Despite Wundt's recognition of the social dimensions of human psychology and behavior, *Völkerpsychologie* did not live up to its social psychological promise. In the final analysis, it remains disappointingly individualistic in content.[24] Wundt's adduced laws of mental development (the laws of *mental growth*, the *heterogony of ends*, and the *development towards opposites*) parallel rather too closely for comfort his laws of individual psychology (the laws of *psychical resultants*, *psychical relations*, and *psychical contrasts*). The laws of mental development themselves bear a very doubtful logical relationship to the body of comparative ethnographic evidence (derived from secondary sources) documented in the 10 volumes of the *Völkerpsychologie*. Despite Wundt's suggestive remarks about differences in linguistic forms providing evidential grounds for inferences about differences in social psychological motives, he never really developed this suggestion in practice, and his own pioneering work on linguistics focused almost exclusively on individual psychological processes in sentence production. Thus Wundt is justly recognized as a founder of psycholinguistics (Blumenthal, 1975; Leahey, 1979) but not of socio- or cultural linguistics.

In his attempt to achieve a grand synthesis of individual and social psychology, along with a general synthesis of the natural and mental sciences grounded in the emergent discipline of psychology, Wundt simply fell back upon the fundamental individualistic assumption that all complex and social psychological processes could be (and had to be) ultimately explained in terms of elemental and individual psychological processes. Indeed, this original commitment is to be found in the opening pages of the first volume of the *Völkerpsychologie*: "Since the individual is basic to the social, and the simple is the foundation of the complex in psychic phenomena, therefore experimental psychology has a more fundamental and general character" (1901, p. 22, quoted in Karpf, 1932, pp. 55–56). As Fay Karpf (1932) diagnosed the problem, "Wundt proceeded on the theory that the fundamental processes of the individual mind are psychologically basic to the processes of the 'collective mind' and that individual psychology is therefore basic to the study of the collective mental life of society" (p. 56).

[24] See Danziger (2001, pp. 88–89) and Kusch (1999, pp. 180–188) for similar versions of this complaint.

Thus, although Wundt did acknowledge the social dimensions of human psychology and behavior and the distinctive subject matter of social psychology, it is doubtful if he consistently maintained (far less demonstrated) that experimental methods cannot be employed in the investigation of social forms of cognition, emotion, and behavior. He seems to have had little interest in the *synchronic* dynamics of forms of human psychology and behavior oriented to social groups or social communities (certainly these matters get virtually no attention in the 10 volumes of the *Völkerpsychologie*), and in this respect it is almost anachronistic to attribute to Wundt any principled objection to an experimental social psychology *conceived as the experimental exploration of the causal dynamics of socially engaged forms of cognition, emotion, and behavior.* Wundt's *Völkerpsychologie* deserves to be recognized in the history of social psychology for its early acknowledgement of the social dimensions of human psychology and behavior as the appropriate subject matter of social psychology but not for its specific theoretical or empirical treatment of them or its methodological prescriptions governing their investigation. Certainly there does not appear to be any good argument offered in the Wundtian corpus against the possibility of an experimental social psychology.

This is not to deny that Wundt's apparent rejection of the possibility of an experimental analysis of the social dimensions of human psychology and behavior may have played a role in the historical neglect of the social dimensions of psychological states and behavior by American social psychologists. Some American practitioners may very well have accepted this attributed Wundtian dictum and pressed on regardless with an experimental psychology of interpersonal psychological states and behavior.[25]

## V

The rejection of Wundt's *Völkerpsychologie* and his (limited) vision of a genuinely social psychology are sometimes also equated with the rejection of a genuinely comparative psychology. Robert Farr (1996), for example, claims that the drive by influential behaviorists such as Floyd Allport (1919, 1924a) to establish an individualistic and experimental social psychology led to "the demise of a truly comparative psychology" (p. 101). Farr's own account of the demise of comparative psychology

[25] See Chapter 9.

is rather narrowly focused on behaviorism. However, it does appear likely that the perceived implications of a comparative psychology in the *Völkerpsychologie* tradition may partially explain the reluctance of some American social psychologists to acknowledge the social dimensions of human psychology and behavior, especially those committed to a narrow empiricist vision of social psychological science.

As noted earlier, Wundt's *Völkerpsychologie* represented the development of a tradition that began with Vico and Herder and included the "Anthropologie" of Kant, the "ethology" of Mill, and the "political ethology" of Herbart. A distinctive feature of these comparative and historical forms of psychology was the recognition of the possibility of distinctive psychologies grounded in the different social formations of different peoples in different places and times. In modern parlance, theorists in this tradition recognized the possibility and probability of distinctive "indigenous psychologies"[26] grounded in different social communities or social groups. Indeed, there might literally be as many distinctive and essentially social psychologies as there are different social communities or peoples – a distinctive social psychology of Catholics, another of families, and another of enlisted soldiers, a social psychology of the ancient Greeks as opposed to medieval Europeans, or a social psychology of the contemporary Chinese as opposed to contemporary Americans.

Such culturally and historically localized social psychologies were presupposed by the comparative studies of different racial and cultural groups to be found in early papers in the *Journal of Social Psychology* (subtitled *Racial, Political and Differential Psychology* until 1949) and by the papers on the different psychologies of "the negro" (Herskovits, 1935), the "red man" (Wissler, 1935), "the white man" (Wallis, 1935b), and the "yellow man" (Harvey, 1935) in Carl Murchison's (1935) *Handbook of Social Psychology*.[27] They were also clearly recognized by Edward B. Titchener,

[26] Although the term "indigenous psychology" is as ambiguous as "folk psychology," since it may be employed to reference the distinctive psychologies of different social groups in different cultures or historical periods or to reference the different theories about psychological states held by different social groups in different cultures or historical periods (to reference the indigenous psychology of the Azande or medieval Christians or the indigenous psychological theories of the Azande or medieval Christians). The term is employed in both senses in Heelas and Lock's *Indigenous Psychologies* (1981), for example.

[27] The presumption of culturally and historically localized indigenous psychologies was developed with great success by American cultural anthropologists, who became known as "cultural determinists" because of their insistence on the sociocultural determination of psychological differences. This is scarcely surprising, since the Columbia school of

despite Titchener's (1916) own doubts about the viability of a comparative social psychology:[28]

> The study of the collective mind gives us a psychology of language, a psychology of myth, a psychology of custom, etc.; it also gives us a differential psychology of the Latin mind, of the Anglo-Saxon mind, of the Oriental mind, etc. (Titchener, 1910, p. 28)

However, the very consideration of the possibility of localized psychologies is often held to be anathema to the notion of a genuinely scientific psychology, including a genuinely scientific and experimental social psychology. The idea of socially indigenous psychologies restricted to different social groups in different times and places (and in the same time and place) is often held to violate two intrinsic principles of scientific thought: that the subject matter of any genuine scientific discipline is invariant in space and time and that scientific explanations are universal in scope (they apply to each and every instance of whatever phenomenon is the object of explanation). On this conception, any psychology that implies the cultural or historical or social restriction of explanatory psychological principles is inherently unscientific. Witness, for example, the extremely negative contemporary response (Kimble, 1989; Spence, 1987; Staats, 1983) to even the suggestion of indigenous psychologies localized to specific social (or cultural or historical) communities (Heelas & Lock, 1981; Moghaddam, 1987).

There is nothing particularly contemporary about this response. Assumptions about the invariance of the subject matter and the universality of explanations in genuinely scientific disciplines have informed American psychology, including American social psychology,[29] from its inception. Such assumptions were essential elements of the behaviorist commitment to a genuinely scientific psychology (behaviorists presumed that all behavior could be explained via relatively simple principles of animal

cultural anthropology was founded by Franz Boas, who was undoubtedly influenced by the program of Wundt's *Völkerpsychologie*.

[28] Although this type of investigation did not form any part of Wundt's *Völkerpsychologie* proper, which was restricted to the general historical laws of social development. For Wundt, the comparative study of different races and nationalities was the province of *ethnic* psychology (Schneider, 1990).

[29] Consider, for example, Murphy, Murphy, and Newcomb's (1937) cautious warning to the effect that the explanatory principles derived from the studies surveyed in their *Experimental Social Psychology* could not be presumed to apply to social groups other than those in the United States in the early decades of the twentieth century. This was dismissed in a review by L. W. Doob (1938) as a "deplorable attitude" that was "almost anti-scientific" (p. 115, cited in MacMartin & Winston, 2000, p. 361).

learning that apply universally).[30] Precisely the same assumptions inform contemporary cognitive psychology, including cognitive social psychology, and were presuppositional for early American functional psychologists such as James R. Angell and structural psychologists such as Edward B. Titchener (whatever his thoughts about the possibility of a comparative *Völkerpsychologie*, Titchener was committed to the principles of invariance and universality with respect to his own restricted vision of experimental psychology). So pervasive are such assumptions that they are also to be found in Wundt's *Völkerpsychologie*,[31] the ultimate goal of which was to develop a *universal* historical theory of social development (Jahoda, 1997, p. 150). The comparative study of the differential psychologies of different peoples in different places and times was parasitic upon this ultimate goal (as it was for Vico and Herder, who also hoped to develop a universal history based on such comparative studies).[32]

Thus it is scarcely surprising that the notion of indigenous psychologies grounded in different social communities, from peoples to professional psychologists, has met with some hostility from those committed to the development of scientific social psychology. It seems quite likely that such hostility played a significant role in the rejection of this type of comparative or ethnic psychology (and seriously impeded the development of a genuinely cross-cultural psychology), as Farr maintains. It also seems quite likely that such hostility played some role in the reluctance of many post-1930 American social psychologists (especially those committed to an experimental social psychology) to accept the genuine theoretical possibility that there might be, quite literally, a social psychology of Catholics distinct from the social psychology of Baptists, a social psychology of engineers as opposed to psychologists, a social psychology of conservatives

[30] See, for example, Clark L. Hull, who maintained in the Preface of *Principles of Behavior* (1943) "that all behavior, individual and social, moral and immoral, normal and psychopathic, is generated from the same primary laws; that the differences in the objective behavioral manifestations are due to the differing conditions under which habits are set up and function" (p. v).

[31] These principles also grounded Wundt's vision of experimental psychology.

[32] Although there is a significant difference between this form of universalism and contemporary versions. For Wundt (as for Vico and Herder), qualitative differences in psychological principles and processes associated with different cultures and historical periods reflected different stages of sociocultural development: qualitative differences in stages were thus held to be consistent with the postulation of a universal developmental process. Contemporary universalists, by contrast, do not recognize fundamental qualitative differences in the psychologies of different social groups (only quantitative differences in their levels of capacities and skills).

as opposed to liberals, a social psychology of teenage gangs as opposed to boy scouts, and so forth, as maintained by many early American social psychologists.

While not surprising, it ought to be stressed that such hostility is completely unwarranted. Neither the principle of invariance nor the principle of the universality of explanation are intrinsic features of scientific thought (Greenwood, 1994). Rather, they are principles of Newtonian physical science that worked well for Newtonian physical science in its day but have been largely abandoned by twentieth century physical science. Commitment to these principles is notably absent in contemporary physics, biology, and cosmology, for example. Evolutionary biologists recognize the variability of species in place and time, and contemporary physicists acknowledge that physical motion can be explained in terms of a variety of forces (gravitational, electromagnetic, and strong and weak nuclear forces).

These principles have doubtful validity with respect to the subject matter and explanatory modes of social and psychological science. However, since the time of Newton they have been mistakenly presumed by psychologists (including most American social psychologists) to be intrinsic components of a scientific approach to any subject matter. David Hume's protopsychology was explicitly modeled on formal features of the Newtonian system;[33] Clark Hull prominently displayed a copy of Newton's *Principia* on his desk and required that his graduate students read it; and Roger Shepard (1987, 1995) recently maintained that Newton's mathematical law of universal gravitation is the standard by which the success or failure of psychology as a science ought to be judged.[34]

To deny that such principles are intrinsic components of a scientific attitude is not to blindly endorse the view that psychological states and processes (including social psychological states and processes) really are variable in (social and cultural) space and time or that there are no universal explanations in social psychological science. Whether such principles apply to any scientific domain, or the degree to which they apply, is a

33 Hume (1739) characterized the "uniting principle" of the "association of ideas" as a "gentle force" (pp. 10–11), the term Newton used to characterize gravity.

34 Shepard himself offers a universal law of stimulus generalization, an exponential decay function "that is invariant across perceptual dimensions, modalities, individuals and species" (1987, p. 1318). Compare Gregory Kimble's (1995) characterization of a proffered hypothetical law of behavior as "psychology's version of Newton's first two laws of motion" (p. 36) and his peculiar (albeit rhetorical) suggestion that a "coherent science that is wrong" is preferable to the "scattered truths" of contemporary psychological science (p. 37).

contingent matter to be determined by empirical research and not by a priori stipulation. Unfortunately, a priori stipulation has been the driving force within psychology from its institutional inception.

That said, it is also worth stressing that the recognition of possible cross-cultural and transhistorical variance in forms of social cognition, emotion, and behavior, or the recognition of distinctive psychological states and behavior oriented to different social groups in any particular place and time, does not itself preclude the investigation or identification of possibly universal mechanisms for the generation, maintenance, and transformation of social psychological phenomena. That is, the recognition of such possible cultural, historical, and intrasocietal differences in social forms of cognition, emotion, and behavior does not rule out the possibility that the social dynamics of cognition, emotion, and behavior may turn out to be largely invariant with respect to different social groups (from professional psychologists to nations) or that it might be possible to explore these experimentally.

Although the contents and objects of psychological states and behaviors that are engaged socially may be radically different from one social group to another, and may need to be taken into account for the purposes of practical prediction and control in everyday life and in particular experimental studies, the basic dynamics of socially engaged forms of cognition, emotion, and behavior may be largely invariant (or much less variable than the variability of their contents and objects might suggest). The basic dynamics of social forms of cognition, for example, could in principle be investigated and identified experimentally by reference to *any* socially engaged beliefs or attitudes, irrespective of their particular contents (e.g., beliefs about baptism among Methodists, beliefs about determinism among psychologists, beliefs about abortion among liberals, and so forth).

This critical point was clearly recognized by Georg Simmel (1894), perhaps the most social psychological of the early sociologists, who maintained that a recognition of local differences in the contents and objects of social psychological states and behavior does not preclude an investigation of "the forms of sociation as such, as distinct from the individual interests and contents in and through which sociation is realized" (p. 272).[35] Or as Thomas Chesterton quaintly put it, "When a man has discovered why men in Bond Street wear black hats, he will at the same moment have discovered why men in Timbuctoo [sic] wear red feathers" (p. 143, quoted in Ross, 1908, p. 98).

[35] Compare Sherif (1948, p. 182).

Unfortunately, this critical point was not recognized by many of those who were committed to a scientific and experimental form of social psychology from the 1930s onward. Comparative psychology was a casualty, but so also was a distinctively social form of social psychology. The fact that contemporary defenders of the possibility of a historically and culturally bounded comparative psychology tend to promote an antiscientific and antiexperimental view, while proponents of a scientific psychology vigorously dismiss the possibility of local differences, does not encourage optimism about the objective investigation of this empirical question (via the comparative experimental analysis of the social dynamics of different forms of socially engaged psychological states and behavior in different cultural and historical contexts). And it does not encourage optimism about the possibility of a return to a distinctively social form of social psychology.[36]

[36] It might be objected that contemporary American social psychology has begun to take social psychological states and behavior seriously, as evidenced by the development of "cultural psychology" (Cole, 1996; D'Andrade, 1981; Shweder & LeVine, 1984; Stigler, Shweder, & Herdt, 1990). This is in fact doubtful, since cultural psychology appears to be a different sort of animal from social psychology (although precisely what sort of animal is a matter of some dispute; i.e., there are serious questions about the disciplinary identity of cultural psychology). Moreover, many of the contemporary advocates of cultural psychology appear to take the antiscientific, antiexperimental line (see, e.g., Schweder, 1990). These and related matters (e.g., other objections based on recent trends in social psychology) are discussed in Chapter 10.

# 3

# Durkheim and Social Facts

Émile Durkheim, often considered the founder of the academic discipline of sociology,[1] is famous for his treatment of social groups or collectives as emergent supraindividual entities. Indeed, among social scientists and philosophers, Durkheim is treated as a paradigm "holist," competing with Plato, Marx, and Hegel for this doubtful honor.

Durkheim is famous for having maintained that social groups or collectives are distinct from mere aggregations of individuals and their individual psychologies: "The whole does not equal the sum of its parts; it is something different, whose properties differ from those displayed by the parts from which it is formed" (1895/1982a, p. 128). Durkheim is also famous for having maintained, apparently dogmatically, that social phenomena can only be explained socially and not psychologically: "Every time a psychological explanation is offered for a social phenomenon, we may rest assured the explanation is false" (1895/1982a, p. 129).

Durkheim's holistic account of social phenomena is opposed by many in the social sciences and philosophy who consider themselves "individualists" or "methodological individualists" (Lukes, 1968). Max Weber, another of the founding fathers of sociology, was one of the earliest to formulate the individualist position in sociology, in apparent opposition to Durkheim. Weber (1922/1978) maintained that references to social groups or collectives are nothing more than references to the potential or actual "social actions" of individual persons, social actions being defined

[1] Although Auguste Comte (1830–1842) and Herbert Spencer (1880–1896) may compete for the title of "founding father" of sociology, Durkheim created the academic discipline of sociology (Lukes, 1973a).

as intentional behaviors "oriented" toward other persons:

> When reference is made in a sociological context to a state, a nation, a corporation, a family, or an army corps, or to similar collectivities, what is meant is... *only* a certain kind of actual or possible social actions of individual persons. (p. 14)

While there are grounds for supposing that Durkheim did believe that social phenomena have emergent properties, his metaphysical position on this matter is largely independent of his account of the social dimensions of cognition, emotion, and behavior and of the nature of social groups. In delineating Durkheim's account of the social dimensions of "social facts," one has to look beyond the artificialities of the holism versus individualism debate, which obscure the fundamental agreement between Durkheim and Weber on the social dimensions of psychological states and behavior.

## I

To understand Durkheim's professed position on these matters, one has to recognize that he was deeply concerned to establish sociology as an *autonomous scientific discipline*: in particular, he was concerned to establish it as a discipline independent of the recently established disciplines of biology and psychology. Durkheim maintained that it is necessary to determine "what are the facts termed 'social'" because "the term is used without much precision, being commonly used to designate almost all the phenomena that occur within society" (1895/1982a, p. 50).

As Durkheim pointed out (1895/1982a), a reference to the general distribution of phenomena within society cannot be employed to demarcate the distinctive subject matter of a scientific sociology, since many biological and psychological phenomena (not to mention physical and chemical phenomena) are also distributed within society:

> Yet under this heading [generality] there is, so to speak, no occurrence that cannot be called social. Every individual drinks, sleeps, eats, or employs his reason, and society has every interest in seeing that these functions are regularly exercised. If therefore these facts were social ones, sociology would possess no subject matter peculiarly its own, and its domain would be confused with that of biology and psychology. (p. 50)

However, Durkheim (1895/1982a) claimed that "there is in every society a clearly determined group of phenomena separable, because of their distinct characteristics, from those that form the subject matter of other sciences of nature" (p. 50).

Although he raised the critical question about the "distinct characteristics" of social phenomena or social facts, Durkheim himself failed to answer it directly. In his formal analysis of social facts, he cited two properties that he held to be common to social facts: they are independent of an individual's consciousness and will, and they exert a causal influence on his or her behavior.[2] He thus defined social facts as "manners of acting, thinking and feeling external to the individual, which are invested with a coercive power by virtue of which they exercise control over him" (1895/1982a, p. 52).

However, these two avowed properties of social facts, externality and causal influence, are not distinctive properties that distinguish social phenomena from nonsocial phenomena, such as merely psychological, biological, and physical phenomena. On the contrary, they are properties shared by *all* the objects of psychological, biological, and physical sciences: by other psychological beings, Golgi bodies, hydrochloric acids, ball bearings, and electromagnetic fields.[3]

Durkheim's failure to answer his own question directly is perhaps best explained in terms of his overriding aim to establish sociology as a legitimate scientific discipline. Thus, the properties he cited in characterizing social facts are the properties that social facts have in common with the objects of other scientific disciplines, not the properties that distinguish social facts from physical, chemical, biological, and psychological facts. Like Wundt, Durkheim was concerned to emphasize the "reality" or "thing-like" nature of social facts rather than their social nature per se: "We do not say that social facts are material things, but that they are things just as material things"(1901/1982c, p. 35).

Durkheim's many holistic or "organicist" references to the thing-like nature of social facts and social groups seem to have been primarily designed to establish sociology as a legitimate scientific discipline distinct from psychology and biology – in the way that biology was held to be a

[2] Durkheim also cited "generality" as a common property of social phenomena but denied that it is a "distinct characteristic," since he recognized that most psychological and biological phenomena also have this property (1895/1982a, p. 55).

[3] Durkheim also characterized social facts as statistical facts about social groups (1895/1982a, p. 55), such as differential rates of suicide between different age, professional, religious, and gender groups – the types of facts documented in his own classic work on *Suicide* (1897/1951a). However, social facts do not appear to be social by virtue of their statistical nature, since there are plenty of nonsocial statistical facts about flora and fauna, the weather, and nuclear reactions. This suggests that statistical facts about certain populations are social by virtue of their being statistical facts about *social* groups, but this of course presupposes some independent characterization of social phenomena.

legitimate scientific discipline distinct from chemistry. It seems to have been this concern that motivated Durkheim's appeal to the supposed analogy between the relation between social facts and individuals and the relation between cells and their chemical constituents:

> Yet what is so readily deemed unacceptable for social facts is freely admitted for other domains of nature. Whenever elements of any kind combine, by virtue of this combination they give rise to new phenomena. One is therefore forced to conceive of these new phenomena as residing, not in the elements, but in the entity formed by the union of these elements. The living cell contains nothing save chemical particles, just as society is made up of nothing except individuals. Yet it is very clearly impossible for the characteristic phenomena of life to reside in atoms of hydrogen, oxygen, carbon, and nitrogen....
>
> Let us apply this principle to sociology. If, as is granted to us, this synthesis *sui generis*, which constitutes every society, gives rise to new phenomena, different from those that occur in consciousness in isolation, one is forced to admit that these specific facts reside in the society itself that produces them and not in its parts – namely its members. In this sense therefore they lie outside the consciousness of individuals as such, in the same way as the distinctive features of life lie outside the chemical substances that make up a living organism. (1901/1982c, pp. 39–40)

That is, Durkheim appears to have thought that social phenomena could only be treated as the distinctive subject matter of sociology if they could be treated as supraindividual entities with emergent causal properties irreducible to the properties of their individual components, similar to the way that biological entities such as cells (part of the distinctive subject matter of biology) supposedly exist as emergent supraindividuals with distinctive causal properties that are not reducible to the properties of the chemical elements that compose them. This belief in a radical ontological discontinuity between the subject matters of the physical, chemical, biological, psychological, and social sciences may also have partly motivated his famous conclusion that social phenomena can be explained only socially and not psychologically:

> There is between psychology and sociology the same break in continuity as there is between biology and the physical and chemical sciences. Consequently every time a social phenomenon is directly explained by a psychological phenomenon, we may rest assured the explanation is false. (1895/1982a, p. 129)

However, although Durkheim does seem to have conceived of social phenomena as supraindividual entities, he appears to have deployed this conception mainly for rhetorical purposes in promoting the new scientific discipline of sociology. Behind his holistic rhetoric may be discerned a conception of the social and psychological – and of the social psychological – that owes virtually nothing to his avowed holistic position on the supraindividuality of social facts.

## II

Durkheim did not practice his own methodological preaching about abjuring nonsocial or psychological explanations of social facts. For example, he directly appealed to the loss of the traditional belief that God punishes suicides in his causal explanation of the correlation between secularization and increased rates of suicide (1895/1982a, pp. 131 ff., cited in Flew, 1985, pp. 46–47). More critically, the examples of social facts that Durkheim employed in arguing that social facts have the properties of externality and constraint make it very clear that he held a distinctively social conception of certain forms of cognition, emotion, and behavior and one very close to the conception held by early American social psychologists. The examples of social facts he cited are the commitments and obligations of family, civic law, and custom and the practices of religion, commerce, and the professions: all forms of human psychology and behavior oriented to the represented psychology and behavior of members of social groups.

Indeed, for Durkheim, it was precisely the orientation of socially engaged psychological states and behavior to the represented psychology and behavior of members of social groups that accounted for the properties of externality and constraint characteristic of social facts (even though such properties are not themselves constitutive of their social nature). It was precisely this feature that Durkheim regularly and forcefully stressed when he distinguished his account of social facts from "the ingenious system of Tarde" (1895/1982a, p. 59), which was based solely upon interpersonal imitation. Durkheim (1895/1982a) insisted that

> imitation does not always express, indeed never expresses, what is essential and characteristic in the social fact. Doubtless every social fact is imitated and has, as we have just shown, a tendency to become generalized, but this is because it is social, i.e., obligatory. Its capacity for expansion is not the cause but the consequence of its sociological character. If social facts were unique in bringing about this effect, imitation might serve, if not to explain them, at least to define them. But an individual state that impacts on others none the less remains individual. (p. 59, n. 3)

The central insight of Durkheim's account of the social dimensions of psychological states and behavior is his recognition that social psychological states and behavior are oriented to the represented psychology and behavior of members of social groups.[4] For Durkheim, it was precisely this feature that explains why social psychological states and behavior are as a matter of fact regularly imitated and generally distributed. As Durkheim put it, social psychological states and behavior tend to be general because they are social, but they are not social just because they are general. A social fact is general "because it is collective (that is, more or less obligatory); but it is very far from being collective because it is general. It is a condition of the group repeated in individuals because it imposes itself upon them" (1895/1982a, p. 56).

Thus, one might reasonably modify Durkheim's formal definition of social facts in the following way, to make explicit their orientation to represented social groups:

> Social facts are ways of thinking, feeling, and acting engaged because and on condition that other members of social groups are represented as engaging these (or other) forms of thinking, feeling, and acting.

This definition accommodates all the illustrative examples of social facts provided by Durkheim: family obligations, legal codes, religious practices, financial instruments, and the like. Such phenomena simply would not exist absent socially engaged forms of cognition, emotion, and behavior. For Durkheim, it was precisely the orientation of social psychological states and behavior to the represented psychology and behavior of members of social groups that distinguished genuinely social forms of thinking, feeling, and acting from merely *general* or *common* forms of thinking, feeling, and acting, such as reasoning, fearing, washing, and eating. This modified definition also takes into account the apparent externality and constraint of social facts, in terms of the represented conditionality of socially engaged psychological states and behavior, and anticipates the conception of social psychological phenomena advanced by early American social

[4] The reason for this orientation is itself an open question for social psychology, albeit a largely neglected one. My own view is close to that of Asch (1952), who traced the social group orientation of socially engaged psychological states and behavior to the role they play in the represented social identity of individuals, or what I have elsewhere called their socially grounded "identity projects" (Greenwood, 1994, following Harré, 1983a, 1983b, & Goffman, 1961).

psychologists. On these grounds alone, Durkheim deserves to be treated as an early proponent of a distinctively social form of social psychology.

In actual fact, Durkheim had a much subtler conception of the relation between the social and the psychological, and of the social psychological, than he is usually given credit for. Unfortunately, he is often misrepresented as demarcating the subject matter of sociology by insisting that social or collective representations, the supposed subject matter of sociology, cannot be reduced to or explicated in terms of the psychological states of individuals, the supposed subject matter of psychology, in much the same fashion as Wundt is often misrepresented as maintaining that social forms of mentality cannot be equated with the psychological states of individual persons (Farr, 1996; Forgas, 1981; Graumann, 1986).

Durkheim (1895/1982a) did maintain that

> yet another reason justifies the distinction we have established... between psychology proper – the study of the individual mind – and sociology. Social facts differ not only in quality from psychical facts; *they have a different substratum*, they do not evolve in the same environment or depend upon the same conditions. This does not mean that they are not in some sense psychical, since they all consist of thinking and acting. But the states of collective consciousness are of a different nature from the states of the individual consciousness; they are representations of another kind. The mentality of groups is not that of individuals: it has its own laws. The two sciences therefore are as sharply distinct as two sciences can be, whatever relationship may otherwise exist between them. (p. 40)

Yet despite his talk about the "mentality of groups," and of social or collective representations having a "different substratum" from individual representations, Durkheim did not distinguish between social and individual representations by maintaining that social representations are held by social groups as opposed to individuals or that social representations are ontologically distinct from the representations of individuals. Rather, Durkheim distinguished between those psychological states of individuals that are held socially (that are oriented to the represented psychological states of members of social groups) and those that are held individually (that are held independently of the represented psychological states of members of social groups).

Durkheim in fact always insisted that the study of social psychological states (or representations) is the study of the psychological (or representational) states of individuals: "I have never said that sociology contains nothing that is psychological and I fully accept... that it is a psychology, but distinct from individual psychology" (1895/1982b, p. 249). Even in the passage quoted earlier, he maintained that social facts are psychological ("psychical"), "since they all consist of thinking and acting." When Durkheim claimed that social facts (including facts about social representations) have a different substratum from individual psychological facts, he did not maintain that they are ontologically grounded independently of the psychology of individuals. Rather, he maintained that the social psychological states of individuals (those socially engaged) are linked to *different causal conditions* (both local and developmental) than the individual psychological states of individuals (those individually engaged): "They do not evolve in the same environment or depend upon the same conditions."

What this amounts to is the claim that the dynamical principles governing socially engaged psychological states and behavior are significantly different from the dynamical principles governing individually engaged psychological states and behavior. The former are oriented to the represented psychology and behavior of members of social groups; the latter are not. It was this fact (or purported fact) that grounded the distinction between social (or group or collective) psychology and individual psychology. Social psychological states and behavior were held to have their own distinctive natures and to be governed by laws distinct from the laws governing individual psychological states and behavior. Yet both social and individual psychological states and behavior were treated quite properly as the (socially and individually engaged) psychological states and behavior of *individual persons*.

All Durkheim seems to have meant in claiming, for example, that the "states of the collective consciousness are of a different nature from the states of the individual consciousness" and that "the mentality of groups is not that of individuals; it has its own laws" (1901/1982c, p. 40) is that the principles of social psychology governing those psychological states and behaviors that are a function of "the way in which individuals associating together are formed in groups" (1897b, p. 171) are very likely distinct from, and certainly cannot be presumed to be equivalent to, the principles of individual psychology governing those psychological states and behaviors that are not a function of "the way in which individuals associating together are formed in groups."

This was an eminently reasonable speculation. It does not seem obvious or likely, for example, that our beliefs or attitudes about the efficacy of psychotherapy, the right to bear arms, or the morality of abortion are solely a function of the dynamical principles of our common cognitive architectures qua human beings and have nothing to do with our membership of social groups, such as the populations of professional psychologists, Republicans, or Catholics. All that Durkheim seems to have reasonably contended is that it cannot be presumed (although it is all too often readily presumed) that the dynamics of our psychology and behavior qua cognitive and social beings are equivalent.

This is precisely the conception of social psychological phenomena that was embraced by early American social psychologists (some of whom were undoubtedly influenced by Durkheim), who distinguished between social and individual psychology in terms of postulated (but prima facie plausible) differences between the dynamical principles of socially as opposed to individually engaged forms of cognition, emotion, and behavior. Durkheim, like early American psychologists, stressed that social psychological explanations analytically reduce to explanations referencing socially engaged psychological states and behavioral dispositions of individuals but do not analytically reduce to explanations referencing individually engaged psychological states or behavioral dispositions of individuals. *Social psychological explanations are psychological explanations but not individual psychological explanations.*

Consequently, Durkheim's oft-quoted pronouncement, that when a psychological explanation of a social phenomenon is offered we may be sure that it is false, is less objectionable when it is taken in its proper context and interpreted not as the statement of a dogma but as a reasonable prediction about the explanatory potential of the individual and atomistic "associationist" psychology of his own day. Durkheim (1901/1982c) actually allowed that the laws governing the combination and development of "collective representations" might in the end turn out to be equivalent to the laws of individual psychology (although he obviously doubted that they would):

> But once this difference in nature is acknowledged one may ask whether individual representations and collective representations do not nevertheless resemble each other, since both are equally representations; and whether, as a consequence of these similarities, certain abstract laws might not be common to the two domains. Myths, popular legends, religious conceptions of every kind, moral beliefs, etc., form a different reality from individual reality. Yet it may be that the manner in which the two attack and repel, join together or separate, is independent of their content and relates solely to their general quality of being representations. While they have been formed in a different way they could well behave in their interrelationships as do feelings, images or ideas in the individual. (p. 41)

What Durkheim reasonably maintained was that until an adequate *social psychology* was developed, the question could not be answered:

> Strictly speaking, in our present state of knowledge, the question posed in this way can receive no categorical answer. Indeed, all that we know, moreover, about the

way in which individual ideas combine together is reduced to those few propositions, very general and very vague, which are commonly termed the laws of the association of ideas. As for the laws of the collective formation of ideas, these are even more completely unknown. Social psychology, whose task it should be to determine them is hardly more than a term which covers all kinds of general questions, various and imprecise, without any defined object.... Now, although the problem is one that is worthy of tempting the curiosity of researchers, one can hardly say that it has been tackled. So long as some of these laws remain undiscovered it will clearly be impossible to know with certainty whether they do or do not repeat those of individual psychology. (pp. 41–42)

As Solomon Asch (1952) later noted, Durkheim's antagonism to psychology and psychological explanation was directed toward the atomistic associationist psychology of his own day, not against psychological explanation – far less social psychological explanation – per se:

When he argues for a sociology independent of psychological principles, he is implicitly assuming the validity of late nineteenth century psychology.... His position acquires force precisely because he adopts without qualification an elementaristic psychology, the contents of which are purged of meaning and therefore of social relevance. Because he accepted this psychology and because he was sensitive to the ordered character of group life and appreciated the enormous hold of group conditions on the fate of individuals, Durkheim saw, correctly, a gulf between it and a science of society. Therefore he perceived a danger in erecting a science of society on a foundation so meager and fragmentary. (pp. 254–255)

Unfortunately, Asch also echoed the popular conception of Durkheim as blinded to a proper recognition of the social nature of social psychological phenomena by his commitment to social groups as supraindividual entities, a conception that treats Durkheim as one of the originators (along with Hegel and Plato) of theories postulating an emergent "group mind" or "collective mind" (*l'ame collective*): "We may say that the group mind theory of Durkheim is a logical consequence of the failure of psychology to develop concepts capable of dealing directly with group facts" (p. 255). This does an injustice to Durkheim, who seems to have had a very clear notion of the social dimensions of human psychology and behavior that owed nothing to his views about the supraindividuality of social groups.

## III

Durkheim's conception of social psychological phenomena, as psychological states and behavior oriented to the represented psychological states and behavior of members of social groups, enabled him to distinguish between genuine social groups and "aggregate groups": that is, populations of individuals that merely have some property or properties in common. Social groups are those populations of individuals whose members are bound by shared social forms of cognition, emotion, and behavior. They are populations of individuals whose members represent themselves as a social group and orient their psychology and behavior to the represented psychology and behavior of members of their social group. Examples of such social groups might include the populations of accountants, gays, historians, Gaelic speakers, the Azande, the Mafia, feminists, Protestants, Democrats, the citizens of the city-state of Singapore (possibly), and citizens of the United Kingdom (doubtfully). In contrast, aggregate groups are those populations of individuals that merely have some property or properties in common, such as the populations of persons with a mole on their left shoulder, who were in the park yesterday between 3.00 p.m. and 3.15 p.m., who are female, who are unemployed, who employ images in abstract thinking, who are afraid of spiders, or who walk with a skip in their step.[5]

This seems to be the distinction Durkheim was driving at when he distinguished between a genuine social group, on the one hand, and a "mere sum of individuals" (1895/1982a, p. 129), on the other, or between

[5] It is an interesting and open question whether "society" itself, conceived as the aggregation of overlapping and intersecting social groups and aggregate groups, itself constitutes a social group: that is, a population constituted as a social group by socially shared forms of cognition, emotion, and behavior. While earlier proponents of the "social mind" or "group mind" tended to focus on whole "societies" or "nation-states," early American social psychologists tended to locate the social dimensions of psychological states and behavior at the level of "local" social groups, such as the family, and religious, occupational, and political groups.

Although an open and empirical question, it may be seriously doubted whether many societies themselves form social groups: that is, whether individuals in any particular society orient their psychology and behavior to the represented psychology and behavior of members of that society as a whole. Hyman (1942), in his pioneering study of what he called "reference groups," noted that most social psychological states and behavior are oriented to the psychology and behavior of members of local social groups and not to the psychology and behavior of members of the general population that constitute society as a whole. He contrasted the "rare occurrence of the total population as a reference group and the great frequency of more intimate reference groups. . . . Individuals operate for the most part in small groups within the total society, and the total population may have little relevance for them. Far more important are their friends, people they work with" (p. 24).

Analogously, in a recent survey, Chang, Lee, and Koh (1996) found that Singaporeans tend to orient their psychology and behavior to the psychology and behavior of members of family, school, and religious groups rather than to members of Singapore society per se. This is despite the fact that Asians are commonly held to be more "collectivist" in their orientation.

social groups and what he called "contingent and provisional aggregates" (p. 108). Further, by means of this distinction, Durkheim provided an answer to what may be termed the "horizontal" as opposed to the "vertical" version of the ontological question at dispute between so-called holists and individualists (Lukes, 1968; A. Rosenberg, 1995): is a social group distinct from a mere sum of individuals? The horizontal version is a question about the difference between those populations of individuals that constitute genuine social groups and those populations that do not. The vertical version is a question about supraindividuality, about whether social groups exist in some sense "over and above" the individuals that compose them:

> There are two apparently opposing views about the ontological status of groups. According to one view – *ontological individualism* – a group is nothing but the individuals who belong to the group. But according to *ontological holism*, a group is something over and above its members. (Schmitt, 1994, pp. 258–259)

Durkheim answered the horizontal version of the question by maintaining that members of a social group are bound by shared social forms of cognition, emotion, and behavior, whereas members of an aggregate group, or "a mere sum of individuals," are not. Durkheim is also well known for promoting an affirmative answer to the vertical version of the question, by insisting on the thing-like nature of social groups. Yet as noted earlier, Durkheim's promotion of the supraindividuality of social groups was rhetorically directed toward emphasizing the scientific status of sociology as an academic discipline and seems to have played no essential role in his characterization of the social dimensions of psychological states and behavior.

It is certainly true that Durkheim's answer to the horizontal version of the question concerning the distinction between a social group and a mere sum of individuals did not depend on his affirmative answer to the vertical version. His distinction between social groups and aggregate groups was not made on the grounds that social groups constitute supraindividuals with emergent properties whereas aggregate groups do not. This is because both social groups and aggregate groups may be said to have properties that the individuals who compose them do not. Thus, for example, both the American Psychological Association and the population of persons who are afraid of spiders have properties that none of the individuals that compose them do: both populations can (physically) occupy six major city hotels even though no individual member can. Yet the American Psychological Association is a social group and the population of persons who are afraid of spiders is not, because the members of the former population are bound by socially shared forms of cognition, emotion, and behavior and the members of the latter are not.[6]

Durkheim's distinction was recognized by American social psychologists who acknowledged the social dimensions of psychological states and behavior. Theodore Newcomb (1951) and Solomon Asch (1952) reprised Durkheim's distinction between social groups and aggregate groups in terms of a distinction between social groups and mere "category" groups:

> For social psychological purposes at least, the distinctive thing about a group is that its members share norms about something. The range covered by the shared norms may be great or small, but at the very least they include whatever it is that is distinctive about the common interests of the group members – whether it be politics or poker. They also include, necessarily, norms concerning the roles of group members – roles which are interlockingly defined in reciprocal terms. Thus an American family is composed of members who share norms concerning their everyday living arrangements, and also concerning the manner in which they behave toward one another. The distinctive features of a group – shared norms and interlocking roles – presuppose more than a transitory relationship of interaction and communication. They serve to distinguish, for social psychological purposes at least, a group from a number of persons at a street intersection at a given moment, and also from a mere category, such as all males in the State of Oklahoma between the ages of 21 and 25. (Newcomb, 1951, p. 38)

Analogously, Asch (1952) distinguished social groups from category groups such as "persons who are five years old or the class of divorced persons" (p. 260).

## IV

The moral of this discussion of Durkheim's position is as follows: the recognition of the social dimensions of human psychology and behavior is not conceptually tied to any commitment to the supraindividuality of social groups. Durkheim's account of the social dimensions of psychological states and behavior and of the distinction between social groups and aggregate groups was not grounded in his views about the supraindividuality of social phenomena.

Another way of making this point is by noting that Max Weber's account of the social dimensions of human psychology and behavior appears

[6] Throughout this work I remain neutral on the question of whether social groups do in fact constitute supraindividuals. Nothing depends upon supposing that they do, and the question appears independent of the distinction between social groups and aggregate groups.

to be in essential agreement with Durkheim's, despite Weber's reputation as an individualist. The apparent difference between Weber's position and Durkheim's position is largely a function of Weber's antiholistic rhetoric and a common (if not entirely unjustified) reading of Weber's account of social action.

Weber (1922/1978) is often classified as an individualist because of his insistence that social groups are not supraindividuals, that social groups do not exist over and above the social actions of the individuals that compose them:

> Even in cases of such forms of social organization as a state, church, association, or marriage, the social relationship consists exclusively in the fact that there has existed, exists, or will exist, a probability of action in some definite way appropriate to this meaning. It is vital to be continually clear about this in order to avoid the "reification" of those concepts. (p. 27)

Weber's regularly expressed doubts about the reality of social groups constituted his negative answer to the vertical version of the question about the distinction between a social group and a mere sum of individuals. Weber maintained that social groups do not constitute supraindividuals over and above the individual persons who compose them:

> There is no such thing as a collective personality which "acts." When reference is made in a sociological context to a state, a nation, a corporation, a family, or an army corps, or to similar collectivities, what is meant is . . . *only* a certain kind of actual or possible social actions of individual persons. (1922, p. 14)

Yet Weber did not deny the conception of social forms of cognition, emotion, and behavior that grounded Durkheim's horizontal distinction between a social group and an aggregate group. On the contrary, Weber appears to have shared it. Weber, like Durkheim, was concerned to delineate the distinctive subject matter of sociology. He maintained that "sociology . . . is a science concerning itself with the interpretative understanding of social action and thereby with a causal explanation of its course and effects" (1922/1978, p. 4). Social action was held to be "its central subject matter, that which may be said to be decisive for its status as a science" (p. 24). Weber defined social action as follows: "Action is 'social' insofar as its subjective meaning takes into account the behavior of others and is thereby oriented in its course" (p. 4).[7]

Now this definition may be read as defining social action as intentional behavior directed toward another person or persons: that is, as defining

[7] Action itself was defined by Weber (1922/1978) as any intentional behavior: "We shall speak of action insofar as the acting individual attaches a subjective meaning to it" (p. 4).

social action in terms of its objects, namely, other persons – as opposed to animals or inanimate objects, for example. On this reading, social action is treated as equivalent to interpersonal behavior. However, on this reading, the definition of social action appears to be too broad and too narrow. Certain interpersonal behaviors that do not appear to have anything social about them, such as some acts of aggression or the action of circumventing another person blocking a doorway or path, turn out to be social actions. Certain prima facie social actions, such as convention-bound forms of solitary genuflection, turn out to be nonsocial actions.

Furthermore, on this reading, Weber's reductive claim that social groups are nothing more than networks of "actual or possible actions of individual persons" is grossly implausible. Networks of actual or possible interpersonal actions do not seem sufficient to constitute social groups or collectives. As Margaret Gilbert (1991, pp. 36–41) has argued, a population of individuals who lived alone in different parts of a forest where they all picked mushrooms for sustenance and who took steps to avoid other individuals if they came across them in the forest would not intuitively constitute a social group, even though there is clearly a network of interpersonal actions.

However, this may not be the best reading of Weber's definition, and his other comments on social action suggest a different position that brings him much closer to Durkheim than is commonly supposed. The manner in which social action "takes into account the behavior of others and is thereby oriented in its course" need not be read as requiring that social action be directed toward another person or persons as the intentional object of the action but instead can be read as requiring that the action be engaged in conformity with the represented behavior of other members of a social group. Weber noted that the individuals toward whom a social action is oriented may be "individual persons, and may be known to the actor as such, or may constitute an indefinite plurality and may be entirely unknown as individuals." Discussing the social value of money as a means of exchange, he commented as follows:

> Thus, money is a means of exchange which the actor accepts in payment because he orients his action to the expectation that a large but unknown number of individuals he is personally unacquainted with will be ready to accept it in exchange on some future occasion. (1922/1978, p. 22)

More significantly, Weber, like Durkheim, was concerned to distinguish between socially engaged and merely imitative behavior, in opposition to theorists such as Gabriel Tarde (1890/1903, 1901/1969). Weber (1922/

1978) denied that actions merely based on imitation, such as learning how to gut a fish by observing another, are social actions:

> Mere "imitation" of the action of others . . . will not be considered a case of specifically social action if it is purely reactive. . . . The mere fact that a person is found to employ some apparently useful procedure which he learned from someone else does not, however, constitute, in the present sense, social action. (pp. 23–24)

Actions based on imitation only count as social actions, according to Weber, when "the action of others is imitated because it is fashionable or traditional or exemplary" or, in other words, when it is based on "a justified expectation on the part of members of a group that a customary rule will be adhered to" (p. 24).[8]

Like Durkheim, Weber insisted that although social actions tend to be common or general, they are not social just because they are common or general:

> Social action is not identical either with the similar actions of many persons or with every action influenced by other persons. Thus, if at the beginning of a shower a number of people on the street put up their umbrellas at the same time, this would not ordinarily be a case of action mutually oriented to that of each other, but rather of all reacting in the same way to the like need of protection from the rain. ( p. 23)

Significantly, this alternative interpretation of Weber's definition of social action, as oriented toward the represented actions of other members of a social group (rather than necessarily directed toward some particular person or persons), does seem sufficient to allow us to characterize social groups in terms of networks of actual and possible social actions of individuals. For example, if M. Gilbert's (1991) mushroom pickers in the forest got together and arranged to pool their resources and take turns at the tasks of picking, storing, and preserving mushrooms and distributing them at agreed times, then they would appear to form a genuine social group, unlike a population of mushroom pickers who simply took care to avoid each other in the forest.[9]

[8] Indeed, it may be argued that Weber went even further than Durkheim in his rejection of Tarde's position. Durkheim was inclined to treat the emotion and behavior of persons in a crowd as social facts (1895/1982a, pp. 52–53), whereas Weber denied that the behavior of persons who simply follow other members of a crowd in the direction that they are heading or who merely imitate the behavior of other persons in a crowd count as instances of social action (1922/1978, p. 23).

[9] While this alternative reading makes better sense of many of Weber's claims, it must be acknowledged that there is some support for the "interpersonal" reading. Weber (1922/

Finally, despite his regular insistence that theoretical references to social groups are no more than intellectual "conveniences," or economical ways of referencing a network of actual or potential social actions, Weber did not deny the essential feature of social collectivity recognized by Durkheim, namely, its potent *psychological* reality: "These concepts of collective entities . . . have a meaning in the minds of individual persons, partly as of something actually existing, partly as something with normative authority" (1922/1978, p. 14).

This last point is important to stress. Durkheim's own conception of a social group as distinct from a mere sum of individuals was essentially *psychological* in nature: that is, Durkheim appealed only to shared psychological states and behavior oriented to the represented psychological states and behavior of members of a social group. On this account, a population constitutes a social group if and only if the members of the population represent themselves as a social group and orient their psychology and behavior to the represented psychology and behavior of other members of the social group.

It was Georg Simmel (1908/1959) who made explicit the role of this shared representation of "unity" in the constitution of social groups:

> Societal unification needs no factors outside its own component elements, the individuals . . . *the consciousness of constituting with the others a unity is actually all there is to that unity*. (p. 7, emphasis added)

As Simmel also noted, however, this consciousness need not be explicit among members of the social group itself, and it often amounts to nothing

1978) did claim that action is nonsocial if "it is oriented solely to the behavior of inanimate objects" and explicitly claimed that "religious behavior is not social if it is simply a matter of contemplation or solitary prayer" (p. 22). However, these comments can be accommodated by the alternative reading.

Weber may have meant that actions oriented to the represented behavior of inanimate objects as opposed to those oriented to the represented behavior of members of social groups are not social rather than maintaining that they are nonsocial just because they are intentionally directed toward inanimate objects. I doubt that he would have denied that a community of farmers working together to build a flood wall to protect their crops from a rising river are engaged in social action *just because* their actions are directed toward inanimate objects (walls and rivers).

Weber may have denied that contemplation or solitary prayer are social actions because he did not conceive of these phenomena as convention bound but as purely private and personal modes of religion expression. Certainly Weber did not appear to think that interpersonal behavior is sufficient for social action: "Not every type of contact of human beings has a social character." As an example he cited "a mere collision of two cyclists" (p. 23).

more than the engagement of psychological states and behavior oriented to the represented psychology and behavior of members of a social group:

> This does not mean, of course, that each member of a society is conscious of such an abstract notion of unity. It means that he is absorbed in innumerable, specific relations and in the feeling and knowledge of determining others and being determined by them. (p. 7)

> To be sure, consciousness of the abstract principle that he is forming society is not present in the individual. Nevertheless, every individual knows that the other is tied to him – however much this knowledge of the other as fellow sociate, this grasp of the whole complex as society, is usually realized only on the basis of particular, concrete contents. (p. 8)

This conception of the essential psychological constitution of social groups was echoed by early American social psychologists. Thus Knight Dunlap (1925) emphasized the constitutive role of social consciousness: "consciousness (in the individual, of course) of *others* in *the group*, and consciousness of them, as *related, in the group*, to oneself; in other words, consciousness of *being a member of the group*" (p. 19). Sherif (1948) noted that "one of the products of group formation is a delineation of 'we' and 'they' – the 'we' thus delineated comes to embody a whole host of qualities and values to be upheld, defended and cherished" (p. 396). Or as Asch (1952) put it, "Social facts are . . . facts of the psychology of individuals who have become social and who act and feel as members of groups" (p. 255).

The conclusion developed in this chapter is perhaps surprising but hopefully instructive. Durkheim and Weber appear to have shared the same basic conception of social psychological phenomena and social groups, despite their rhetorical commitments to holism and individualism respectively. Moreover, they appear to have shared the same basic conception of the social dimensions of human psychology and behavior as early American social psychologists. This demonstrates that the conception of social psychological phenomena embraced by Durkheim, Weber, Simmel, and early American social psychologists is not essentially tied to any holist position on the supraindividuality of social groups. This is important to stress, because although the historical association of claims about the social dimensions of psychological states and behavior and holistic claims about supraindividuality did in fact play a significant role in the historical neglect and eventual abandonment of the original conception of the social, this need not have been the case. Theories about the supraindividuality of the group mind could have been abandoned and the original conception of the social dimensions of human psychology and behavior maintained (as indeed was done by a minority of American social psychologists). Consequently, contemporary social psychologists who might be tempted to reembrace the original conception of the social dimensions of human psychology and behavior need not feel bound to endorse theories about the supraindividuality of social groups.

# 4

# The Social and the Psychological

In the last chapter it was suggested that, despite their apparent differences, Durkheim and Weber were in basic agreement on the nature of social psychological phenomena: both grasped the social dimensions of human psychology and behavior as conceived by early American psychologists. If this is correct, it demonstrates the irrelevance of the holist versus individualist debate with respect to the delineation of the social dimensions of cognition, emotion, and behavior. It does not, however, demonstrate the irrelevance of this debate (or the conceptual distortions produced by it) to our understanding of the historical neglect of the social in American social psychology. The significant role played by the historical association of a social conception of human psychology and behavior with supraindividual theories of the "social mind" or "group mind" is documented in Chapter 5.

However, the aim of the discussion thus far has not been to demonstrate that early American social psychologists were especially influenced by Durkheim (or by Weber or Simmel). Although some no doubt were, others were influenced by European theorists such as Gustav Le Bon (1895/1896) and Gabriel Tarde (1890/1903). The significant role played by the work of such crowd theorists in shaping the later asocial tradition of American social psychology is documented in Chapter 7.

The main aim of the discussion so far has simply been to establish that a good many early American social psychologists shared the same conception of social forms of human psychology and behavior as Durkheim and Weber, irrespective of Durkheim's avowed commitment to holism and Weber's to individualism. The main point of the exercise has been to more clearly demarcate the conception of the social that was shared by these classical social theorists and early American social psychologists but was abandoned by later generations of American social psychologists.

## I

Following Durkheim and many early American social psychologists, social groups have been characterized as populations whose members share social forms of cognition, emotion, and behavior. These social groups, and the shared social psychology and behavior of the populations that constitute them, may reasonably be characterized as *intrinsically social* (following Asch, 1952)[1] and may usefully be regarded as the fundamental *social constituents* of the social world: they are the elemental "building blocks"[2] from which the complex fabric of particular societies are constituted and historically constructed. These intrinsically social groups also deserve to be designated as the fundamental subject matter of the social sciences, including sociology and social psychology.

However, this appears to raise a problem concerning the demarcation of sociology and social psychology, especially given Durkheim's claim that sociology is a special form of psychology (1895/1982b, p. 251). One might wonder whether sociology is an autonomous discipline at all, that is, whether sociology has any subject matter distinct from the subject matter of social psychology.

However, the subject matter of sociology (and other social sciences) is not restricted to the intrinsically social. Sociology is also concerned with relations between intrinsically social groups, such as relations of domination, economic inequality, and differential marriage and suicide rates, and with a variety of phenomena related to intrinsically social groups and socially engaged forms of cognition, emotion, and behavior. Some populations, and some forms of human psychology and behavior, are characterized as social *derivatively* but are nonetheless legitimate objects of sociology and social science. Thus some aggregate groups, such as the populations of unemployed persons, divorced persons, disabled persons, persons with AIDs, blacks, and women, are characterized as social derivatively[3] because the properties of being unemployed, divorced,

[1] See Asch (1952), who talked of "intrinsically social attitudes": those attitudes central to the identity of social group members *as* social group members.

[2] While elemental, there is nothing logically atomistic about these social constituents. For a discussion of the difference, see Greenwood, 1989, 1994.

[3] It is of course an open and empirical question whether these populations do in fact constitute mere aggregate groups that can only be characterized as social derivatively. These

disabled, a person with AIDs, black, or a woman are represented as socially significant by members of intrinsically social groups.[4] These derivatively social groups, the relations between them, and the relations between intrinsically and derivatively social groups are proper objects of sociological analysis. Analogously, some forms of human psychology and behavior are characterized as social derivatively, not because they are socially engaged, but because they are directed toward (intrinsically or derivatively) "social objects," such as psychologists or Democrats or the unemployed or the disabled.[5]

Moreover, the explanation of such phenomena is not restricted to social psychological explanation. Although some social "structural" explanations in sociology are undoubtedly social psychological in nature, such as explanations of the behavior of nations bound by treaties or partners bound by marriage, others are clearly not, such as explanations citing differential economic and power relations based on differential access to water, oil, or raw materials. While some explanations in sociology cite social psychological factors, others do not, such as the individual psychological explanation of differential depression rates between men and women in terms of differences in biochemistry. As noted earlier, most rational-choice explanations of social behavior are individual psychological explanations, as are sociobiological explanations (Dawkins, 1976; Wilson, 1975). Some explanations in sociology do not even cite psychological factors but appeal to material or environmental conditions responsible for certain social relations (as in Marx's account of capitalism); local conditions that directly constrain or promote certain forms of

examples are employed for illustrative purposes only, since it appears doubtful (to the author at least) that any of these populations are bound by shared social forms of cognition, emotion, and behavior (i.e., it is doubtful if they constitute intrinsically social groups). It is also worth stressing that this question is independent of the question about the degree of empathy that disabled persons (or blacks or women) may feel for each other, which may of course be very high.

4 Of course, the parties to conventions about the social significance of being a woman or unemployed may, by virtue of their membership of different social groups, be parties to *different* conventions about the social significance of being a woman or unemployed. For instance, the social significance of being a woman or unemployed may be contested by psychologists, biologists, feminists, Marxists, and Catholics. However, this increases rather than diminishes the social significance of being a woman or unemployed.

5 The main thesis of this work may consequently be articulated in these terms. Early American social psychologists were primarily concerned with intrinsically social forms of cognition, emotion, and behavior. Contemporary American social psychologists are almost exclusively concerned with derivatively social forms of cognition, emotion, and behavior: with cognition, emotion, and behavior directed toward "social objects."

thought and behavior (such as employment opportunities for women or blacks); or the "function" of certain social practices and relations, as in so-called structural-functional explanations (Malinowski, 1944; Merton, 1963; Parsons, 1951; Radcliffe-Brown, 1958).[6]

## II

The moral of all this is as follows. The primary subject matter of social psychology is relatively clear and unambiguous: it is socially engaged forms of cognition, emotion, and behavior. These form a legitimate domain of inquiry for both sociologists and psychologists, but sociologists need not feel threatened by "psychological" forms of social psychology (social psychology as practiced within the academic discipline of psychology), since sociologists are concerned not just with the social psychological but with social relations, social structures, and their functional roles. There is thus no danger that sociology will find itself reduced to social psychology. Neither is there any danger that individual psychology will find itself reduced to social psychology (or vice versa), given the real distinction between social and individual psychology.

There is no special problem generated by the investigation of social psychological phenomena by both sociologists and psychologists. According to the present account, the social psychological represents the proper intersection of the social and the psychological and is thus the proper common domain of sociology and psychology (in much the same way that molecular biology represents the proper intersection and proper common domain of chemistry and biology). As Bogardus (1918) put it,

> The new science of social psychology must develop its own methodology and speak from its own vantage ground. Its sector of the field of the social sciences is that important territory where the activities of psychology and sociology overlap. (p. 11)

There is thus no essential difference between sociological and psychological forms of social psychology. A social psychology concerned with

[6] Thus, with respect to the so-called explanatory dispute between holists and individualists as to whether social explanations can be reduced to psychological explanations (Lukes, 1968; A. Rosenberg, 1995), there does not appear to be any straightforward answer. Some social explanations reduce to social psychological explanations but not to individual psychological explanations, some reduce to individual psychological explanations, and some do not reduce to any form of psychological explanation at all (as in the examples given). For a detailed development of these points, see Greenwood (2003).

the social dimensions of human psychology and behavior ought to look pretty much the same whether practiced within departments of sociology or psychology (plus or minus a bit to allow for different areas of emphasis and interest).

Thus claims (or hopes) that American social psychology would be more social if it inclined more to sociological social psychology than psychological social psychology (Backman, 1983; House, 1977; Quinn, Robinson, & Balkwell, 1980; Stephen & Stephen, 1991; Stryker, 1983) are misdirected and redundant. They are misdirected because it is far from obvious that sociological social psychology is any more social than psychological social psychology (and usually unhelpful, because what is held to be distinctively social about sociological social psychology is rarely specified, other than vague references to trans- or supraindividual "structures" or "social interaction"). They are redundant because what would make psychological social psychology more social can be independently specified, as indeed it was specified by early American social psychologists working in both departments of psychology and sociology (and economics, anthropology, and so forth).

Moreover, such claims promote the misguided idea that what is required of a genuine social psychology is the integration of the "individual psychological" approach of psychology with the "social structural" approach of sociology – as if the structure of intrinsically social groups and the social psychology of individuals were ontologically distinct. Yet as Asch (1952) astutely noted, it is precisely this external (or extrinsic) conception of the relation between the social and the psychological that vitiates most "integrative" accounts (vitiates them as much as the usual extremes of holism and individualism). This was not an error made by Durkheim or Weber or by early American social psychologists, who were acutely aware of the internal (or intrinsic) relation between social groups and the socially engaged psychological states and behaviors of individual persons.

Furthermore, such claims only tend to reinforce the historically sedimented association of theories of the social dimensions of human psychology and behavior with discredited theories of supraindividual social minds, an association that played a significant role in the historical abandonment of the original conception of the social dimensions of human psychology and behavior by American social psychologists.[7] In the chapters that follow, I restrict the use of the terms "psychological social

[7] See Chapter 5.

psychology" and "sociological social psychology" to the forms of social psychology actually practiced within departments of psychology and sociology, without prejudice to the question of whether either form of social psychology represents a genuinely social form of social psychology.

## III

It remains a separate question whether psychological or sociological forms of social psychology have maintained the original conception of the social dimensions of cognition, emotion, and behavior embraced by early American social psychologists who worked in both academic settings (and thus whether psychological and sociological forms of social psychology are similar or different as a matter of contingent fact). In this work, I argue that the original conception of the social dimensions of human psychology and behavior was abandoned by psychological social psychologists from the 1930s onward while leaving it a largely open question whether it was maintained by sociological social psychologists. However, some doubts are expressed in the remainder of this chapter.[8]

Without too much simplification, one can roughly characterize the twentieth century development of individual (or general) psychology, social psychology, and sociology in America as follows. American individual (or general) psychology has remained individualistic since its institutionalization at the end of the nineteenth century: that is, it has remained concerned with individually engaged psychological states and behavior (from Titchener and Angell to Miller and Neisser). While originally concerned with socially engaged psychological states and behavior in the early decades of the twentieth century, psychological social psychology began to abandon this conception from the late 1920s and 1930s onward, increasingly focusing (like individual or general psychology) on individually engaged psychological states and behavior.

Sociological social psychology was also concerned with socially engaged psychological states and behavior in the early decades of the twentieth century[9] and was indistinguishable from early forms of psychological social psychology. In the early days, sociology itself amounted to little

[8] See also the final section of Chapter 9.

[9] That is, many early sociologists such as Emory Bogardus, Luther L. Bernard, Charles Ellwood, Ellsworth Faris, W. I. Thomas, and Kimball Young maintained the same conception of social forms of cognition, emotion, and behavior as psychologists such as Knight Dunlap, J. R. Kantor, Daniel Katz, William McDougall, Richard Schanck, and Wilson D. Wallis.

more than sociological social psychology, but as the century developed, it became increasingly focused on social relations and structures within and between social groups and their functional roles, notably in what became known as the "structural-functionalist" tradition (Merton 1963; Parsons 1968). Explanations in terms of social relations and structures (and the functions of such relations and structures) were advanced in large part independently of explicit (or empirically supported) theories of the psychological or social psychological mechanisms implicated. During the early decades of the twentieth century, the position that came to be known as "symbolic interactionism" (Blumer, 1937, 1969, 1984) developed out of the work of John Dewey (1917, 1927), James Mark Baldwin (1895, 1897), Charles Horton Cooley (1902, 1909), and George Herbert Mead (1934).

Within sociology, this position is often characterized as a form of individualistic "micro-sociology" and has remained a minority position. However, it is also often characterized as a distinctively sociological form of social psychology and even as the form of genuinely social psychology to which contemporary psychological social psychology ought to aspire (Farr, 1996; J. M. Jackson, 1988): "Mead's social behaviourism, which only later came to be called symbolic interactionism by Blumer (1969), is one of the very few genuinely social psychological theories" (Forgas, 1981, p. 9).

In the very early days (i.e., in the late nineteenth century), American sociology was hardly more social than American psychology. While social groups were recognized, they were generally conceived as mere aggregate groups: "The group was, however, primarily viewed in an additive, aggregational, individualistic or nominal sense. The group was seen more as a by-product, and not considered to be influential in determining behavior" (Hayes & Petras, 1974, p. 391). According to Hayes and Petras (1974), the development of the position that came to be known as symbolic interactionism introduced the notion of "group determinism" to sociology: "With the rise of symbolic interactionism, the group was seen primarily in terms of associations and interactions between individuals with the emphasis on the role of perceived expectations in behavior" (p. 391).

It is certainly true that some of the early twentieth century psychologists, sociologists, and philosophers who later came to be identified as the originators of the symbolic interactionist tradition in sociology did recognize the social dimensions of cognition, emotion, and behavior. However, as the tradition developed, it increasingly focused on *interpersonal* cognition, emotion, and behavior, or "interaction," to the point that its advocates have come to recommend the study of the interpersonal as the defining mark of a properly social psychology.

Thus J. M. Jackson (1988), for example, responding to the "crisis" in social psychology, recommends an "integrative orientation," conceiving of "human social conduct as occurring within meaningful, bounded social contexts, or social acts," following Mead's (1934) definition of a "social act" as a "cooperative process" that involves "two or more persons who have, for the purpose of the action, a common end" (p. 119). Analogously, Collier, Minton, and Reynolds (1991) claim to have begun their historical study of the development of American social psychology as an attempt to say something about (and for) the "neglected area" of "interpersonal relations," noting that this area was not neglected by earlier social theorists such as Mead.

## IV

To understand how this came about, it is worth looking briefly at the historical background to the early development of social psychology in America in relation to the early development of the disciplines of psychology and sociology. European theorists played a significant role in shaping the development of American social psychology. American social psychologists were influenced by (and reacted to) the social psychological positions of Durkheim, Weber, and Simmel and the theories of crowd suggestion and imitation advanced by Gustav Le Bon and Gabriel Tarde.[10] The influence of the German "folk psychology" tradition[11] was probably weaker and

[10] The historical significance of this latter influence is discussed in detail in Chapter 7.

[11] Wundt's *Völkerpsychologie* had little direct influence on the development of American social psychology, with the exception of Judd (1925, 1926), Dewey (1886/1967, 1896) and Mead (1904, 1906) (see also n. 25). In Chapter 2 it was noted that Wundt had little to say about the social *dynamics* of human psychology and behavior. However, independently of Wundt, theories of the social dynamics of human psychology and behavior were developed in Germany around the turn of the century by scholars such as Albert Schäffle, Ludwig Gumplowitz, and Gustav Ratzenhofer. Schäffle's *Bau und Leben des Socialen Körpers* (1875–1878) distinguished between social psychology, social morphology, and social physiology, and it contained a long section entitled "Outline of Social Psychology: The Psychic Facts of Social Life in Their General Bearing on the General Phenomena of the Folk Mind." It was primarily Schäffle's restricted use of the term "social psychology" to reference higher cultural phenomena that disinclined Wundt to accept "social psychology" as a translation of *Völkerpsychologie* (1916, p. 4). Gumplowitz (1885) developed a "group-conflict" theory and Ratzenhofer (1898) a theory of "social interests." These theories were discussed by American sociologists and social psychologists such as Bogardus (1922), Small (1905), Small and Vincent (1894), and Ward (1883).

more indirect, particularly as filtered though the interpretative lenses of popularizers such as Small and Vincent (1894).

Other significant influences came from Great Britain, notably the individualistic associationist psychology and utilitarian social theory of Thomas Hobbes, David Hume, David Hartley, Adam Smith, Jeremy Bentham, James Mill, John Stuart Mill, and Alexander Bain and the evolutionary theories of Charles Darwin and Herbert Spencer. Fay Karpf (1932) claimed that "American social psychology emerged, as might be reasonably expected, more or less as the natural outgrowth of its English background, and particularly the evolutionary standpoint as it was defined by Darwin and Spencer" (p. 147). The associationist and hedonistic psychology[12] of Hobbes, Hume, Hartley, J. Mill, J. S. Mill, and Bain was atomistic and individualistic, referencing only individually engaged psychological states and behavior, and was employed to furnish the primary explanatory principles of utilitarian moral and political theory and "laissez-faire" economics.

Individual psychology in America was instituted and initially developed by Wundt's American students, but aside from Edward B. Titchener's idiosyncratic British empiricist reinterpretation of Wundt's psychology (Leahey, 1981), Wundt's American students carried back only the experimental skeleton (and institutional academic program) of Wundt's psychology. They reapplied the principles of resemblance, contiguity, and repetition (and the pleasure principle) of associationist psychology to the study of animal and human behavior, as the functionalism of James R. Angell and Harvey Carr displaced the structuralism of Titchener and developed into the behaviorism of John B. Watson, Clark L. Hull, Edward C. Tolman, and B. F. Skinner.

Darwin's theory of evolution had an enormous influence on American psychology and sociology, especially in promoting a generally naturalistic and functionalist approach. In psychology it promoted the development of comparative psychology, functionalism, and behaviorism; in sociology it promoted theories of the evolution of social forms and societies and the functionalist approach to the explanation of the persistence of social structures (in terms of their social function). Yet although Darwin's theory provided an initial naturalistic impetus to the development of American

[12] By "hedonistic psychology" I mean any psychology based on some version of the "pleasure principle," according to which behaviors are performed (or avoided) because of their association with past pleasure [or pain]. Early anticipations of the pleasure principle (and of the "law of effect") are to be found in Hobbes, Locke, Hartley, and Bain.

psychology, sociology, and social psychology, much of American psychology, sociology and social psychology originally developed in reaction to a central explanatory component of Darwinian theory, namely, the explanation of human and animal psychology in terms of inherited instincts, especially *social* instincts (conceived by Darwin and Spencer as naturally selected psychological and behavioral repertoires).

Much of the early debate in sociology concerned the question of whether the evolved "higher" forms of mentality and sociality found in humans are fixed or subject to rational "social engineering" (usually conceived in line with utilitarian principles). Spencer's own laissez-faire version of what came to be known as "social Darwinism" originally dominated the intellectual scene and was maintained by William Graham Sumner (1906), who founded the Department of Sociology at Yale University. However, other social theorists, notably Walter Bagehot (1884), Wilfred Trotter (1916), Graham Wallas (1914, 1921), and Leonard T. Hobhouse (1904, 1913) maintained that evolved human intelligence enables humans to surmount the limitations of their biological heritage and, via the use of social science, to rationally direct and control the process of social evolution.[13]

Hobhouse (1913), for example, talked of "a purpose slowly working itself out under limiting conditions which it brings successively under control" (p. xxvi). According to Hobhouse, qualitative as well as quantitative psychological differences between humans and animals enable humans to surmount their biological inheritance, so that humankind

> ceases to be limited by the conditions of its genesis. It becomes self-determining, is guided, that is to say, by values which belong to its own world, and finally it begins to master the very conditions which first engendered it. (pp. 11–12)

[13] Therein lay the fundamental difference between the theories of Darwin (1859) and Spencer (1855, 1870–1872) and their influence on the development of American social psychology (and American psychology in general). Darwin maintained that natural selection is *sufficient* to explain the evolution of biological traits. Darwin consequently rejected both the notion that biological organisms have an innate tendency to perfect themselves in relation to their environment and the notion that the process of evolution is inherently progressive (notions accepted by Lamarck and Spencer). It was because later American social psychologists (and psychologists in general) adopted Darwin's position rather than Spencer's that they felt no inclination to adopt a hands-off approach to psychological and social evolution. If the process of psychological and social evolution did not itself guarantee progression (far less perfection), they felt morally obliged to rationally deploy the principles of the new sciences of psychology, sociology, and social psychology to improve the human psychological and social condition. Only Spencer's theory of evolution prescribed the laissez-faire position that came to be known as social Darwinism.

This was also the position developed in Lester Ward's *Dynamic Sociology* (1883), often held to set the stage for the development of sociology as in autonomous scientific discipline in America (in much the same way as James's *Principles*, 1890, is often held to have set the stage for the development of psychology as an autonomous scientific discipline).[14] Ward (1883) maintained that a properly scientific sociology could only be developed on the assumption that social phenomena are "capable of intelligent control by society itself in its own interest" (pp. vi), appealing to Galtonian "artificial modification" as "the only practical value that science has for man" (p. vi). The same basic position was promoted by Albion Small (1905), who founded the Department of Sociology at Chicago[15] (which attracted such luminaries as W. I. Thomas and Robert Park), and Franklin Giddings, who founded the Department of Sociology at Columbia.[16]

Edward Ross's *Social Psychology* (1908) and William McDougall's *Introduction to Social Psychology* (1908) are generally recognized as the first textbooks on social psychology, written by a sociologist and psychologist respectively (although Ross was an economist and McDougall a bit of a philosopher). Ross's *Social Psychology* followed the Bagehot and Ward position that social forces, or "social control," could be surmounted and exploited by human reason based on scientific social understanding.[17]

[14] Although James's influence on the development of social psychology was probably as great as Ward's.

[15] Small and Vincent's pioneering work *Introduction to the Study of Society* (1894) was largely responsible for the popularization of German social thinkers such as Schäffle, Ratzenhover, and Simmel. It followed Schäffle's (1875–1878) text by including a section specifically on "social psychology," one of the first American uses of the term in an academic context.

[16] Franz Samelson (1974) has noted how historians have distorted the theoretical position of Auguste Comte (1830–1842, 1851–1854), the French positivist who coined the term "sociology" and who vies with Durkheim and Spencer as the founding father of sociology. Historians of social psychology (notably G. W. Allport, 1954) have tended to focus on Comte's positivist stage of science, based on the description and corelation of observables (Comte was undoubtedly the founder of "positivism"), but have tended to ignore Comte's transcendent "moral science" (the final stage of Comte's developmental "law of stages"), generating an "origin myth" about sociology and social psychology. Still, it would be fair to say that many early sociologists and social psychologists followed Comte rather than Spencer in supposing that social science could transcend the limitations of biologically grounded social evolution.

[17] Like many other works of this period, Ross's text was influenced by, and acknowledged its debt to, the work of Gabriel Tarde. Yet while Ross did appeal to principles of "suggestion" and "imitation," he did not blindly follow Tarde's interpersonal account of imitation. Like many of those who appealed to imitation as an explanatory principle (such as Bagehot, Giddings, and Ellwood), Ross recognized (with Durkheim) that social forms of imitation are oriented to the represented psychology and behavior of members of social groups.

McDougall's *Introduction to Social Psychology* is often rather unfairly represented as advocating an individualistic biological and psychological position. This work was actually intended as "an indispensable preliminary of all social psychology" (1908, pp. 17–18), that is, a preliminary to the study of socially engaged forms of cognition, emotion, and behavior, which McDougall considered to be the distinctive subject matter of social psychology (or collective or group psychology).

Thus, although *Introduction to Social Psychology* was largely restricted to instincts, McDougall followed Ross in maintaining the critical importance of the "social environment" in the determination of human psychology and behavior, rejecting both individualistic associationist psychology and biological determinism. He claimed that the

> very important advance in psychology toward usefulness is due to the increasing recognition that the adult human mind is the product of the moulding influence exerted by the social environment, and of the fact that the strictly individual human mind, with which alone the older introspective and descriptive psychology concerned itself, is an abstraction merely and has no real existence. (p. 16)

McDougall's *Introduction to Social Psychology* dealt with only the first part of what he considered to be the "fundamental problem of social psychology":

> For social psychology has to show how, given the native propensities and capacities of the individual human mind, all the complex mental life of societies is shaped by them and in turn reacts upon the course of their development and operation in the individual. (p. 18)

The second part was dealt with by McDougall in *The Group Mind* (1920).[18]

American social psychology, as much as American psychology and sociology, developed through its rejection of instinctual explanations of human psychology and behavior. In sociology, this form of explanation was challenged by writers such as Bagehot and Hobhouse. In individual

[18] Although *The Group Mind* (1920) focused on the metatheoretical explication and justification of the principles of group or collective psychology as they were held to apply to the nation-state, which McDougall considered to be the "most interesting, most complex and most important kind of group mind." Much of this work involved a comparative survey of the evolution of the nation-state somewhat analogous to Bagehot's *Physics and Politics* (1884) and Wundt's *Elements of Folk Psychology* (1916). McDougall planned to deal in more detail with the influence of less inclusive social groups (such as families, trade unions, and religious groups) in a projected third volume, which he never completed (McDougall, 1930).

or general psychology, it was famously challenged by Kuo (1921) and Watson (1919). In social psychology, it was targeted by Dunlap (1919) and Bernard (1921, 1924), leading to a rather acrimonious dispute within social psychology (Faris, 1921; Kantor, 1923; McDougall, 1921a). McDougall's *Introduction to Social Psychology* was usually identified as the main offender, despite the fact that most critics acknowledged an explanatory role to instincts, albeit under a different designation, such as Dunlap's "drives" (1925), Bogardus's "original human nature" (1924b), and Floyd Allport's "prepotent reflexes" (1924a). In the end, most theorists in psychology and sociology by and large followed McDougall's position, allowing some role for instincts while attributing a greater role to the "social environment."[19]

## V

Early American social psychologists, including those who came to represent the "social interactionist" tradition, emphasized the plasticity of the human biological inheritance and the learned nature of much of human psychology and behavior. In this respect they were no different from their counterparts in individual psychology. What distinguished early American social psychologists from their counterparts in individual psychology was not their focus on learned as opposed to instinctual behavior but their focus upon socially engaged – and socially learned – forms of cognition, emotion, and behavior as opposed to individually engaged forms of cognition, emotion, and behavior, grounded in inherited biological propensities and nonsocial learning.

James, Dewey, Baldwin, Cooley, and Mead did recognize the social dimensions of human psychology and behavior (the orientation of some psychological states and behavior to the represented psychology and behavior of members of social groups). However, their advocacy of social psychological as opposed to individual psychological forms of explanation too often confounded the distinction between the social and the individual psychological with the distinction between the innate and the learned. In consequence, references to the "social environment" by these (and other) theorists are frequently ambiguous, sometimes denoting the represented social groups to which socially engaged psychological states and behavior are oriented and sometimes denoting other persons as the

[19] This was true of a good many avowed behaviorists as well, for example, Thorndike (1913) and Tolman (1922, 1923). See also Woodworth (1918).

focus of interpersonally learned behaviors or "habits," conditioning and conditioned by other persons.[20]

Thus, while James, for example, did recognize the social dimensions of human psychology and behavior (notably in his discussion of "social selves"), his *Principles* (1890) was also instrumental in promoting the idea that learned habits, notably those interpersonally learned habits promoted by education, form the "fly-wheel" of society.[21] As mentioned in Chapter 1, interpersonally learned forms of cognition, emotion, and behavior, such as those based on interpersonal imitation, are not necessarily social forms of cognition, emotion, and behavior. Yet Dewey, Baldwin, and Cooley were too often inclined to equate social learning with interpersonal learning.

Baldwin (1895, 1897), for example (following James and Dewey), promoted the Hegelian idea that human individuality develops through the systems of social relations afforded by family, school, play groups, and religious, occupational, and political groups:

> The consciousness of the self, thus developed, carries with it that of the "alter" selves, the other "socii" who are also determinations of the same social matter. The bond, therefore, that binds the members of the group together is reflected in the self-consciousness of each member. The external social organization in which each has a certain status is reinstated in the thought of the individual. It becomes for each a psychological situation constituted by selves or agents, in which each shares the duties and rights common to the group. Upon the background of commonness of nature and community of interests the specific motives of reflective individuality – self-assertion, rivalry, altruism – are projected; but they are the fruits of self-consciousness, they are not the motives that exclusively determine its form. All through its history, individualism is tempered by the collective conditions of its origin. (1897, p. 133)

However, Baldwin emphasized interpersonal learning and interaction as much as the social dimensions of human psychology and behavior. While he objected (with some justice) to being lumped together with Gabriel Tarde, he did often talk of child development in terms of interpersonal

[20] A similar ambiguity infects common talk about "mental interstimulation" and "interconditioning," which sometimes refers to psychological states and behavior oriented to the represented psychology and behavior of members of social groups (e.g., in Dunlap, 1925 and Ellwood, 1925) and sometimes to interpersonal psychological states and behavior, with other persons serving as the object or stimulus (e.g., Gault, 1921 and F. H. Allport, 1924a).

[21] Although James, like just about every other theorist, maintained a limited explanatory role for instincts.

imitation and interaction, characterizing the child as a veritable "copying machine."

Cooley (1902, 1909) recognized the social engagement of much of human psychology and behavior (its orientation to the represented psychology and behavior of members of "primary" and "secondary" social groups), but he also tended to emphasize the interpersonal and interactive nature of the "social environment," to the point where the social environment came to be virtually equated with the interpersonal. Thus Cooley, like Baldwin, suggested that, aside from biology, the idea of a purely individual psychology was an illusion, grounded in the fiction of an interpersonally "isolated" individual. As Cooley (1902) famously maintained, "A separate individual is an abstraction unknown to experience" (p. 1).

Yet early American social psychologists such as McDougall, Wallis, Dunlap, Thomas, Bernard, and Bogardus did not contrast social psychology with an individual psychology conceived as the study of individuals *in isolation* from other individuals (whatever exactly that would amount to). According to the conception of the social dimensions of human psychology and behavior advocated by McDougall, Wallis, Dunlap, Thomas, Bernard, and Bogardus, for example, social psychological states and behavior can be engaged by individuals in physical isolation from others (e.g., the solitary golfer or genuflector), and individual psychological states and behavior can be engaged by individuals in physical proximity to others, including other members of a social group (the Baptists coming out of church may all individually believe that it is raining given the liquid evidence, or the trade unionists may hear about the hurricane from other trade unionists and individually follow their example in quitting the meeting early to stock up on supplies). For these social psychologists, the social could not be equated with the interpersonal. Nonetheless, the regular putative contrast of the social with the thought, feeling, and behavior of individuals "in isolation" led many to conceive of the contrast between social and individual psychological states and behavior in terms of the contrast between interpersonally learned psychological states and behavior and purely instinctual psychological states and behavior, which an individual could in principle have or develop independently of any contact with other persons.

This was the basis of Dewey's (1917) hugely misleading claim that *all* psychology (that is, all psychology that is not biological) is social psychology: "From the point of view of the psychology of behavior all psychology is either biological or social psychology" (p. 276). For Dewey, the contrast between social and individual psychological states and behavior, and thus between the subject matter of social and individual (biological) psychology, became a matter of the contrast between interpersonally learned versus innate or instinctual forms of cognition, emotion, and behavior:

> A consideration of the dependence in infancy of the organization of the native activities into intelligence upon the presence of others, upon sharing in joint activities and upon language, make it obvious that the sort of mind capable of development through the operation of native endowment in a non-social environment is of the moron order, and is practically, if not theoretically negligible. (p. 272)

Thus for Dewey, aside for the study of some biologically fixed forms of sensation and drive, "all that is left of our mental life, our beliefs, ideas and desires, falls within the scope of social psychology" (p. 267).

To be fair, Dewey, like Baldwin and Cooley and later Mead and Blumer, did often emphasize that some learned forms of cognition, emotion, and behavior are oriented to the represented psychology and behavior of members of distinctive social groups, such as the family, the local community, religious and professional groups, and so forth. However, these theorists also often neglected the social group orientation of social learning, virtually equating social learning with any form of *interpersonal learning* regularly provided by family, school, or religious community. By the time this form of sociological social psychology came to be developed as symbolic interactionism (Blumer, 1937, 1969, 1984), it focused almost exclusively on the interpersonal (albeit the interpersonal construction of meaning).

Mead (1934), for example, did link social psychological states and behavior to social groups:

> The individual experiences himself as such, not directly, but only indirectly, from the particular standpoints of other individual members of the same social group, or from the generalized standpoint of the social group as a whole to which he belongs. (p. 138)

Yet the emphasis for Mead was usually on the interpersonal development of consciousness and the self through interpersonal interaction rather than (as, e.g., in Baldwin and Cooley) the development of individuality within social groups.[22] For Mead, the "social act" was the primary unit, and it

[22] As Merton and Kitt (1952) later complained,

> The terms "another," "the other" and "others" turn up on literally hundreds of occasions in Mead's exposition of the thesis that the development of the social self entails response to attitudes of "another" or of "others." But the varied status of "these others" presumably

was defined interpersonally as any act "in which one individual serves in his action as a stimulus to a response from another individual" (1910, p. 397). Mead's concern with the interpersonal rather than the social dimensions of human psychology and behavior (as conceived by social psychologists such as McDougall, Wallis, Dunlap, Thomas, Bogardus, and Bernard) is clearly demonstrated in his (1909) characterization of social psychology as the "counterpart" to physiological (individual) psychology:

> The important character of the social organization of conduct or behavior through instincts is not that one form in a social group does what the others do, but that the conduct of one form is a stimulus to another to a certain act, and that this act again becomes a stimulus to the first to a certain reaction, and so on in ceaseless interaction. (p. 406)[23]

This definition of a social act is indistinguishable from Floyd Allport's (1924a) definition of what he called "circular" social behavior:[24]

> When individuals respond to one another in a direct, face-to-face manner, a social stimulus, given, for example, by the behavior of individual A, is likely to evoke from individual B a response which serves in turn as a stimulus to A causing him to react further. The direction of the stimuli and of their effects is thus circular. (pp. 148–149)

Allport cited ordinary conversation as an illustrative example of this form of circular social behavior and noted that such a succession of responses was aptly characterized by Mead as a "conversation of attitudes" (p. 149).

Indeed it would be an overstatement, but not much of an overstatement, to claim that this sociological version of social psychology, with its emphasis on interpersonally learned forms of human psychology and behavior,

taken as frames of self-reference is glossed over, except for the repeated statement that they are members of "the" group. (p. 435)

Too often Mead appears to be talk about the interpersonal development of the self in relation to "the group" conceived as society as a whole.

23 Again it is worth stressing that although interpersonal behavior cannot be equated with social behavior, many social forms of human psychology and behavior are causal products of interpersonal behavior, especially the reciprocating interactions of the sort described by Mead, and many interpersonal behaviors and interactions are grounded in social forms of human psychology and behavior (e.g., oral examinations, pickups, arguments, dancing, and gang fights). Thus, many of the phenomena described by Mead and others representing the social interactionist tradition would naturally form part of the subject matter of social psychology (irrespective of whether it is practiced in departments of sociology or psychology).

24 Allport (1924a) defined social behavior as "behavior in which the responses either serve as social stimuli or are evoked by social stimuli" (p. 148).

is much closer to Floyd Allport's individualist brand of social psychology than the social form of social psychology advocated by McDougall, Wallis, Dunlap, Thomas, Bernard, and Bogardus. The social psychologies of Dewey, Mead, and Floyd Allport were fundamentally behaviorist and functional in spirit.[25] These three theorists rejected instinctual accounts of human psychology and behavior of the sort advocated in McDougall's *Introduction to Social Psychology* (1908) in favor of accounts in terms of interpersonally learned habits (despite the fact that they were obliged to grant some role to the instincts and that McDougall also called himself a behaviorist).[26]

While others talked of "imitation" and the "social environment" as merely the imitation of cognition, emotion, and behavior in the interpersonal environment, early American social psychologists such as McDougall, Wallis, Dunlap, Thomas, Bogardus, and Bernard displaced

[25] There are some interesting links between Wundt and the functionalist and behaviorist traditions that originated at the University of Chicago. Mead's (1904, 1906) interactionist position derived from his reflections on Wundt's theory of "gestures" and language development in the early volumes of *Völkerpsychologie*. Mead spent a semester at the University of Leipzig in the winter of 1888–1889 and enrolled in one of Wundt's classes (Collier, Minton, & Reynolds, 1991). Dewey's "reflex arc" paper of 1896, often treated as one of the defining statements of functionalist psychology at the University of Chicago, owed much of its inspiration to Wundt's voluntarism and teleology (Shook, 1995). Mead's own conception of social interaction, as a sequence in which reactions to social stimuli come to be treated as stimuli for further reactions, seems to have been inspired by Dewey's reflex arc paper. Watson's brand of behaviorism, which inspired Floyd Allport, developed out of Chicago functionalism but largely in reaction to its perceived limitations (at least for Watson, who spent the initial years of his career at Chicago, with Angell and Carr as colleagues and sometime mentors).

In later years, Floyd Allport (1955) became critical of his earlier approach as excessively mechanical and endorsed the more fluid and functional position of Dewey's reflex arc paper (Collier, Minton, & Reynolds, 1991), which he had employed as a model for his earlier treatment of circular social behavior (1924a). Allport (1919) acknowledged Ernst Meumann (1907) as one of the anticipators of his "social facilitation" research. Meumann was a student of Wundt who developed the discipline of "experimental pedagogy" and associated himself with the progressive educational reforms of G. Stanley Hall and Dewey. Allport was introduced to this work by his Harvard doctoral supervisor Hugo Münsterberg, another of Wundt's students.

[26] Although McDougall's (1912) "purposive" behaviorism was diametrically opposed to the "objective" behaviorism of John B. Watson (Watson & McDougall, 1928). It is also interesting to note that although Knight Dunlap (1919) was one of the most vigorous and dismissive critics of McDougall's appeal to instincts and seemed to have despised McDougall personally (in announcing McDougall's death at an American Psychological Association convention in 1938, Dunlap claimed that McDougall had done a great service to social psychology by dying! [M. B. Smith, 1989]), he clearly shared McDougall's conception of the social dimensions of human psychology and behavior (Dunlap, 1925).

instinctual explanations of human psychology and behavior in favor of explanations in terms of socially engaged (and learned) forms of cognition, emotion, and behavior *while still maintaining the critical distinction between socially and individually engaged psychological states and behavior*. Therefore, while Dunlap, for example, was critical of many instinctual explanations of the sort advanced by McDougall (although he did not abandon such explanations altogether), he clearly distinguished between social and individual forms of human psychology and behavior in *Social Psychology* (1925), on virtually the same grounds as McDougall (see Chapter 5). Analogously, within sociology, Bogardus (1924b) and Bernard (1926a) displaced explanations in term of instincts (without abandoning them altogether) in favor of explanations in terms of socially as opposed to individually engaged forms of cognition, emotion, and behavior. Thus Bogardus (1924b), for example, claimed that

> Social psychology is more than an application of the psychology of the individual to collective behavior. It is more than an imitation theory, an instinct theory, a herd theory, or a conflict theory of social life. It is developing its own approach, concepts, and laws. It treats of the processes of intersocial stimulation and their products in the form of social attitudes and values. (p. xi)
>
> Group life is the medium in which all intersocial stimulation occurs. Human nature, personal attributes, and social values emerge only out of group life. Groups provide all social contacts and stimuli. Once formed the group is prior to the individual. Into groups all individuals are born; up through them personality emerges; and in turn persons dominate and create groups. . . . Group environment is the matrix of all intersocial stimulation. (p. 24)

Luther Bernard (1926a) identified himself as a behaviorist and maintained that "the influence of environment is cumulative in our lives and the decline of the influence of instinct is progressive" (1924, p. 524). Yet while Bernard (1926a) focused upon learned habits as opposed to instincts, he was careful to distinguish between socially and individually learned habits and between the merely physical and interpersonal environment and the "psycho-social" environment of social groups:

> Social psychology studies the behavior of individuals in a psycho-social situation. This behavior is valid subject matter for social psychology whether it conditions or is conditioned by other social behavior or responses. It is also concerned with all collective responses, that is, responses of individuals which mutually and reciprocally condition each other and those which are uniform throughout the group, regardless of what environment they arise from. Of course the chief source of stimuli of which social psychology takes cognizance is the psycho-social environment, and the chief type of behavior in which it is interested is collective behavior. (p. 18)

Thus for Bernard the goal of social psychology "is to find out how men behave in groups, or, in other words, to study the reactions of individuals to the psycho-social environment and the consequent building up of collective adjustment patterns in the individuals in response to social stimuli" (p. 589).

Of course, not all psychological social psychologists maintained the distinction between socially and individually engaged forms of human psychology and behavior, while replacing instinct explanations with explanations in terms of environmental learning or conditioning. Floyd Allport and his followers, including those who developed his experimental program of social psychology (Dashiell, 1930, 1935; Gurnee, 1936; Murphy & Murphy, 1931; Murphy et al., 1937),[27] made no distinction between socially and individually engaged forms of human psychology and behavior and equated the "social environment" with the interpersonal environment.

Of special interest historically is the position of Charles Ellwood, who originally held Floyd Allport's *later* position that there are no distinctively social forms of explanation. In his *Introduction to Social Psychology* (1917), Ellwood maintained that all social explanation is just the application of psychological or biological (or geographical) principles of explanation:

> Whatever may be thought of the doctrine of the unity of nature, it is evident that "the social" is no distinct realm in itself, but is evidently a certain combination of biological and psychological factors. Every social situation is made up of, and may be analyzed into, geographical, biological, and psychological elements. Ascertained truths in biology and psychology may be used directly, therefore, to explain certain social phenomena. From this it follows that the chief method of social psychology, or psychological sociology, must be to take ascertained laws and principles of the mental life and apply them to the explanation of phases of the social life in which these laws and principles are manifestly at work. Deduction

[27] It may seem a little unfair to characterize the Murphys as "followers" of Floyd Allport and the asocial experimental paradigm he established, given their frequently expressed concerns about the artificiality of experiments in social psychology, the limited generalizability of experimental results in social psychology, and the general applicability of natural scientific models of social psychological science. As Katherine Pandora (1997) has recently argued, Louis and Gardner Murphy (along with Gordon Allport and others) were "rebels within the ranks" who dissented from precisely the overly narrow conception of science appealed to by mainstream social psychologists such as Floyd Allport. Nonetheless, despite their reservations, Murphy and Murphy's *Experimental Social Psychology* (1931; Murphy et al., 1937) both endorsed and promoted the individualistic and experimental program initiated by Floyd Allport. See Chapter 7.

from ascertained laws and principles of antecedent sciences must then be the prime method of social psychology. (p. 12)[28]

Yet in the *Psychology of Human Society* (1925), Ellwood carefully distinguished between social and individual psychology, maintaining that group habits distinguish human groups from animal groups (in which uniformities of psychology and behavior are based on instinct or mere suggestion and imitation):

Thus the social life presents itself as a process, but a process made up both of individual psychic elements and of social psychic or cultural, elements; that is, of elements of interstimulation and response among individuals – such as communication, suggestion, imitation, sympathy, conflict – and of cultural elements – such as custom, tradition, conventions, and institutions. All of these processes enter into, and determine, the form of group behavior. Some of them are individual psychic, others are social psychic. The social psychic, or the cultural, however, can operate only through the individual and hence the individual has a chance to modify it. On the other hand, the individual's psychic life itself is largely determined by the social psychic, or the cultural. Individual behavior, in other words, comes largely from group culture; but culture in the last analysis, as we have said, comes from the individual mind. (pp. 465)

Accordingly, Ellwood (1924) defined social psychology as "the psychology of group behavior":

Social psychology, in the sense of the psychology of group behavior, is accordingly a part of sociology. *It is the study of the psychic factors involved in the origin, development, structure and functioning of social groups.* (p. 9)

Nonetheless, it was Ellwood's earlier and more individualistic characterization of social psychology as "psychological sociology" (in *Introduction to Social Psychology*, 1917) that was targeted by Floyd Allport (1924a) when he complained:

This seems to the present writer to minimize unjustly the claims of the psychologist. It is surely a legitimate interest to consider social behavior and consciousness merely as a phase of the psychology of the individual, in relation to a certain portion of his environment, without being concerned about the formation or character of groups resulting from these reactions. (p. 11)

Allport largely abandoned instinctual explanations in favor of explanations referencing the social environment, conceived in terms of individually engaged (and learned) forms of cognition, emotion, and behavior

[28] Compare Allport (1924a, pp. 154–168).

relating to other persons, no different in form from the individual psychological explanations of individual psychology. Thus for Allport, as much as for Dewey and Mead, all psychology was social psychology. Allport held that there is no fundamental distinction between social and individual psychology because he believed that most forms of human psychology and behavior are interpersonally learned, directed, and related.

# 5

## Social Psychology and the "Social Mind"

One reason for the neglect of the social in post-1930 American social psychology appears to have been ultimately misguided but not entirely unjustified fears concerning the illegitimate "reification" or "personification" of social psychological phenomena. Such fears were a reaction to some of the more extreme claims made by social theorists such as Durkheim (1895/1982a), Espinas (1877), Fouillée (1885), Hegel (1807/1910), Hobhouse (1913), Le Bon (1895/1896), Lévi-Bruhl (1923), McDougall (1920), Martin (1920), Schäffle (1875–1878), Sighele (1892), Tarde (1890/1903), Wallis (1925, 1935a), and Wundt (1916, 1900–1920), who sometimes talked of social groups as emergent supraindividuals or organisms. However, in rejecting the (albeit sometimes rather extreme) positions of such theorists tout court, many American social psychologists effectively threw out the baby with the bathwater.

The notion that social groups, or societies themselves (or states or nations), form emergent supraindividuals or organisms has been popular with social theorists since at least the time of Plato and was particularly prominent among idealist social theorists such as Hegel (1807/1910), Green (1900), and Bosanquet (1899). The notion of a social "mind" or "spirit" or "soul," usually but not invariably associated with a nation or state, became the common intellectual currency of such idealist thinkers and was imported into social scientific disciplines such as history, sociology, and the new German discipline of *Völkerpsychologie*. Such notions were popularized and applied to smaller social units by European social theorists such as Espinas (1877), Tarde (1890/1903), and Le Bon (1895/1896), who were particularly impressed by (and concerned about) the behavior of crowds and mobs but who also extended the notion of a social or group mind to all forms of social groups (large and small, disorganized and organized, voluntary and involuntary, and so forth).

For example, Le Bon (1895/1896) maintained that "psychological crowds" form "a single being" subject to "the *law of the mental unity of crowds*":

> Under certain given circumstances, and only under these circumstances, an agglomeration of men presents new characteristics very different from those of the individuals composing it. The sentiments and ideas of all the persons in the gathering take one and the same direction, and their conscious personality vanishes. A collective mind is formed, doubtless transitory, but presenting very clearly defined characteristics. (p. 1)

Analogously, Durkheim (1895/1982a) characterized social groups as "greater than the sum of their parts," Wundt (1916) talked about the "social mind," and even Titchener (1910) maintained the existence of some form of "collective consciousness."

Many early American social psychologists also articulated their theoretical positions in terms of a social or group mind. Perhaps the most famous example of a commitment to the notion of a social or group mind in early American social psychology is William McDougall's *The Group Mind* (1920). In this work, McDougall stated the following:

> Since, then, the social aggregate has a collective mental life, which is not merely the sum of the mental lives of its units, it may be contended that a society not only enjoys a collective mental life but also has a collective mind or, as some prefer to say, a collective soul. (p. 7)

Some theorists were led to personify groups, attributing to them not only social or group minds but social or group wills and purposes as well: "Permanent groups are themselves persons, group persons, with a group-will of their own and a permanent character of their own" (Barker, 1915, p. 175, quoted with approval in McDougall, 1920, p. 19).

However, during the 1920s and 1930s, the notion of a social or group mind, along with the associated notion of a legitimate form of social psychology distinct from individual psychology, was soundly rejected by most American social psychologists, although perhaps most vigorously and famously by Floyd Allport (1924a, 1924b). Allport (1924b) characterized as the "group fallacy" the supposed error of attempting to explain "social phenomena in terms of the group as a whole, whereas the true explanation is to be found only in its component parts, the individuals" (p. 60). He railed against the idea that social groups constitute supraindividuals with emergent powers, properties, or personalities and maintained that all talk about the actions of social groups is just talk about the actions of individuals:

> When we read that a certain army captured a city, or a certain football team defeated a rival team, the language, though not precise, is not misunderstood. It is clear that it is solely the individual soldiers or players who combined their efforts and accomplished the feat described. (p. 60)[1]

In his critique of the reification and personification of social phenomena, Allport echoed the complaints of individualist sociologists such as Weber (1922/1978), who famously maintained:

> When reference is made in a sociological context to a state, a nation, a corporation, a family, or an army corps, or to similar collectivities, what is meant is . . . *only* a certain kind of development of actual or possible actions of individual persons. (p. 14)

> Even in cases of such forms of social organization as a state, church, association, or marriage, the social relationship consists exclusively in the fact that there has existed, exists, or will exist a probability of action in some definite way appropriate to this meaning. It is vital to be continually clear about this in order to avoid the "reification" of those concepts. (p. 27)[2]

[1] Compare Jarvie (1959), a well-known methodological individualist: "Army is merely the plural of soldier and all statements about the army can be reduced to statements about the particular soldiers comprising the army" (p. 57).

[2] The parallels to Weber are even more striking in comments Allport made in 1961 (when invited to respond to Bernard's 1926b review of his 1924 *Social Psychology* text):

> To say that the "*team*" runs down the field, though useful, does, however, imply a personification and a specious singularity.
>
> Suppose now that a forward pass occurs. Instead of saying the team runs down the field, we say that the team "executes a forward pass." . . .
>
> Since we are unable to describe the forward pass episode as an act of an individual, suppose that we call it an act of the group. The "team" carries out the play. Here, at the social level, we have again invented through the term *team* a useful singularity. . . . The "corporate fiction" is something without which our economic, political and organizational life in general could hardly go on. But the trouble here, from the standpoint of objective science, is that the term for the agency which is said to "execute the forward pass" is devoid of any unambiguously denotable referent. When we try to touch or speak to the "team," we are encountering or addressing only individuals. The corporate fiction, though a useful orienting device for perceiving and handling a situation in a certain way, is still a fiction. (p. 195)

Compare Weber (1922/1978):

> For still other cognitive purposes – for instance, juristic ones – or for practical ends, it may on the other hand be convenient or even indispensable to treat social collectivities, such as states, associations, business corporations, foundations, as if they were individual persons. Thus they may be treated as the subjects of rights and duties or as the performers of legally significant actions. But for the subjective interpretation of action in sociological work these collectivities must be treated as *solely* the resultants and modes of organization of the particular acts of individual persons, since these alone can be treated as agents in a

Analogously, Gordon Allport (1954) complained about the notion of a "collective consciousness" or "group mind":

> Probably it is regrettable that the concept was ever used by anyone. We see now that it has unnecessarily imposed metaphysical blocks in the path of constructive conceptualization. (p. 40)

According to Gordon Allport, the ascription of psychological properties to groups is an illegitimate reification of the abstract concept of psychological states *common* to many individuals, "of certain attitudes and beliefs from the personal mental life of individuals" (p. 39).[3]

Many other social psychologists followed the Allports in rejecting the notion of a social or group mind and insisted that a scientific social psychology should only concern itself with the psychological states and behavior of individuals. Thus R. H. Gault (1921), for example, maintained that

> there is something about group life, about co-operation among individuals, that leads us to talk about a crowd mind, a group mind, or a social mind. The individual in a crowd or in a highly organized group does not behave as does the same individual in isolation. In highly organized groups there grows up a keen feeling of unity and of identity from day to day, and there is a conviction that progress has been made and will be made in the future by the whole group, and that it stands for a certain set of ideals and purposes. These are probably the phenomena above all that prompt some students to postulate the social mind – however it may be phrased. To do so perhaps satisfies a desire for completeness and system when the term is used in the sense in which we employ it as applied to an individual. No objection can be made to the term when it implies *only* the facts, theories, ideas, purposes, traditions, etc., that *are held in common by the members of the group*. On another hand it is described as a super-individual mind – one that is in addition to the minds of individuals but of the same stuff as they are, and that co-ordinates and unifies them. It is beyond the methods of science to arrive at such a mind. (p. 43, emphasis added)

However, although these social psychologists may have shared Weber's distaste for the reification and personification of social groups, they did not acknowledge Weber's own critical distinction between social and merely common forms of cognition, emotion, and behavior. In rejecting the notion of an emergent or supraindividual social mind, such theorists also rejected the original conception of social forms of cognition,

course of subjectively understandable action ... for sociological purposes there is no such thing as a collective personality which "acts." (pp. 13–14)

[3] Compare Floyd Allport (1924b): "The intangibility of the phenomena combines with the collective or abstract use of language to produce an error" – the error in question being the "group fallacy" (p. 60).

emotion, and behavior as psychological states and behaviors oriented to the represented psychology and behavior of members of social groups.

## I

It is worth distinguishing the different sorts of reasons that early social theorists and social psychologists had for talking about the social or group mind and social or group mentality. Often enough it was because such theorists held that social phenomena are constituted as supra-individual entities with emergent properties and powers based on a presumed analogy between social entities and emergent biological entities such as cells or organisms. Thus, for example, Le Bon (1895/1896) maintained that

> there are certain ideas and feelings which do not come into being, or do not transform themselves into acts except in the case of individuals forming a crowd. The psychological crowd is a provisional being formed of heterogeneous elements, which for a moment are combined, exactly as the cells that constitute a living body form by their reunion a living body which displays characteristics very different from those possessed by each of the cells singly. (p. 2)

Likewise, Durkheim (1895/1982a) famously maintained that social groups or collectives are greater than the sum of their parts and based his claim on the supposed analogy between cells and their chemical constituents (1901/1982c, pp. 39–40).[4] Precisely the same analogy was taken up in social psychology by enthusiastic proponents of the notion of a social mind such as Wilson Wallis (1935a):

> The living organism is an entity because it embodies a system of persistent relations which enables the constituent parts to function as a unity.... Similarly a group of people is an entity and not a mere manifold when the members function as an interrelated unity. (pp. 367–368)

Another common reason for talking about the social or group mind was the avowed fact that social groups maintain their identity over time despite changes in their component members:

> It is maintained that a society, when it enjoys a long life and becomes highly organized, acquires a structure and qualities which are largely independent of the qualities of the individuals who enter into its composition and take part for a brief time in its life. (McDougall, 1920, p. 9)

[4] Although, as noted in Chapter 3, Durkheim's main objective in emphasizing the thing-like nature of social phenomena was to establish their legitimacy as objects of sociological science.

A social group has the power "of perpetuating itself as a self-identical system, subject only to slow and gradual change" (McDougall, 1920, p. 9; cf. Wallis, 1935a, p. 367). Thus social "entities" such as the Roman Catholic Church, City College, and the American Psychological Association are held to maintain their identity over time as the Roman Catholic Church, City College, and the American Psychological Association despite (eventually complete) changes in their membership.

However, it is not clear whether this latter consideration involves any commitment to the existence of supraindividual entities with emergent powers or properties. It merely alludes to different criteria for the numerical identity of certain populations over time. Thus the American Psychological Association today counts as the same association as the one that was formed by Granville Stanley Hall in 1892 despite complete changes in its membership, whereas the crowd on the street corner this Friday (or the population of unemployed persons in 1999) is a different crowd from the crowd on the street corner last Friday (or different from the population of unemployed persons in 1929) if it is composed of different persons. Such differences are recognized by the most rigorous individualists or methodological individualists (see, e.g., Flew, 1985, p. 44; Popper, 1957, p. 17) and are not in fact universal with respect to social groups as opposed to aggregate groups. Social groups such as government cabinets and supreme courts become different cabinets and supreme courts when their memberships change substantially.

The third and perhaps most common reason for talking about the social or group mind (or social or group mentality) is the most interesting and important for present purposes. For some theorists, the social dimensions of human psychology and behavior provided a reason for talking about the social or group mind. The notion of a social or group mind was held to derive from the purported fact that individuals in social groups think, feel, and act in a different fashion from individuals in isolation.

Thus Le Bon (1895/1896) claimed that individuals "transformed into a crowd" are possessed by a "sort of collective mind" that "makes them feel, think, and act in a manner quite different from that in which each individual of them would feel, think or act if he were in a state of isolation" (p. 2). McDougall's (1920) fundamental reason for talking about the social mind was the avowed fact that "under any given circumstances the actions of the society are, or may be, very different from the mere sum of the actions with which its several members would react to the situation in the absence of the system of relations that render them a society" (p. 9). Analogously, Dunlap's (1925) reason for endorsing talk about the social mind was the avowed fact that individuals think, feel, and act in different ways within social groups than in isolation from them: "The social group... is not the sum of the individuals as they would be if isolated from the group, but as they actually are in the group" (p. 17).

Since this reason is critical for understanding what came to be neglected in American social psychology, it is important to distinguish between the positions of crowd theorists such as Le Bon, on the one hand, and early American social psychologists such as McDougall and Dunlap, on the other. Le Bon and other crowd theorists maintained that individuals think, feel, and act differently in the *physical presence* of other individuals than they do in *physical isolation* from other individuals. The mechanism of influence postulated is interpersonal and perceptual, grounded in "imitation" or "contagion." In contrast, McDougall, Dunlap, and many other early American social psychologists maintained that social psychological states and behavior (oriented to the represented psychology and behavior of members of social groups) are different from individual psychological states and behavior (not oriented to the represented psychology and behavior of members of represented social groups) *independently of the physical presence or absence of group members*. The mechanism of influence postulated by early American social psychologists was *social psychological* rather than interpersonal. As Dunlap (1925) put it,

> "social consciousness"... is the consciousness (in the individual, of course) of *others* in *the group*, and consciousness of them as *related, in the group*, to oneself; in other words, consciousness of *being a member of the group*. The consciousness of others may be perceptual, or it may be ideational. One may be conscious of one's membership in the Lutheran Church, or in the group of atheists, when physically alone; and this group consciousness *may* be as important and as vivid under such circumstances as when one is physically surrounded by members of the group. (p. 19)[5]

However, critics of the notion of social mentality and of a distinctive social psychology were blind to these subtle but significant differences. Whatever the reasons offered for talking about a social or group mind (or social or group mentality), they were soundly rejected by many in the developing social psychology community, most notably and effectively by Floyd and Gordon Allport.

[5] The significance of this distinction between social influence and merely interpersonal influence, and the role played by their association in the historical neglect of the social, will be discussed in greater detail in Chapter 7. As noted in Chapter 4, the later development of the tradition that came to be known as "symbolic interactionism" failed to maintain the critical distinction between the social and the interpersonal.

## II

Floyd Allport was perhaps the most vigorous and vitriolic critic of the notion of a social or group or crowd mind and thus of the notion of any form of social psychology distinct from individual psychology. Allport was critical of the usual arguments offered for the postulation of a distinctive form of social psychology, with one notable exception. He scarcely considered what was held by many to be the primary reason for postulating a distinctive form of social psychology: namely, the fact that certain psychological states and behavior are engaged socially as opposed to individually (the third reason noted above). Virtually all of Allport's arguments against a distinctive social psychology were directed against versions of the supraindividual "social mind" thesis.

Allport (1924b) used the term "group fallacy" to characterize the supposed error of "*substituting the group as a whole as a principle of explanation in place of the individuals in the group*" (p. 62, original emphasis). The version of this fallacy of particular relevance to this work (and the one on which Allport tended to focus) is the version that mistakenly supposes "that it is possible to have a 'group psychology' as distinct from the psychology of individuals" (p. 62).

Allport (1924a) rejected the idea that psychological properties such as consciousness or impulsivity can be ascribed to crowds, because crowds, unlike the individuals who compose them, do not have a nervous system or introspective faculty:

> Psychologists agree in regarding consciousness as dependent upon the functioning of neural structure. Nervous systems are possessed by individuals; but there is no nervous system of the crowd. Secondly, the passing emotion or impulse common to the members of a crowd is not to be isolated introspectively from the sensations and feelings peculiar to the individual himself. (pp. 4–5)

Talk about a crowd emotion, for example, is most naturally analyzed as talk about the emotions of the individuals who compose it: "When we say that the crowd is excited, impulsive, and irrational, we mean that the individuals in it are excited, impulsive, and irrational" (p. 5).

No doubt one could say the same thing about the ascription of beliefs or attitudes to social groups such as Methodists or professional psychologists. No belief or attitude can literally be ascribed to a social group: our ascription of beliefs or attitudes to social groups is just shorthand for the ascription of beliefs or attitudes to many or most of its members. Yet this acknowledged fact does not demonstrate that socially engaged beliefs and attitudes, the putative subject matter of a distinctive social psychology, are equivalent to individually engaged beliefs and attitudes, the subject matter of individual psychology.

Allport (1924a) seems to have mistakenly supposed that theorists such as McDougall and Dunlap were led to postulate social mentality or a social mind just because of the commonality of beliefs or attitudes held by members of a social group:

> Another sense in which the group is sometimes said to possess a consciousness and behavior of its own is in the *sameness* of thought and action among the members of such a body as an army, a political party, or a trade union. In these groupings the uniformities of mind are considered as elevated to the position of a separate entity participated in by all. . . . A particular segment of the individual's life is picked out because of its similarity with the corresponding segments in other individuals, and is set up as a separate psychological entity. (pp. 5–6)

Yet none of the early American social psychologists who held a distinctively social conception of psychological states and behavior (as oriented to the represented psychology and behavior of members of social groups) treated merely common beliefs or behaviors as equivalent to social beliefs or behaviors, since common beliefs and behaviors (even those common to members of the same social group) may be held individually (most of the Methodists leaving the church may individually believe it is raining, given the liquid evidence, and individually raise their umbrellas to protect themselves from the rain). Allport ignored the critical distinction between social and merely common forms of cognition, emotion, and behavior recognized by early American social psychologists and insisted upon by early social theorists such as Durkheim and Weber.

Allport rejected the familiar argument in favor of the conception of crowds or social groups as emergent supraindividual entities, with distinctive properties and powers, that was based on their purported analogy with biological entities. Such an analogy, certainly favored by Durkheim, Le Bon, and others, treated the relation between social groups and the individuals who compose them as analogous to the relation between biological entities such as cells and the chemical elements that compose them. Just as cells were held to have properties and powers that cannot be reduced to or explained in terms of the properties and powers of the individual chemical elements that compose them, so too social groups were held to have properties or powers that cannot be reduced to or explained in terms of the properties and powers of the individual persons who compose them.

This analogy was frequently based on essentially "vitalist" conceptions of biological phenomena, which had been largely abandoned by biologists by the end of the nineteenth century. Allport rejected the general analogy between social groups and living systems but accepted the general hierarchical conception of ontology and scientific explanation upon which the analogy was based – he merely reversed its implications. In the traditional reductive fashion of many empiricists, he maintained that entities (and the properties of entities) on a "higher" ontological stratum can only be explained in terms of the entities (and the properties of entities) on the lower ontological stratum that compose them. For Allport, just as biological phenomena and their properties can only be explained in terms of the chemical properties of the chemical elements that compose them, so too social phenomena and their properties can only be explained in terms of the psychological properties of the individual persons who compose them.

Thus, in opposition to Durkheim's apparent dogma that social phenomena can be explained only in terms of other social phenomena and not in terms of psychological phenomena, Allport maintained that social phenomena can be explained only in terms of psychological phenomena. For Allport, to advance a social explanation of social phenomena is to advance a pseudo-explanation that remains on the level of *description* only. To offer a genuine explanation of social phenomena, one has to move to the lower compositional level of individuals and their psychologies. This reductive conception of explanation was entirely general, according to Allport, and applied to all the various sciences. Thus, in highlighting the group fallacy, Allport (1924b) conceived of himself as articulating an entirely general principle and applying it to social psychology:

> The phenomena studied by any science are approachable from two different viewpoints. The first is that of description, the second is explanation. A complete program for any science embodies both these forms of approach. Now the essential fact is that in the hierarchy of sciences the field of description of one science becomes the field of explanation for the science immediately above it. (p. 69)

Thus, for example, the physiological properties of the reflex arc must be explained in physicochemical terms:

> First, the physiologist notes that when a stimulus, such as a pin prick, is applied to a sensory nerve ending a certain muscle contraction follows. He notes also certain properties of this event, such as latent time, refractory phase, and inhibition of other reflexes by this one. These are descriptive aspects of that physiological unit called the reflex arc. By the aid of the microscope he is also able to describe such minute features as the synapse. But the physiologist must not be content with mere description; he must explain. In order to do this he must borrow from the lower sciences, physics and chemistry. Thus to account for neural transmission and the action of the synapse he employs the laws of electro-chemical change, polarization, and combustion. The existence of an intermediate science, such as organic

chemistry, proves how closely the organic is dependent upon the inorganic for its causal principles. Description is thus carried in physiological terms, explanation in physico-chemical terms. But we call this science *physiology*. (p. 69)

Precisely the same is true for the psychologist, who must resort to physiology for his explanations:

The psychologist in turn is attracted by the field of human behavior. He observes the higher integrations of response, such as emotions, habits, and thought, their speed of operation, and ability to inhibit or reinforce one another. He is interested not so much in the reflex as a detached physiological unit as he is in what response (involving usually a pattern of reflexes) is linked up through synaptic functions with a particular stimulus. The realm of phenomena described by the psychologist thus transcends in scope and complexity that of its lower constituent science, physiology. But how about explanation? It will be seen that for principles of causation in the study of behavior we must descend directly to the reflex arc level, and accept as explanatory its conditions and characteristics as described by the physiologist. (pp. 69–70)

Finally, the sociologist, in like manner, must explain the social in terms of the psychology of individuals:

Turning now to the sociologist, we find that the data which he describes reach the highest point of breadth and complexity. They embrace collections of individuals in organized societies, the products of such organizations, and the changes which they undergo. This is indeed a vast field for descriptive analysis. Yet for explanation sociology is in turn dependent upon the descriptive formulae of the science just below it, namely psychology. Just as psychology has to seek its causation within the units (reflex arcs) of which its material, individual behavior, is composed; so sociology must find its explanatory principles in the units (individuals) of which society is composed. (pp. 70–71)

Such an analysis is of course question begging and was disputed by Wallis (1925) and other proponents of a distinctive social psychology:

Those who find that individual psychology is the only psychology, unwittingly are paying homage to the superior reality of (assumed) analysis, a superiority which the social psychologists do not admit. The latter would not admit that cells possess a higher order of reality than does a living organism composed of these selfsame cells. The existence and the reality of the cells, it might be argued, depend as much upon the existence and reality of the organism, as the organism depends upon the cells. . . . Individual psychology, then, cannot give us group behavior, any more than a study of cells can explain the difference in behavior of two individuals, A & B. (pp. 148–150)

Moreover, Allport's account appears paradoxical and open to an obvious reductio ad absurdum argument. For this account implies not only that there can be no legitimate level of genuine social explanation but also that either the only real or ultimate explanation of anything is in terms of the "lowest" science on the "lowest" ontological level (presumably microphysics) or that there are no genuine explanations at all (since putative explanations in microphysics cannot themselves be reductively explained, they would appear to remain merely descriptive).[6]

Even if one accepted the dubious principle that explanatory adequacy always parallels ontological dependency, this sort of familiar reductive argument ignores important differences between forms of explanation on the same ontological level. Both gases and organic acids are composed of molecules, and the properties of gases and organic acids are explained in terms of the properties of the molecules that compose them. Yet the *aggregative* form of the statistical mechanical explanation of the properties of gases is quite different from the *structural* form of the electrovalent bonding explanation of the properties of organic acids. Analogously, the explanation of differential rates of suicide between Protestants and Catholics and between men and women may both be in terms of the psychological states of individuals, yet an explanation in terms of socially engaged psychological states (such as different social beliefs or attitudes about the morality of suicide embraced by Catholics and Protestants) is quite different from an explanation in terms of individually engaged psychological states (such as different levels of depression among men and women that are a function of differences in their learning history or biochemistry). Allport's argument to the effect that all social phenomena require some form of psychological explanation simply ignores the critical distinction between social psychological and individual psychological forms of explanation – the distinction upon which earlier conceptions of a distinctive social psychology were based. Since Allport did not acknowledge social as opposed to individual psychological forms of explanation, he saw no need for a social psychology distinct from individual psychology.[7]

[6] It is also very hard to imagine the justification for this account, since the concept of causal explanation employed by Allport is clearly Humean, in terms of descriptions of empirical invariance. Yet it is not clear why one level of description of empirical invariance ought to be privileged over any other.

[7] Allport (1924b) made a legitimate point about the limitations of putative social psychological explanations in terms of the "intolerance, emotionality, and irrationality of crowds" (p. 62). Such putative explanations fail to explain why one crowd behaves in one way and another crowd differently:

> There is at hand no means for explaining the differences in the behavior of different crowds, since all are emotional, irrational, and the like. Why should the excitability of one crowd express itself in whipping non-church-going farmers, that of another crowd in

Yet for many of the empiricist and positivist psychologists who forged the science of psychology and social psychology in the early twentieth century, this type of reductive account was far more appealing than the Durkheimian supraindividual account, whose appeal to emergent entities and properties had obvious affinities with vitalistic forms of explanation in biology. After all, had not the new discipline of psychology itself been modeled upon the discipline of physiology, which developed in the nineteenth century precisely via its rejection of vitalist explanations in favor of reductive physicochemical explanations,[8] by maintaining, as Emil Du Bois-Reymond and Ernst Brücke put it, that "no other forces than the common physical-chemical ones are active within the organism" (1842, p. 19, letter from Du Bois-Reymond to Karl Ludwig, cited in Lowry, 1987, p. 75).

As noted earlier, Allport's reductive individualist position repeated Weber's critique of Durkheim's treatment of social groups as supraindividuals, and indeed often echoed it in remarkable ways. In particular, Allport followed Weber in regularly complaining about the "reification" and "personification" of social phenomena, although unfortunately he did not follow Weber in distinguishing between socially and individually engaged psychological states and behavior. In maintaining this reductive position, Allport was simply following the general empiricist and behaviorist trend that came to dominate academic psychology in America at this time and remains dominant to the present day (Toulmin & Leary, 1992).

That is, the reactionary fear of the reification of social phenomena expressed by Allport and many of his followers is just one aspect of the more general fear, still prevalent among many contemporary psychologists, of the illegitimate reification of unobservable "theoretical posits" or "hypothetical constructs" (Kimble, 1989). The rejection of such suspect theoretical phenomena was part of the general behaviorist and neobehaviorist metatheoretical position in psychology that developed from

looting grocery stores, and that of still another in lynching negros? These questions throw into relief the necessity of delving deeper for our notions of cause than terms which describe the crowd as a whole. We must seek our mechanisms of explanation in the individuals of whom the crowd is composed. (p. 62)

However, it is not clear that explanations in terms of the reflex arc or its physiology would fare any better, and it is in fact doubtful if the adequacy of forms of explanation in social psychology depends on their ability to answer such essentially *historical* questions.

[8] Nonetheless, many biologists who were committed to vitalism, such as Johannes Müller, Claude Bernard, and Louis Pasteur, made substantive contributions to the discipline.

the 1920s onwards:[9] social minds and social mentality went out with imageless thoughts, Oedipal complexes, and cognitive maps. It is probably no accident, for example, that Howard Kendler (1952), in a famous paper in which he argued that psychologists ought to reject putative theoretical references to "hypothetical constructs" such as "cognitive maps" (Tolman, 1948) as "reified fictions," employed the example of conceiving the "family" as an emergent supraindividual to illustrate the muddle generated by the reification of theoretical concepts. Kendler cited the proverbial error of the foreigner who complained that although he had met all the individual Smiths while staying in their lovely home, he had never observed the "Smith family." According to Kendler (1952), like the theoretical construct "cognitive map," the construct "the Smith family" does not refer to any "supra-sensible 'entity'." To suppose otherwise is to commit "the fallacy of reification or hypostatization" (p. 271).

Regrettably, this is how the situation continues to be perceived. Floyd Allport's contribution has come to be seen by mainstream social psychologists as "liberating" social psychology from its commitment to metaphysically and morally dubious entities such as the social mind (Cartwright & Zander, 1968, p. 12) and enabling social psychologists to employ the experimental methods of individual psychology in the development of their discipline. Allport is often represented as having established that social psychological phenomena are nothing more than aggregations of individual psychological phenomena and thus legitimate objects of individual psychological explanation, amenable to the experimental methods of individual psychology, whose application to social phenomena was pioneered by social psychologists like Allport himself and J. F. Dashiell (1930, 1935), H. Gurnee (1936), and Lois and Gardner Murphy (1931; Murphy et al., 1937). Thus J. M. Levine and Moreland, for example, writing in the 1998 *Handbook of Social Psychology*, claim the following:

> Many early observers (e.g., Durkheim ... Le Bon ... McDougall ... ), impressed by how differently people can act when they are together than alone, claimed that groups possess "emergent qualities" that cannot be fully understood by studying their members. Analyzing group phenomena may thus require the development of new research methods and theories. Other observers (e.g., Allport ... ), however,

[9] Despite the "cognitive revolution," psychologists in general remain ultra-cautious about the postulation of (unobservable) theoretical psychological states. Most psychologists, including cognitive and social psychologists, remain committed to "methodological behaviorism" (Cahan & White, 1992): they believe that such hypothetical constructs must be operationally defined and are legitimate only insofar as they facilitate the prediction and control of observable behavior.

claimed that groups are simply sets of individuals, without any special qualities beyond those already possessed by their members. . . . If groups do not really exist apart from their members, then group phenomena *can be analyzed using the same research methods and theories that are used to analyze individual phenomena.* (p. 417, emphasis added)

However, as was demonstrated in the previous chapters, a distinctive social psychology concerned with the social dimensions of psychological states and behavior does not depend on or entail any commitment to the existence of emergent supraindividual entities such as social minds that somehow exist "apart from their members."[10] Yet Floyd Allport and his followers rejected this distinctive form of social psychology along with the views of the social mind theorists.

That is, Floyd Allport did not merely reject the notion of supraindividual social minds. In rejecting the notion of a "'collective mind' or 'group consciousness' as separate from the minds of the individuals of whom the group is composed" (1924a, p. 4), he also rejected the possibility of a distinctive social psychology concerned with socially engaged psychological states and behavior (oriented to the represented psychology and behavior of members of social groups). He rejected the conception of a distinctive social psychology that had been reasonably maintained by theorists as diverse as Wundt, Durkheim, Weber, Simmel, Titchener, Külpe, Angell, McDougall, Ross, Wallis, Dunlap, Schanck, Thomas, Katz, Bernard, and Bogardus.

Allport (1924a) famously maintained that there is no form of group psychology distinct from individual psychology:

There is no psychology of groups which is not essentially and entirely a psychology of individuals. Social psychology must not be placed in contradistinction to the psychology of the individual; *it is a part of the psychology of the individual*, whose behavior it studies in relation to that sector of his environment comprised by his fellows. (p. 4)

Allport was thus one of the first to clearly and categorically deny that social psychological states and behavior are distinct from individual psychological states and behavior. He maintained, in contrast to Durkheim,

[10] It also ought to be stressed that social psychological states and behavior do not need to be reduced to individual psychological states and behavior in order to be analyzed according to the same research methods – including experimental methods – employed in the analysis of individual psychological states and behavior, although many social psychologists appear to have followed Allport in supposing this to be the case. That is, both socially and individually engaged psychological states and behaviors can be investigated via the same empirical and experimental methods.

Weber, and many early American social psychologists, that social psychological states and behavior are *nothing more* than psychological states and behavior that are common among individuals:

Collective consciousness and behavior are simply the aggregation of those states and reactions of individuals which, owing to similarities of constitution, training, and common stimulations, are possessed of a similar character. (p. 6)

Allport essentially redefined social consciousness and social behavior as interpersonal forms of consciousness and behavior, as forms of consciousness and behavior directed toward social objects, namely, other persons (in contrast to the earlier conception in which social consciousness and behavior were defined in terms of their orientation to the consciousness and behavior of members of a social group).[11] Thus, in contrast to earlier social psychologists who recognized that one could have socially engaged attitudes and behavior directed toward nonsocial objects, such as colors and rivers (e.g., Thomas & Znaniecki, 1918, pp. 30–31), Allport (1924a) denied that social psychological states and behavior could be directed toward such objects:

Social behavior comprises the stimulations and reactions arising between an individual and the *social* portion of his environment; that is, between the individual and his fellows. Examples of such behavior would be the reaction to language, gestures, and other movements of our fellow men, *in contrast with our reactions towards non-social objects, such as plants, minerals, tools, and inclement weather.* (p. 3, second italics added)

In consequence, Allport provided a definition of social psychology (familiar and acceptable to most practitioners of social psychology from this point onward) that effectively excluded socially engaged psychological states and behavior:

Social psychology is the science which studies the behavior of the individual in so far as his behavior stimulates other individuals, or is itself a reaction to their behavior; and which describes the consciousness of the individual in so far as it is a consciousness of social objects and social reactions. (p. 12)

No trace of the original concept of the social psychological remains in this definition, which essentially redefines social psychology as the psychology

[11] It is worth noting that Allport's definitions of social psychological states and behavior as merely common psychological states and behaviors and as psychological states and behaviors directed toward other persons are not equivalent. Common beliefs about the composition of the moon are not interpersonal, and interpersonal acts of rape and aggression are not common behaviors (at least in most places and times).

of interpersonal consciousness and behavior and avoids any reference to the consciousness and behavior of members of social groups.[12]

Florien Znaniecki, the joint author of *The Polish Peasant* with W. I. Thomas, similarly redefined the domain of social psychology in line with "the demands put by the ideal of scientific exactness":

> There is one historically important conception of social psychology which must be rejected at once: that which gives this science the task of studying *collective* consciousness or *collective* behavior as against individual consciousness or individual behavior, which are left to "individual psychology" to investigate. (1925, p. 52)

For Znaniecki, social experiences are not linked in any way to social groups but are simply individual or collective (in the sense of plural) experiences of "social objects," namely other persons or groups of persons: "They are simply *individual or collective experiences of human beings – separate persons or groups – and of their behavior*" (p. 55). Social action is likewise defined as interpersonal action – as individual or collective (in the sense of plural) behavior directed toward other persons or groups of persons: "*Social acts are specifically those individual or collective acts whose purpose is to influence human beings, i.e., to modify persons or groups in a certain definite way*" (p. 57).

A similar reconceptualization and redefinition of social psychological states and behavior, and of the discipline of social psychology, occurred in the work of Gordon Allport. Allport (1935) maintained that social attitudes are *nothing more* than attitudes common to a number of individuals and that theories of the so-called social mind are nothing more than illegitimate reifications of common or similar attitudes:

> Many theories of the "social mind" are reducible essentially to the fact that men have similar attitudes. These "common segments" of mental life, when regarded apart from the personalities which contain them, and when viewed in relation to

[12] Samelson (2000) notes that Floyd Allport later affirmed that he never denied the reality of social influence or of socially engaged psychological states and behaviors. In his contribution to Boring and Lindzey's *History of Psychology in Autobiography* (1974, Vol. 6), Allport claimed, "I was not attacking the evident truth they contain [doctrines of social causality] but the failure to give them more specific meaning" (p. 15). However, as noted in the following chapter, although it is true that Allport was forced to recognize socially engaged psychological states and behaviors (however grudgingly), he generally denied or disparaged them in his published works. Moreover, by the 1970s the damage was already done, and his later endorsement of "social causality" remained half-hearted. In describing his general theory of "structure in nature," Allport acknowledged emergent causal properties as a function of emergent physical, chemical, and biological structures but refused to recognize them in the case of "social structures" (pp. 18–19).

the corresponding segments of the mental life of others, give rise to an impression that the group itself has a mental life of its own. But, psychologically considered, the "group mind" can mean nothing more than the possession by a group of people of common attitudes. (p. 827)

Allport maintained this position despite the fact that he documented clear examples of social attitudes that were restricted to members of specific social groups, noting that such attitudes are often uncritically adopted (in the absence of any evidence) and rigidly maintained (in the face of contrary evidence):

Through the imitation of parents, teachers or playmates, they [attitudes] are sometimes adopted *ready-made*.... thousands of such attitudes and beliefs are adopted ready-made and tenaciously held against all evidence to the contrary. (1935, p. 811)

He cited a variety of studies that indicated that such socially engaged attitudes form the basis of racial prejudices and stereotypes (D. Katz & Braly, 1933; Lasker, 1929), but he still maintained that such attitudes are nothing more than *common* attitudes.

In his 1954 paper on the history of social psychology, Gordon Allport offered his well-known interpersonal definition of social psychology:

With few exceptions, social psychologists regard their discipline as *an attempt to understand and explain how the thought, feeling, and behavior of individuals are influenced by the actual, imagined, or implied presence of other human beings.* (p. 5)

Again, no trace of the original conception of the social psychological remains in this definition.[13]

## III

It is worth stressing that the rejection by the Allports and their followers of the earlier social conception of psychological states and behavior – and of a distinctive social psychology – was unwarranted. For they failed to address the central and critical argument offered by early American psychologists such as McDougall, Wallis, and Dunlap in favor of a distinctive

[13] It is true that in the very same paragraph Allport went on to talk about "complex social structure" and "membership in a cultural group" (p. 5), but as the rest of his discussion makes clear, this amounts to no more than a reference to common attitudes and behavior. Moreover, most social psychologists who cite this definition approvingly (e.g., S. T. Fiske & Goodwyn, 1996, p. xiv) usually ignore this addendum.

social psychology. According to such theorists, certain psychological states and behavior require explanation as socially engaged psychological states and behavior (in terms of their orientation to the represented psychology and behavior of members of social groups). These psychological states and behavior cannot be explained or predicted individually: that is, they cannot be explained or predicted by reference to the principles of an individual psychology concerned with those psychological states and behaviors that are the product of genetics or nonsocial learning or "habit formation" – states and behaviors engaged independently of the representation of the psychology and behavior of members of social groups.

Thus, for example, McDougall (1920) maintained that the recognition of socially engaged psychological states and behavior was the primary reason for developing a "group psychology":

> For the collective actions which constitute the history of any such society are conditioned by an organization which can only be described in terms of mind, and which yet is not comprised within the mind of any individual; the society is rather constituted by the system of relations obtaining between the individual minds which are its units of composition. Under any given circumstances *the actions of the society are, or may be, very different from the mere sum of the actions with which its several members would react to the situation in the absence of the system of relations which render them a society; or, in other words, the thinking and acting of each man, insofar as he thinks and acts as a member of a society, are very different from his thinking and acting as an isolated individual.* (pp. 9–10, emphasis added)[14]

Given that he believed that some psychological states and behaviors are engaged socially and can thus only be explained and predicted socially, McDougall maintained that the goal of group psychology, as distinct from individual psychology, is "to display the general principles of collective mental life which are incapable of being deduced from the laws of the mental life of isolated individuals" (pp. 7–8): that is, to delineate those explanatory psychological principles that are incapable of being deduced from the laws governing individually engaged psychological states and behavior.

[14] By "society" McDougall meant any form of social group as well as the composite of social groups and other aggregates that make up society in general. This seems clear from his answer to Maciver's (1917) objection that if nations can be said to have a collective mind, then so can smaller social collectivities: "If a nation has a collective mind, so also have a church and a trade union" (p. 76). McDougall was quite prepared to accept this implication, or, as he put it, "My withers are quite unwrung" (1920, p. 11).

This justification for the recognition of a distinctive form of social psychology was also clearly expressed in Wallis (1925):

> In this paper we shall maintain . . . that there is a social dimension to mind; that individual psychology is not a guide to the behavior of the group. Our justification for these conclusions will be our ability to predict social phenomena by the study of social phenomena, and the corresponding inability to predict them by a study of individual psychology. (p. 147)

> The scientific basis for social psychology lies in the fact that by means of it one can predict the action of the group or the individuals in it insofar as they act *as members of the group*. If I am acquainted with the concepts current in the tribe, I can predict what a Dakota Indian will do when face to face with a will-o'-the-wisp, with an enemy, with a buffalo, with a spider. We may safely defy any individual psychologist who does not take account of the social atmosphere to make that prediction. (p. 149, emphasis added)

This seems to get things exactly right. One needs to know what psychological states and behaviors are engaged socially in order to explain and predict the distinctive psychological states and behaviors characteristic of members of a social group. An individual psychology that ignores the orientation of certain psychological states and behavior to the represented psychology and behavior of members of social groups is incapable of explaining and predicting these psychological states and behaviors.[15] However, this central argument was simply ignored by the Allports and most other critics of a distinctive social psychology, who effectively rejected this conception of the social dimensions of human psychology and behavior and of a distinctive social psychology along with the notion of the social mind as a supraindividual entity with distinctive properties and powers.

However, to be fair, it is perhaps not that surprising that many social psychologists rejected both, for many of the advocates of a distinctive social psychology did maintain that the reasons for recognizing a distinctive social psychology were also reasons for postulating social minds as distinct supraindividual realities. Thus, over and above his appeal to socially engaged psychological states and behavior as the primary reason for recognizing a distinctively social psychology, McDougall (1920) went on to maintain that

> we must nevertheless recognize the existence of over-individual or collective minds. We may fairly define a mind as an organized system of mental or purposive

[15] Compare Wundt (1908, pp. 227–228, cited in Kusch, 1999, p. 169): "the conditions of mental reciprocity produce new and specific expressions of general mental forces, expressions that cannot be predicted on the basis of knowledge of the properties of the individual consciousness."

forces; and, in the sense so defined every highly organized human society may properly be said to possess a collective mind. (p. 9)

Analogously Wallis (1925), after offering his articulate statement of the theoretical rationale for a distinctive group psychology, went on to maintain that

> the group is a reality over and above the individuals that compose it....
>
> A group is an entity when it shows persistence of qualities and when the parts of which it is composed are mutually interdependent. In such a case it functions as a unity, and so is a reality.
>
> These are, in general character, though not in details of quality, the traits which enable us to recognize an aggregate as also an individual. (p. 147)

Indeed, in the first paragraph of this paper, Wallis claimed that the aim of the paper was not merely to maintain that there is a "social dimension to mind" and thus the need for a distinctive social psychology but also to maintain "the thesis that the group is a reality over and above the individuals who compose it" (p. 147). By the time he wrote "The Social Group as an Entity" (1935a), Wallis virtually abandoned the "social dimensions of mind" argument altogether and focused almost exclusively on the supraindividual reality thesis.

Thus, although there is no intrinsic connection between the argument for a distinctive social psychology based on a recognition of the social dimensions of human psychology and behavior and the notion of a supraindividual social mind, it is scarcely surprising that they were associated by the Allports and others. It is also little wonder that McDougall came to recognize that he had made a "tactical error" in employing the phrase "group mind" in conjunction with his advocacy of a social psychology concerned with the social dimensions of human psychology and behavior.

While the form of social psychology recognized by McDougall and Wallis did not require the maintenance of the supraindividual social mind thesis, one can have at least some sympathy with Floyd Allport's complaint (1924b) that this form of social psychology would "abolish the individual; and... therefore abolish the services of psychology" (p. 63), especially when faced with such rhetorical overstatements of the "social dimension of mind" argument as the following (which rivals Durkheim[16]

[16] Compare Durkheim's (1895/1982a) claim that a social theorist who identifies the social causes of suicide by statistics relating age, sex, social class, and religion with suicide rates can ignore the contribution of individuals:

> Since each one of these statistics includes without distinction all individual cases, the individual circumstances which may have played some part in producing the phenomenon

in its apparent devaluation of the individual):

> The fact is that the student of group activities can disregard individuals, can predict behavior of the group and members of the group without acquaintance with the respective individuals. In a word, we need to know only their social attributes, their place in the group, and we then can disregard their individual attributes.... Here, in group life, the social attributes override the individual attributes.... Given the circumstance of place and person you can predict that the man of our culture will wear a white collar. You do not have to wait until you investigate his tastes in the manner of collars or of colors.
>
> Where individual differences, and time and place, are relatively unimportant... there we have the operation of a reality that transcends individual psychology. This is the field of social psychology. In that world the individual as an individual is relatively unimportant, for it is not his individual attributes which determine his behavior, but rather the attributes of the group. (Wallis, 1925, p. 150)

It should be noted, however, that not all those who advocated a distinctive social psychology based on socially engaged psychological states and behavior also advocated a supraindividual theory of the social mind. E. A. Ross (1908) and Kantor (1922) both clearly acknowledged the social dimensions of human psychology and behavior but had no truck with the notion of a supraindividual social mind. Dunlap (1925) defended the notion of a distinctive social psychology while explicitly rejecting the notion of a supraindividual social mind:

> Since psychology is the study of the mind, it has been customary to speak of social psychology as the study of the *social mind*, or of the *group mind*. To this manner of speaking there is no objection, if we do not forget that the only mind of which we can legitimately speak is the mind of an individual. Unfortunately, the phrases have in the past led to the supposition of a "group mind" which either is distinct from the individual minds in the group, or which contains some mental factors over and above the individual minds. To avoid the possibility of such confusion, we must clearly understand that the social mind or group mind, in any concrete case, is the mind of an individual in a social group; and that "social tendencies" or other social mental factors are those factors in individual minds which make social association possible or which result from social association. (pp. 11–12)[17]

cancel each other out and consequently do not contribute to determining the nature of the phenomenon. What it expresses is a certain state of the collective mind. (p. 55)

[17] Analogously, D. Katz and Schanck (1938), while clearly recognizing the social dimensions of psychological states and behavior (their orientation to distinctive social groups or "in-groups" [p.35]), maintained:

> We have carried on our theme without resorting to such mysterious principles as the group mind or the superorganic. We have not attempted to develop the problems of

Richard T. La Piere (1938) condemned the notion of a supraindividual social mind as a product of the mistaken belief that

> from the interaction of people in social groups, there arises a collective ethos – variously termed as mind, spirit or soul. This mind is supposed to be the directive force of collective activities. To express it otherwise: when men interact collectively, they are thought to lose their identities as individuals and to merge into a whole, the spirit or force of which then determines the behavior of the individual members. (pp. 4–5)

However, he also emphasized that there are objective patterns of collective (social) behavior to be studied by a distinctive social psychology:

> Actually, the group-mind is nothing but the personification of patterns of collective behavior. (p. 5)

> It is now rather generally recognized that there are patterns of collective behavior and that these patterns are subject to objective study without their being personified. Collective behavior must be considered a special type of behavior. It involves what might be described as its own "laws." (p. 7)

La Piere naturally associated the personified group-mind view with that of McDougall: "here in the United States the idea was most dogmatically and effectively expressed by the philosopher-psychologist William McDougall, when he published his *Group Mind* in 1920" (p. 5). Yet La Piere was equally critical of theorists such as Floyd Allport, who he claimed were "guilty of an equal error – that of denying the very existence of collective behavior" (p. 6). He likened this error to that of someone observing a Virginia reel who maintained "that all he had actually seen was an aggregate of moving individuals and that there had been no pattern of the Virginia reel – all but the individuals being a figment of his imagination" (pp. 6–7).

However, the Allports won the day, clearly helped by McDougall's tactical error in talking about the group mind and the regular joint advocacy of theories of the social dimensions of human psychology and behavior and theories of the supraindividuality of the social mind. Thus, in reviewing the 1928 edition of *The Group Mind*, Gordon Allport (1929, p. 124) could afford to be personally as well as intellectually accommodating.

social psychology in terms of mysterious forces over and above the problem of individuals adjusting in a world that offers things which satisfy human need. (p. 683)

Schanck's (1932) own Elm Hollow study (the fictitious name of an Upstate New York community), which included the data on the socially and individually held attitudes of Methodists and Baptists, was carefully titled "A Study of a Community and Its Groups and Institutions Conceived as Behaviors of Individuals."

Allport praised the work, claiming that McDougall's conditions for "the formation of a group mind" were "already recognized by the most austere of individualists" (such as his brother Floyd). This was because Allport simply reinterpreted McDougall's conditions in terms of common sentiments and "common segment" behavior. Thus, by 1954 Allport could talk about the "common thread in the thinking of McDougall and of F. H. Allport. What is called the group mind, they say, is essentially the abstraction of certain [common] attitudes and beliefs from the personal mental life of individuals" (p. 39).[18]

## IV

Much has been made of the role played by Floyd Allport's supposed commitment to behaviorism in his rejection of a distinctively social psychology (Farr, 1996). However, the sense in which he may be properly characterized as a behaviorist is a matter of some dispute (Parkovnick, 2000), and Allport's rejection of theories of the social mind and his commitment to behaviorism appear to have been *joint* products of the restrictive empiricist conception of science that he held in common (and possibly socially) with many early American psychologists. Nonetheless, his specifically behaviorist commitments did play some role in his rejection of theories of the social dimensions of psychological states and behavior as well as his rejection of theories of a supraindividual social mind.

At first sight, Allport's behaviorism appears fairly innocuous and not of the radical Watsonian stripe. Allport (1924a) did not maintain that all references to consciousness and cognition should be eliminated from a science of behavior, and he in fact criticized radical behaviorists such as Watson for claiming that they should. In contrast, Allport maintained that the study of (social) consciousness forms an integral part of social psychological science:

> There are a few psychologists who maintain that . . . consciousness . . . has no place in the science which studies behavior. This is a serious mistake. No scientist can afford to ignore the circumstances attendant upon the events he is observing. Introspection on conscious states is both interesting in itself and necessary for a complete account. The consciousness accompanying reactions which are not readily observable also furnishes us with valuable evidence and information of

[18] Throughout this work I remain neutral on the question of whether any population of individual persons can in fact instantiate a social mind, conceived as a form of mentality not reducible to the socially or individually engaged psychological states of individual persons. However, for a recent spirited defense of the possibility, see Brooks (1986).

these reactions, and thus aids us in our selection of explanatory principles within the mechanistic field. (p. 3)

This sounds eminently reasonable and much more like a broad-minded functionalist such as Angell than a radical behaviorist such as Watson. However, the differences between Watson and Allport are more apparent than real. While Allport recognized the evidential role of consciousness and cognition (and their potential utility as an *exploratory* resource), he was as adamant as Watson in maintaining that references to consciousness and cognition play no role in the *explanation* of behavior. Consciousness and cognition are just one of the many possible end points in stimulus-response chains:[19]

It is clear that consciousness stands in some intimate relation to the biological need and the behavior which satisfies it. Just what this relation is still constitutes an unsolved and perplexing problem. One negative conclusion, however, seems both justified and necessary as a working principle: namely, that consciousness is in no way a *cause* of the bodily reactions through which the needs are fulfilled. Explanation is not derived from desire, feeling, will, or purpose, however compelling these may seem to our immediate awareness, but from the sequence of stimulation–neural transmission–and reaction. Consciousness often accompanies this chain of events; but it never forms a link in the chain itself. (p. 2)

As an example, Allport claimed that, although salient, our consciousness of hunger does not explain our behavior of going to the dining hall, sitting down, and eating, which he claimed would occur even in the absence of consciousness:

The man himself experiences hunger-pangs, and considers these sufficient reason for his eating. Actually, however, "hunger sensations" are only a *description* of the consciousness *accompanying* the behavior. The cause of going to the table lies in the sequence stomach-stimulation–nerve transmission–reaction. The act would be equally well-explained if the subject had no consciousness whatsoever. (p. 2, n. 1)

Analogously, our consciousness of our beliefs about the injustice of the war, or shame over our drunken behavior, and our beliefs and shame themselves, could likewise be presumed to play no role in the causal explanation of our consequent protesting and apologetic behavior.

[19] Watson never denied the existence of consciousness or cognitive states. What he denied was the existence of behaviors generated by "centrally initiated" conscious or cognitive states. His point in equating thought with movements of the larynx was not primarily directed toward the specification of its anatomic location but rather was intended to emphasize its status as a form of motor response that could be "integrated into systems which respond in serial order" (1913, p. 174).

Whatever the prima facie plausibility of this claim (and it is not great), it seems clear that it played a significant role in Allport's thought and does partially account for his denial of the social dimensions of human psychology and behavior, since Allport effectively denied the possibility of socially engaged psychological states and behavior. Allport had no problem recognizing individually engaged psychological states and behavior, for these could be conceived simply as reflexive responses to nonconscious and noncognitive stimuli (independently of any relation to a represented social group). However, Allport had a real problem recognizing socially engaged psychological states and behavior, since, as many theorists stressed, these were held to be products of the cognitive – and often conscious[20] – representation of the psychology and behavior of members of a social group, often in the physical absence of actual members of the relevant group. Allport's behaviorist commitment precluded him from treating social psychological states and behavior as representational products, and he defined social psychological states and behavior in terms of the "social stimuli" (i.e., other persons and social groups) to which they were held to be directed.

Allport's behaviorist commitment led him to officially deny the social dimensions of human psychology and behavior.[21] One expression of this denial was Allport's confident claim that social forms of behavior can be found in the animal kingdom down to the lowest levels. According to Allport (1924a), although social forms of behavior can be attributed to lowly biological forms such as parasites and bacteria "only in the most general sense" (p. 154), other organisms such as insects, birds, dogs, cats, monkeys, and apes engage in social behavior because they respond reflexively (through innate predispositions and learned habits) to behavioral stimuli from other insects, birds, dogs, cats, monkeys, and apes.[22] This claim followed naturally from his behaviorist commitment to the continuity of reflex arc explanations (in terms of environmental "adjustments").

In contrast, although those who recognized socially engaged psychological states and behavior acknowledged the possibility of socially

[20] It is not necessary that the representational states required for socially engaged psychological states and behavior be conscious (see Chapter 1), although often enough they are in practice.

[21] However, as noted in Chapter 6, he was in practice forced to acknowledge them, albeit grudgingly.

[22] Thus social behavior for insects, birds, dogs, cats, monkeys, and apes was defined as interinsect, interbird, interdog, intercat, intermonkey, and interape behavior. Compare Zajonc, Heingartner, and Herman (1969) on the "social behavior" of the cockroach: that is, on intercockroach behavior.

engaged psychological states and behavior among nonhuman animals, notably primates, they were properly cautious about such attributions. This was because they recognized that the viability of such attributions depends on the ascription to animals of representations of psychological states and behavior shared by other members of a discriminated social group (as noted in Chapter 1, it involves the ascription of a fairly sophisticated "theory of mind"). The viability of such attributions depends on the ascription of forms of "mental interconnection" that doubtfully apply to most nonhuman animals.[23]

As Ellwood (1924) put it,

> Living in groups is not peculiar to man, but characterizes many plants and animals as well. Nor is living in groups in itself "social life." A clump of grasses, a forest of trees, a colony of bacteria, or a group of protozoa may show interdependence in the life activities of their separate units; but we do not usually call such groups "societies," because so far as we know, no conscious relations of "comradeship" are involved in such forms of collective life. The relations between their units seem to be purely physical or physiological. Such groups, it is true, show the first mark of social life in that they share a common life; but since they are lacking in conscious relations they cannot be regarded as having social life.
>
> As soon as mentality appears in the world of animal life another sort of interdependence is possible. This new interdependence takes the form of mental interaction, or, as we might more accurately say, of mental interstimulation and response. In other words, more or less conscious relations arise among the members of animal groups, and the group activities begin to be carried on by means of more or less conscious interactions or mutual adjustments between the members of such a group. In this case, the association of the members of a group is guided and controlled by conscious or mental processes, giving rise to what we may call, properly speaking, collective or group behavior. (p. 4)

[23] It is worth pointing out that the proper emphasis in such debates was on mental or cognitive capacity rather than consciousness per se. We are liable to be confused by the fact that many writers during this period used the terms "conscious" and "mental" (or "psychological") interchangeably because of the long tradition (beginning with Descartes and Locke) of treating consciousness as an essential mark of the mental or psychological. Yet often enough in such discussions, "consciousness" meant little more than "discriminatory awareness," and many early American social psychologists were willing to grant that socially engaged psychological states and behavior might be unconsciously engaged (see Chapter 1).

# 6

# Individualism and the Social

In the last chapter it was suggested that Floyd and Gordon Allport and their followers were committed to a restrictive empiricist form of ontological individualism that blinded them to the social dimensions of human psychology and behavior and to the possibility of a distinctively social form of social psychology. They rejected the social dimensions of human psychology and behavior because they rejected everything associated with the notion of a social mind.

However, this cannot be the whole story. For on one level at least, both Floyd and Gordon Allport did recognize a fundamental difference between socially and individually engaged psychological states and behavior. In "The J-Curve Hypothesis of Conforming Behavior," Floyd Allport (1934) cited numerous examples of "conforming behavior" whose statistical distribution was highly asymmetrical. The statistical distributions of these forms of behavior were often J-shaped, with the mode on the terminal step, in contrast to the distributions of behaviors expressive of random personal differences, which tended to be normal and symmetrical. One of the interesting features of Allport's examples of conforming behavior is that they included social forms of consciousness and behavior that violated Allport's own interpersonal definition of social consciousness and behavior. For example, they included the practice of bowing in silent prayer before church service among Episcopalians, the practice of stopping before a red light among motorists, and belief in the deity (as a personal creator and ruler) among Catholics[1] (D. Katz & Allport,

[1] Strictly speaking, among Catholic men.

1931).[2] Analogously, Gordon Allport (1935) cited Schanck's finding that social or institutional attitudes, such as the Methodists' preference qua Methodists for sprinkling as a form of baptism, also displayed this skewed or J-shaped distribution, whereas individual (or "private") attitudes tended to be more normally and symmetrically distributed: "The private attitudes of the members of these groups quite often resemble in moderateness and variability the attitudes of outsiders who are not members" (p. 825).

Now such differences in statistical distributions are not terribly reliable markers of socially as opposed to individually engaged psychological states and behavior. This is because conforming behavior may be distributed asymmetrically in accord with the J-curve hypothesis but may be a product of nonsocial factors (as in the case of the Methodists raising their umbrellas individually because it is raining and likely in some of Allport's other examples, such as punching a time-clock in a factory, for which every worker would have had good individual reasons). Conversely, beliefs, emotions, and behaviors may be engaged socially by any proportion (including a small proportion) of the members of a social group (and by nonmembers, for whom the social group is a reference group but not a membership group).[3] Whether for these or different reasons, the J-curve hypothesis was rejected by social psychologists.[4] Nevertheless, it does seem to indicate that both Floyd and Gordon Allport had some grasp of the difference between socially and individually engaged psychological states and behavior.

[2] Moreover, in his later work with Chiang Lin Woo on "structural dynamics," Allport noted how the degree of conformity behavior (as measured by the steepness of J-curve distributions) appeared to be a function of the degree of "subject involvement" with the social groups for which the conformity behavior was appropriate (belief in God as a supreme deity for Catholics, for example). Unfortunately, the data were lost and the experiments never published (Allport, 1974).

[3] Although some psychological states and behaviors must be shared socially for a collection of individuals to form a social group, once such a social group is established, other psychological states and behaviors may be engaged socially by only a minority of the members, based on the false assumption that most others engage these psychological states and behaviors.

[4] Allport's mathematical definition of the J-curve as requiring at least 50 percent of cases falling in the region of maximum conformity was criticized as arbitrary (Dudycha, 1937), and later empirical studies suggested that the J-curve distribution covered only a limited number of social and conforming behaviors (Fearing & Krise, 1941; Waters, 1941; both cited in Gorman, 1981).

## I

Although Floyd and Gordon Allport may have recognized socially engaged psychological states and behaviors, both refused to take them seriously as distinctive forms of human psychology and behavior. Floyd Allport (1934) explained the J-curve distribution in terms of individually engaged interpersonal "conformity producing agencies," such as conditioning, punishment, and the like (p. 168). Gordon Allport (1935) treated social or institutional attitudes as merely "uniform 'common segment' attitudes" (p. 825). The reasons for this take us beyond the merely methodological individualist objections to the reification or personification of the social mind. Other forms of individualism, more subtle and pervasive perhaps, led them to deny and disparage the social dimensions of human psychology and behavior whenever they found themselves forced to acknowledge them (however grudgingly). For the recognition of genuinely social psychological states and behavior also posed a threat to their cherished ideals of moral individualism: to their ideals of personal autonomy and responsibility.

Perhaps one of the best ways of illustrating this prima facie threat is by considering the implications of socially engaged beliefs and attitudes for our scientific and moral judgments. Consider, for example, the implications of one of our earlier examples. An individual Catholic might hold that abortion is wrong individually (based on a consideration of arguments and evidence) and independently of whether other Catholics are represented as holding that abortion is wrong. Alternatively, an individual Catholic might hold that abortion is wrong socially – that is, because and on condition that other Catholics are represented as holding that abortion is wrong. Yet the very idea that some Catholics might hold that abortion is wrong socially casts doubt on our image of ourselves as autonomous moral thinkers, and many would be reluctant to admit that their most cherished moral beliefs or attitudes are in fact held socially, even when they honestly avow a high valuation of their membership of the relevant social groups. Some might reasonably treat this as a threat to morality itself.[5]

Consider the implications of the idea that some of the theoretical beliefs of scientists are held socially. There has, of course, been much recent

[5] It would, for example, pose a serious threat to the Kantian theory of morality, the main point of which is that morality is (or is at least supposed to be) *unconditional*: one ought to do one's duty for its own sake in any circumstance, independently of whether any others are represented as having done their duty in similar circumstances.

discussion of the social dimensions of science, the social construction of scientific knowledge, and the threats consequently posed to the rationality of science (Barnes, 1977; Bloor, 1976; Brown, 1984; Collins, 1985; Fuller, 1988; Hronszky, Feher, & Dajka, 1984; Kuhn, 1970; Latour & Woolgar, 1979; Longino, 1990; McMullin, 1992, Pickering, 1984). However, many social aspects of scientific activity do not appear to pose any intrinsic threat to the rationality of science. For example, the fact that experimental tasks are often apportioned to different members of research teams, or the fact that different research programs are apportioned to different research groups within a scientific community, seems to pose no intrinsic threat to the rationality of science. On the contrary, these social practices secure a division of labor that is generally recognized as eminently rational in light of the exploratory goals of science, even if, in actual practice, the constitution of such groups sometimes contingently leads to forms of bias. Analogously, the various institutional and organizational structures of scientific communities and their systems of professional rewards and gatekeeping also appear to generally promote the exploratory goals of science, although again it may be a contingent fact that some particular structures do pose particular threats: some grant-awarding committees, for example, may happen to be biased in various ways.

The fact that theoretical constructs are themselves socially constructed and constituted as meaningful by social convention does not appear to pose any intrinsic threat to the rationality of science, unless one holds an extremely radical view about the construction of reality by our concepts of it. Any problems about the rationality of science tend to relate to the employment of such constructs in accord with empirical standards of adjudication – or not, as the case may be. The fact that experimental practice is itself based on social conventions about procedures, controls, and forms of experimental reporting does not appear to pose any intrinsic threat to the rationality of science, so long as these procedures, controls, and forms of experimental reporting are capable of independent rational justification and are properly followed by most practitioners. It is perhaps necessary that some scientists accept these procedures individually on rational grounds, since it is important that there are recognized rational grounds for these procedures. Yet what matters most of the time is that scientists actually follow these conventions.

Much ink has been spilled on these questions, yet it is not difficult to remain unmoved by most recent claims about intrinsic threats to the rationality of science posed by these social dimensions of scientific activity while accepting particular and empirically supported claims about the individual frailties and frauds of particular scientists and the biases of some grant committees and scientific institutions. What would, however, pose an intrinsic threat to the rationality of science would be the discovery that many or most scientists hold their theoretical positions *socially*, that is, because and on condition that other members of their scientific community are represented as holding these theoretical positions: that biologists, for example, generally accept the neo-Darwinian theory of evolution because and on condition that other biologists are represented as holding this theory.

It was this form of social adoption of scientific theories that Bacon (1620) was concerned about when he referred to the (social) dangers of the "idols of the theater" (the blind allegiance to theoretical positions), and it was this, too, that lay behind Max Planck's pessimistic aphorism that "a new scientific truth does not triumph by convincing its opponents and making them see the light, but rather because its opponents eventually die, and a new generation grows up that is familiar with it" (1949, pp. 33–34). This form of social adoption of scientific theories was also the threatening implication of Thomas Kuhn's (1970) treatment of a scientific revolution as analogous to a "religious conversion."

It would certainly pose a serious threat to the self-image of many scientists if these supposed paradigms of individual rationality acknowledged that they generally adopted their theories socially.[6] Most scientists represent themselves as autonomous rational agents who can and do – and ought to – adjust their theoretical beliefs according to the available evidence and arguments: that is, as persons who adopt and abandon theoretical beliefs individually. For many scientists, to acknowledge that they accept scientific theories socially would be tantamount to admitting that they are not really scientists. This point, of course, applies equally to social psychologists themselves. Social psychologists would in general be extremely reluctant to acknowledge that they accept social psychological theories socially – because and on condition that other social psychologists are represented as accepting these theories.

The issue has in fact been scarcely explored, and it remains an open and empirical question whether practicing scientists, including social psychologists, accept scientific theories socially or individually. Some evidence supports pessimism, such as B. Barber's (1961) study of the inflexible

[6] To claim that theoretical beliefs are held socially is to suggest that they are held (at least in part) fashionably, that is, held (at least in part) in the manner of fashionable beliefs, attitudes, and behaviors.

reaction of scientists to Lister's theory of antisepsis, Mendel's theory of genetic inheritance, and Arrhenius's theory of electrolytic dissociation, and Krantz's (1972) study of the insularity of scientific schools such as "radical behaviorism"[7] (cited in Bar-Tel, 1990). Other evidence gives reason for cautious optimism, such as Klotz's (1980) study of the N-ray affair. Blondlot's "discovery" of N-rays in 1903 was "replicated" by about 50 French scientists in 1904, but belief in N-rays was abandoned by all but Blondlot when the American professor of optics R. W. Woods documented (in *Nature*) the insufficiency of the evidence after his visit to Blondlot's laboratory in 1905, despite obvious French national pride in Blondlot's "achievement."

Whatever the empirical reality, the idea that our scientific theories, moral viewpoints, and other personally cherished beliefs and attitudes are held socially is not one that any scientist, including any social psychological scientist, is going to rush to embrace. We believe that we are to a significant degree autonomous and rational agents who adjust our scientific beliefs and moral viewpoints according to available evidence and arguments. The suggestion that things might be otherwise, that our theories, beliefs, and attitudes might be determined socially (i.e., independently of evidence and argument), strikes deep at the heart of our "folk psychological" conception of ourselves.

The point is made by Jackson and Pettit (1992) in their recent analysis of "subversive" theories that treat social structures as external causes determining belief, attitude, and desire independently of the assessment of evidence and argument. Jackson and Pettit note that acceptance of the view that our beliefs, attitudes, and desires are caused in this way

> would undermine our folk psychology more radically than may at first appear.... It is part of the folk-psychological notion of belief that a belief is not formed or held in a manner that is entirely insensitive, at least under suitable exposure, to matters of evidence and consistency.... our image of ourselves, and more generally of human beings, as belief-driven agents is going to be put under strain by the view that structural factors have direct and unrecognized causal influences on our psychological make-up. (pp. 110–111)

Nevertheless, it seems an open and empirical question whether any particular set of beliefs, attitudes, or desires attributed to individuals actually are held socially. The fact that Jackson and Pettit clearly abhor such a state of affairs is no argument against its empirical possibility. If any beliefs,

[7] The "operant conditioning" school associated with B. F. Skinner.

attitudes, or desires really are held only socially by some individuals, the beliefs, attitudes, and desires of such individuals would not be sensitive to "matters of evidence and consistency."[8]

## II

These points were not lost on the early pioneers of American social psychology. For example, Edward Ross, the author of one of the first introductory texts on social psychology (Ross, 1908), clearly recognized the moral significance of socially engaged forms of cognition, emotion, and behavior. Ross was as much a moral individualist as Floyd or Gordon Allport: he was clearly committed to the ideals of personal autonomy and responsibility. However, he was also perhaps more of a realist about the social psychological world than either Allport. It was precisely Ross's commitment to the ideals of personal autonomy and responsibility that led him to focus on the threats to moral individualism posed by socially engaged psychological states and behavior in the hope that a science of social psychology would enable individuals to overcome such subversive influences.[9]

Ross maintained that the goal of social psychology is to reveal the sources of social influence so that enlightened individuals can surmount them. He clearly recognized the real threat posed to moral individualism by social influence:

> [Social psychology] seeks to enlarge our knowledge of the *individual* by ascertaining how much of his mental content and choice is derived from his social surroundings. Each of us loves to think himself unique, self-made, moving in a path all his own. To be sure, he finds his feet in worn paths, but he imagines he follows the path because it is the right one, not because it is trodden. Thus Cooley observes: "The more thoroughly American a man is, the less he can perceive Americanism. He will embody it: all he does, says, or writes will be full of it; but he can never truly see it, simply because he has no exterior point of view from which to look at it." Now, by demonstrating everywhere in our lives the unsuspected presence of social factors, social psychology spurs us to push on and build up a genuine individuality, to become a voice and not an echo, a person and not a parrot. The realization of how pitiful is the contribution we have made to what we are, how few of our ideas are our own, how rarely we have thought out a belief for ourselves, how little our feelings arise naturally out of our situation,

[8] Although, as noted in Chapter 1, it is probably rare for any belief to be held only socially. Most socially held beliefs are also held individually, which makes it easy for agents to rationalize socially held beliefs.

[9] It is no accident that Ross's first work was entitled *Social Control* (1906).

how poorly our choices express the real cravings of our nature, first mortifies, then arouses, us to break out of our prison of custom and conventionality and live an open-air life close to reality. Only by emancipation from the spell of numbers and age and social eminence and personality can ciphers become integers. (p. 4)[10]

A stronger expression of moral individualism would be hard to find, although it came from a sociologist (strictly speaking, an economist) who clearly acknowledged the social dimensions of human psychology and behavior. However, the same moral individualism led Floyd and Gordon Allport to deny and depreciate the social dimensions of human psychology and behavior precisely because of the threat posed to their cherished notions of autonomy and responsibility.

A good number of historians of psychology and social psychology have claimed that the Western and characteristically American value of individualism played an important role in shaping the development of American social psychology (Farr, 1996; Graumann, 1986; Pepitone, 1981). In maintaining this familiar position, Farr (1996) treats individualism as a product of the Renaissance and Reformation, stressing the developing traditions of religious and political dissension and liberalism:

At least since the time of the Renaissance individualism has been a key component in the Western intellectual tradition. It is the tap root of modern social psychology, at least in its psychological forms....

Within the Western branch of Christendom the Reformation quickened the cause of individualism. The invention of the printing press and the spread of literacy further promoted it by producing active minorities who could read holy writ for themselves rather than accept the word of others, who protested at the propaganda, who dissented from the former consensus, and who failed, generally, to conform. The spokesmen (for they were men) for the pious majority represented these deviants as Protestants, Dissenters and Nonconformists. Once the representation had been formed, individuals could be identified and then persecuted. Persecution in the Old World led to selective migration to the New. This, in turn, led to individualism being a more central value in North America than, say, in Central Europe....

[10] Compare Baldwin (1897):

Opinion is formed on social models, social authority preceded logical validity, private judgement is never really private. (p. 134)

And Faris (1925):

Thus we are often entirely unaware of the influence of social attitudes upon our own judgments and activity; our subjective feeling is that we make up our own minds and freely choose our activity, even though detached observation shows that we are clearly dominated by collective definitions. (p. 205)

> The roots of individualism lie buried in the soil of the whole Western intellectual tradition but its flowering is a characteristically American phenomenon. Its roots are to be found in Cartesian dualism, right at the start of modern philosophy. If individualism is a core value within a particular culture then it should be possible to detect its effect in the history of the social sciences. I believe this to be true in the case of the history of social psychology in America. (pp. 103–104)[11]

However, it is important to recognize that the individualization of the social by the Allports and their followers was a function of their commitment to a very specific form of moral individualism, one that had its roots in European thought but which was also a distinctively American product. It was also a function of their special way of dealing with the threat posed by the social dimensions of psychological states and behavior, which they generally denied or disparaged, rather than developing strategies to surmount social influences (as recommended by Ross).

Farr (1996) appeals to the form of moral individualism that stresses the distinction between the individual and the social. However, this was not the only form of European individualism. In medieval thought, the term "individual" actually meant "inseparable" and was generally employed to individuate "a member of some group, kind, or species" (R. Williams, 1961). Persons were defined as individuals by reference to the groups of which they were members: group membership defined their very identity as individual "units." On this conception, group membership was treated as intrinsic to the identity of individuals *as* individuals:[12] "to describe an individual was to give an example of the group of which he was a member, and so to offer a particular description of that group and of the relationships within it" (R. Williams, 1961, p. 91).

Partly as a result of the Renaissance and Reformation and the rise of capitalism, the concept of the individual came to be divorced from its original intrinsic connection with social community (Lyons, 1978; Morris, 1972). In the tradition of social thought from Hobbes, Locke, and Smith to Bentham and Mill, individuals came to be treated in isolation and

[11] In locating the roots of individualism in Cartesian dualism, Farr expresses a common but ungrounded assumption about Descartes's philosophical position. Dualists are no more prone to individualism than materialists (most materialists are also individualists), and Descartes himself embraced a decidedly holistic and relational conception of the cosmos and the social world. See Wee (2002).

[12] The concept of the individual originally developed in the context of medieval debates about the nature of the Holy Trinity, in which the individual natures of the Father, Son, and Holy Ghost were treated as determined by their relation to an indivisible whole (Morris, 1972).

abstraction from social community, as "self-contained" individuals. The liberal political tradition that built upon this conception emphasized the "bare" individual as bearer of absolute personal rights, similar to the way that the parallel tradition of laissez-faire economics emphasized the purely egotistical rational agent.[13]

Yet it is important to stress that the original medieval conception of individuality as embedded in social community was maintained and continued to play a significant role in moral and political thought. It was developed in particular by Rousseau and Hegel and by English idealists such as Bosanquet (1899) and Green (1900). It was precisely this conception of individuality that William McDougall embraced in *The Group Mind* (1920). According to this conception, a person's moral worth and individuality (or identity) are grounded in social community rather than independent of it or opposed to it. What Ross and Floyd and Gordon Allport saw as a threat to individuality, McDougall (1920) viewed as the *source* of individuality and personal integrity, namely,

> the vital relation between the life of the individual and the life of the community, which alone gives the individual worth and significance, because it alone gives him the power of full moral development; the dependence of the individual, for all his rights and for all his liberty, on his membership of the community. (p. 17)

McDougall thought that an asocial being, a being whose psychology and behavior is engaged only individually, independent of or in opposition to social community, was not only an illusion but a morally undesirable ideal.

These differences between McDougall and the Allports on the nature of moral individualism bordered on the political and were intimately related to the Allports' objection to the notion of a social mind. Floyd and Gordon Allport objected not only to the reification and personification of the social mind associated with the advocacy of a distinctive social psychology. They also objected to the implied commitment of such theories to collectivist political ideals associated with the philosophy of Hegel. Gordon Allport (1954) clearly articulated the implied association with totalitarian forms

[13] R. Williams (1961) astutely notes that around the same time as the individual was reconceived as independent of social community, or as opposed to it, the social came to be treated as an abstract individual, or supraindividual, to which individual persons were conceived as externally or extrinsically related. As Asch (1952) later noted, therein lay the ultimate source of the historical neglect of the social in twentieth century American social psychology, which also came to treat the psychological and the social as externally or extrinsically related.

of government:

According to Hegel's idealistic philosophy there is only one Mind (1807). It is absolute, all-embracing, divine. It works itself out in the course of history. Individual men are but its agents. Its principal focus is in the state, which is therefore the chief agent of divine life on earth. Each state has, in fact it *is*, a group mind. It has its own laws of growth and development (the dialectic) and while it makes much use of individuals, it is by no means reducible to their transitory mental life. Marx, as well as Hitler, was among the sinister spiritual children of Hegel. Like him, they equated personal freedom with obedience to the group, morality with discipline, personal growth with the prosperity of the party, class, or state. *Du bist nichts: dein Volk ist alles* was the Nazi rallying cry.

We can trace Hegel's psychological influence in several directions. As we have said, it underlay Karl Marx's exaltation of social class as a superindividual entity. All contemporary Soviet psychology reflects this view (Bauer, 1952). In Britain, Bosanquet (1899) and Green (1900) were among the political philosophers who, following Hegel, viewed the state as an organic mind transcending the component minds of individuals, and demanding "sober daily loyalty." It is hardly necessary to point out that psychological apologists for racism and nationalism (e.g., Jaensch, 1938) tend no less than Hegel to apotheosize the group mind, as represented by the state, race, folk, or *Kultur*. (pp. 34–35)

This implied association was also the basis of Popper's famous (if overstated) critical attack on the notion of social collectivity in *The Open Society and its Enemies* (1945).

However, as Adrian Brock (1992) has pointed out, this is grossly unfair to the advocates of social or group psychology from Wundt to McDougall. The Nazis, for example, burnt copies of Wundt's *Völkerpsychologie* (1900–1920) precisely because Nazi apologists such as von Eickstedt (1933) insisted that a race is defined by blood rather than by social community (as Wundt had maintained), and it was the historical disciples of Wundt, notably the social anthropologist Franz Boas (1934) and his followers, who were among the staunchest critics of the racist policies of Nazi Germany.

The work of William McDougall did come to be associated with the policies of some totalitarian states. McDougall was a committed eugenicist and author of the elitist and positively racist *Is America Safe for Democracy?* (1921b). His practical policy recommendations were followed by the Nazis, who implemented subsidies for Aryan mothers and weekend spa retreats for the breeding of the SS elite (positive genetics) and developed sterilization programs and (eventually) extermination camps (negative genetics). However, these unsavory positions were a product of McDougall's hereditarian beliefs, not a product of his alleged commitment to an objective *Völksgeist*.

In actual fact, McDougall (1920) was quite specific in his rejection of Hegelian notions of supraindividuality, for reasons not far removed from Gordon Allport's own:

I may . . . remind the reader that the conception of the State as a super-individual, a superhuman quasi-divine personality, is the central conception of the political philosophy of German "idealism." That conception has, no doubt, played a considerable part in bringing upon Europe its present disaster. It was an instance of one of those philosophical ideas which claim to be the product of pure reason, yet in reality are adopted for the purpose of justifying and furthering some already existing interest or institution. In this case the institution in question was the Prussian state and those, Hegel and the rest, who set up this doctrine were servants of that state. They made their doctrine an instrument for the suppression of individuality which greatly aided in producing the servile condition of the German people.

I would say at once that the crucial point of difference between my own view of the group mind and that of the German "idealist" school (at least in its more extreme representatives) is that I repudiate, provisionally at least, as an unverifiable hypothesis the conception of a collective or super-individual consciousness, somehow comprising the consciousness of the individuals composing the group. (p. 35)

Nonetheless, despite his explicit rejection of supraindividuality and totalitarianism, McDougall endorsed what he called the "grain of truth" in the idealist philosophy, as represented by "the more enlightened British disciples of this school," such as T. H. Green (cited disapprovingly by Gordon Allport) and E. Barker (quoted approvingly by McDougall), who winnowed "the wheat from the chaff." This was the conception, noted earlier, of the "vital relation between the life of the individual and the life of the community," which McDougall traced back to the Greek city-state. McDougall's conception was directly opposed to the utilitarian and laissez-faire conception of Bentham, Mill, and Spencer, in which the individual was conceived in terms of his or her *independence* from social influence and interference. The virtue of this alternative "communitarian" conception of individuality, according to McDougall (1920), quoting Barker (1915, p. 11), was that it

could satisfy the new needs of social progress, because it refused to worship a supposed individual liberty which was proving destructive of the real liberty of the vast majority, and preferred to emphasize the moral well-being and betterment of the whole community, and to conceive of each of its members as attaining his own well-being and betterment in and through the community. (p. 17)

It was to this communitiarian conception of individuality, this supposed grain of truth in the Hegelian (and Greek and medieval) tradition, to which the Allports objected, morally and politically as much as methodologically. In rejecting this communitarian conception of individuality, the Allports appealed to the distinctively American conception of "autonomy" that had evolved in the preceding centuries. Bellah et al. (1985) articulate this familiar American conception in the following fashion:

> We believe in the dignity, indeed the sacredness of the individual. Anything that would violate our right to think for ourselves, judge for ourselves, make our own decisions, live our lives as we see fit, is not only morally wrong, it is sacrilegious. (p. 142)

However, although it is common to suppose (with Farr, 1996) that the originating ideas of American society were tied to "notions of individual autonomy" (Jehlen, 1986, p. 5) and that commitment to personal autonomy has remained a constant of American life from its inception to the present day (Dolbeare, 1984), the original colonists and revolutionary founders did not completely abandon European notions of community (Shain, 1994). Although "the animating idea of the American founding was individual liberty" (Himmelfarb, 1988, p. 117), it was a "liberty characterized by a voluntary submission to a life of righteousness that accorded with objective moral standards as understood by family, by congregation, and by local communal institutions" (Shain, 1994, p. 4).

The form of individualism embraced by early American colonists was decidedly communitarian in nature. The "independent citizens" of colonial times embraced a form of individualism bound to local cohesive communities of family and church. According to Shain (1994), until at least the end of the revolutionary era, most Americans believed that "human flourishing" was "to be achieved in group life rather than individually" (p. 6). Isolative, selfish, and licentious forms of individualism were perceived as a threat to morality and democracy by Puritan moralists such as John Winthrop, the first governor of the Massachusetts Bay Colony, and political theorists such as Thomas Jefferson, the primary author of the Declaration of Independence. The psychological and behavioral constraints imposed by community were held to sustain democracy by restraining the potential excesses of material greed and capitalist exploitation, since it was believed that a mass of mutually antagonistic individuals could degenerate into anarchy and become easy prey to despotism (Bellah et al., 1985).

This concern was famously echoed by Tocqueville (1830/1969), who identified American individualism with the goal of "material betterment" that emerged with the geographic and economic expansion of the nineteenth century. Tocqueville identified distinctively American "habits of the heart" – such as commitment to family life and religion and participation in local politics – that encouraged citizens to maintain links with the broader social and political community and served to sustain the free democratic institutions of the new country. However, Tocqueville also warned that the individualism of material betterment could undermine these conditions of freedom and democracy and that any form of individualism that rejected all civic, religious, and communitarian traditions rejected the social meaning and significance that is the very basis of human individuality and dignity.[14]

Yet this form of communitarian individualism could not survive the intellectual, economic, and political developments of nineteenth century America. A newer and more distinctively American form of individualism emerged and began to develop independently of the older religious and civic forms that were European in origin. The humanistic Unitarianism that eventually came to displace the Calvinistic Protestantism of New England emphasized the capacity of the individual soul for growth and perfectability, and all social and political forms came to be seen as subservient and secondary to the perfection of the soul. This new individualist credo was expressed in the religious teachings of W. H. Channing, which many identified as the basis of the new religion for the new democracy, but it found its purest expression in the writings of Ralph Waldo Emerson. For Emerson, self-reliance and self-determination formed the true basis for democracy: "The root and seed of democracy is the doctrine, Judge for yourself" (1834/1912, p. 369). Emerson and later theorists such as Josiah Warren, whose doctrine of the "sovereignty of the individual" greatly influenced John Stuart Mill (1873/1924, pp. 216–217, cited in Arieli, 1964), rejected any form of social organization that would constrain the autonomy of the individual, including the new socialist forms of civil organization. In doing so, they set the autonomous individual against society, whether in the form of local community or national government (Arieli,

[14] A concern also expressed in a variety of sociological works of the 1920s and 1930s, such as *Middletown* (Lynd & Lynd, 1929) and *Middletown in Transition* (Lynd & Lynd, 1937), which lamented the loss of the "independent citizen" in the face of industrial expansion. It is also expressed in postmodern angst about the "emptiness of a life without sustaining social commitments":

> If this is the danger, perhaps only the civic and biblical forms of individualism – forms that see the individual in relation to a larger whole, a community and a tradition – are capable of sustaining genuine individuality and nurturing both public and private life. (Bellah et al., 1985, p. 143)

1964). As Emerson put it, "Society is everywhere in conspiracy against the manhood of every one of its members" (1841, cited in Bellah et al., 1985).

It is not difficult to imagine how such an conception of individuality might come to be represented as a threat to society itself, undermining the communitarian basis of democracy and human dignity, as indeed it later was by McDougall. Unfortunately for the communitarian conception of individuality, Southern apologists for slavery such as John C. Calhoun (1838) and George Fitzhugh (1854) regularly promoted it before the Civil War and contrasted the Southern culture of harmonious "small communities" with the Northern culture of individuals "eagerly pursuing... [their] own selfish welfare." The acrimonious national debate this generated, and the victory of the North in the Civil War that followed, served to focus, distill, and enshrine the ideal of autonomy or self-determination as both the basis and the ultimate goal of a democratic free society (Arieli, 1964). The work of European theorists such as John Stuart Mill and Herbert Spencer reinforced this distinctively American ideal of autonomy individualism, which became deeply entrenched by the end of the nineteenth century and provided a link to the "evolving ideology of private enterprise and laissez-faire, postulating absolute equality of opportunity and the claim that private accumulation leads to public welfare" (Lukes, 1973b, p. 30).

It is clear that Floyd and Gordon Allport were committed to this distinctively American form of autonomy individualism. In his contribution to Boring and Lindzey's (1974) *History of Psychology in Autobiography*, Floyd Allport characterized himself as someone who as a young man resolved "to accept nothing on faith, but to carry forward an analysis and examination of every point to the bitter end" (Vol. 6, p. 5). He proudly acknowledged that he had "failed to show even a decent regard for traditional beliefs and convictions" and was "reclusive by nature," someone "who did not mingle with colleagues and associates" (p. 5). He claimed that in his work on institutional psychology he was "pleading no cause nor advocating a change of institutions, but was in effect questioning the efficacy of institutions as such" (p. 17), and he suggested that his position anticipated the assault on institutions advanced by the 1960s generation. In his own contribution to Boring and Lindzey's (1967) *History of Psychology in Autobiography*, Gordon Allport described himself as a "political liberal" (Vol. 5, p. 17) with a strong belief in the "integrity of each individual human life" (p. 11), and he expressed his conviction "that any adequate psychology of personality must deal with the essential uniqueness of every personal structure" (p. 16). Like his brother Floyd, Gordon Allport obviously prided himself on his pursuit of his "own personal idea in the face of contrary fashion" (p. 13), and he emphasized that his own work had been consistently "critical of prevailing psychological idols" and "fashionable explanatory principles" (p. 22).[15]

It is also possibly not accidental that some of those who defended a distinctively social conception of psychological states and behavior, such as Daniel Katz, Muzafer Sherif, and Solomon Asch, were European socialists committed to social causes,[16] as were a good many of those who later embraced Kurt Lewin's vision of a socially useful psychological science (or "action research"). However, there is no intrinsic connection between a commitment to socialism or social causes and recognition of the social dimensions of human psychology and behavior. McDougall was no socialist, and many individualistically oriented social psychologists, including the Allports, were deeply committed to social causes. Moreover, as noted earlier, most social psychologists (like most psychologists in general) rejected the laissez-faire brand of social Darwinism promoted by Spencer in favor of scientifically based interventionist strategies of social improvement. Both Floyd and Gordon Allport were as opposed to the uniform economic "units" of institutionalized capitalism as they were to the leveled uniformity of socialism and communism (Pandora, 1997). Their own advocacy of autonomy individualism was largely based on their distinctive emphasis on the particularity of individual character and personality.[17]

Nonetheless, McDougall had a point when he complained that autonomy individualists tended to caricature all forms of socially engaged human psychology and behavior as involuntary and regimented when in fact many social forms of individuality are freely engaged by members of social groups. As D. Katz and Schanck (1938) noted, when attitudes

[15] In Gordon Allport's case, this included opposition to the then fashionable commitment to a narrowly behaviorist and experimentally oriented conception of social psychological science (Pandora, 1997) – the conception favored by his brother Floyd.

[16] Daniel Katz (1991) described his socialist background as the source of one of the few conflicts he had with his teacher Floyd Allport:

> The first of these occurred when I told Floyd that I believed that his characterization of radicals as personality misfits was incorrect and reactionary. I came from a socialist home and I thought I knew more about radicalism than Floyd did with his elitist background. His response to that challenge was to turn his advanced social psychology course over to me for two sessions to document my opinions. This turned out to be an easy assignment since the students were on my side and I was knowledgeable on the subject. In the end, Floyd graciously admitted he might have been wrong. (p. 126)

[17] This was especially true of Gordon Allport (1939), a pioneer of American personality theory and research.

and behaviors are engaged socially and (in consequence) uniformly by members of a social group, it does not follow that they are engaged in an involuntary and regimented fashion, since "the uniform activities of a group of people may represent their fundamental wishes very well" (p. 9). Conversely, attitudes and behaviors that are engaged individually, out of fear of punishment or death, may be also be uniform and regimented, as in the case of "good citizens" frightened into conformity with totalitarian ideology and practices.

It is instructive to compare D. Katz and Schanck's (1938) and Floyd Allport's (1934) interpretation of conformity behavior as represented by the J-curve distribution. Allport (1934) explained such behavior in terms of individually engaged interpersonal "conformity producing influences," such as conditioning, punishment, and the like (p. 169), and he treated such behavior as psychologically as well as statistically abnormal, as pathological forms of "crowd-like subservience" (Allport, 1924a, p. 396). In contrast, D. Katz and Schanck (1938) maintained:

> By itself the J distribution is a measure of uniform behavior, but it is not an indication of whether the uniform behavior represents voluntaristic or regimented behavior. By definition, regimented conformity is a complete compliance in a specific situation through the inhibition of opposed reaction tendencies. (p. 48)

As D. Katz and Schanck observed, individuals may willingly engage in such forms of conforming behavior because their *identity* is oriented to a particular social group (such as Methodists or lawyers):

> Through identification with a larger group we enhance our egos. We conform because in so doing we become part of a great university, a great church, a great nation. We can claim as our own the accomplishments of our fellow members and our leaders. (p. 174)

One of the casualties of the post-1930s neglect of the social dimensions of human psychology and behavior was the consequent neglect of this social conception of identity, according to which a person's sense of individuality is grounded in social community. This social conception of identity was a distinctive feature of the social developmental theories of Baldwin (1897), Cooley (1902), and Mead (1934) and later Goffman (1959, 1961).[18] Baldwin (1911), for example, claimed "the most remarkable outcome of modern social theory" to be "recognition of the fact that the individual's normal growth lands him in essential solidarity with his

[18] Although, as noted in Chapter 4, it came to be attenuated via the increasingly restricted focus on interpersonal interaction and learning within the social interactionist tradition.

fellows, while on the other hand the exercise of his social duties and privileges advances his highest and purest individuality" (p. 16). Only Sherif (1948) and Asch (1952) and later Pepitone (1976) and Sampson (1977) preserved this social conception of identity as the basis of socially engaged forms of human psychology and behavior.

Finally, D. Katz and Schanck (1938) astutely noted that different forms of individualism, such as the communitarian and autonomy forms of individualism advocated by McDougall and the Allports, may themselves be adopted socially:[19]

> On a more fundamental level, moral and political positions on these different forms of individualism may themselves be held either socially or politically, willingly or reluctantly, by their advocates as well as their objects. (p. 9)

## III

The Allports were not completely blinded (methodologically) to the social dimensions of human psychology and behavior. They reluctantly and grudgingly recognized them, but abhorred them on moral and political grounds. Consequently, they refused to grant the study of social psychological states and behavior a central position in the developing scientific discipline of social psychology.

It is clear enough that Gordon Allport, for example, recognized both the possibility and actuality of socially engaged beliefs and attitudes, despite his regular characterization of social beliefs and attitudes as nothing more than common beliefs and attitudes. In his 1935 paper on attitudes, for example, Allport described one of the studies of D. Katz and F. H. Allport (1931) on the attitudes of student fraternity members:

> The attitudes of fraternity members were investigated. Two-thirds of the students turned out to be "institutionalists" who believed that their fraternities were of the order of super-individual Beings; the remaining third were "individualists" who believed that their fraternities were mere collections of individuals. (p. 829)

[19] There is of course no inconsistency in recognizing that principles of autonomy individualism may be socially held. Indeed, some theorists have maintained that the social adoption of principles such as autonomous rationality would be positively virtuous:

> If rationality were once to become really respectable, if we feared the entertaining of an unverifiable opinion with the warmth with which we fear using the wrong implement at the dinner table, if the thought of holding a prejudice disgusted us as does a foul disease, then the dangers of man's suggestibility would be turned into advantages. (Trotter, 1916, p. 45)

It may be doubted whether Syracuse students really were committed to a Hegelian conception of their fraternities as "super-individual Beings," although it is reasonable to suppose that they represented themselves as members of a genuine social group as opposed to a mere collection of individuals and that some of their attitudes were in consequence held socially. However, Allport went on to stress that the individualists, that is, those who represented their fraternities as mere collections of individuals (and who, it might also be suggested, likely did not see themselves as members of a distinctive social group), did not adopt the beliefs or attitudes of the fraternity members who were committed to the fraternity as an institution:

> This latter group, as contrasted to the fraternity institutionalists, believed that varsity teams should not jeopardize individualized athletics ... that professors should have considerable freedom of expression in the classroom, that some restrictions should be placed upon the privileges of fraternities, and that eccentric and unpopular student types should be admissible to fraternities. (p. 829)

Allport had a legitimate point to make about the "generality of attitudes":

> Individualists in one situation are individualists in another; and institutionalists in one are institutionalists in another. (p. 829)

What is of interest for present purposes is the conclusion Allport drew from the study. He noted that although all the students sampled were members of fraternities, the individualists did not appear to be strongly affected in their beliefs and attitudes by their membership of the fraternities:

> It will be noted that all of these students are members of fraternities, and yet not all are equally affected by such membership. One-third of them refuse to accept the cultural pattern at its face value. Their attitudes, therefore, *are not determined, as some sociologists would argue, exclusively by the social influence to which they are exposed.* (p. 829, emphasis added)

Allport dismissed the idea that student attitudes are socially determined as a false doctrine propagated by "some sociologists."[20] However, the acknowledged fact remains that two thirds of the fraternity members do appear to have been affected by their membership of the fraternities: they appear to have held certain attitudes (at least in part) socially rather than individually. This was not something that Allport supposed deserved special discussion or investigation, although it is very clear that he considered the individualists as the true representatives of the American ideal of autonomy individualism.

[20] Allport seems to have had Faris (1925) in mind.

Since Allport equated sociality with uniformity, and uniformity with the involuntary, even his own desocialized conception of social attitudes as merely common attitudes posed a threat to his cherished conception of autonomy individualism and to the viability of his own vision of social psychology:

> If all attitudes were common attitudes, it would be possible to construct acceptable social laws, for in such a case all attitudes would be the same for all people, and human nature would thus become a "constant." Since, however, common attitudes are not the only type, social laws are merely the statement of tendencies based upon the resemblance between *some* of the attitudes of *some* of the individuals within any group. Prediction is virtually impossible in social psychology largely because common attitudes are not sufficiently universal to be depended upon. Social psychology cannot afford to overlook the "individual perturbations" in social life. The differences between human beings, in spite of all standardizing influences, are more noteworthy than their resemblance. Personality everywhere intrudes itself. (p. 827)

He might just as easily have said that the autonomous individual intrudes himself in a social psychology devoted to "individual perturbations."

Floyd Allport (1933) also grudgingly recognized socially engaged psychological states and behavior but thought that their social engagement was morally and politically reprehensible and grounded in illusion. In maintaining this position, he abandoned his strict behaviorism and acknowledged that socially engaged beliefs and attitudes (albeit grounded in illusion) causally influence people's behavior:

> It is true that students and teachers sometimes *think* of their university as having a life and a continuity of its own. Thinking in this way, moreover, may modify, to some extent, their academic behavior and their relationships. But the fact that they think and act upon the assumption that their university is a super-individual reality does not prove that it is such a reality. The early Greek conception of lightning as the thunder-bolt of Zeus was widely accepted and transmitted through succeeding generations; it also entered into the emotional life and the activities of the people in a profound way. Yet these facts do not constitute the slightest proof that the imagined Being called Zeus existed. (p. 6)

It is true that thinking of a university as a supraindividual no more constitutes it as a supraindividual than thinking of lightning as the thunderbolt of Zeus establishes the reality of Zeus. Yet as Simmel (1908/1959) constantly stressed, if many or most members of a university come to conceive of themselves as forming a unity and thus come to think, feel, or act in certain ways because and on condition that other members are represented as thinking, feeling, and acting in these ways, then that is sufficient to constitute them as a genuinely social group – a "real" social group as opposed to a "mere sum of individuals" (Durkheim, 1895/1982a, p. 129). Indeed, as Simmel also often stressed, it is not strictly necessary that they explicitly represent themselves as a social unity to constitute a social unity: it is sufficient that they share socially engaged psychological states and behavior. Moreover, although their representation of themselves as a unity by itself neither constitutes them as a social group nor as a supraindividual, it may nonetheless exert a powerful (social) influence on their thought, emotion, and behavior, just as the ancient Greek belief that lightning was the thunderbolt of Zeus "entered into the emotional life and the activities of the people in a profound way."

Allport (1933) acknowledged this reluctantly, but he clearly held that it represented a threat to personal autonomy and moral and political well-being:

> This collection of segmental habits, which we call the institution, is something we envisage as belonging more properly to Society than to particular individuals. But Society, if conceived as a being like a human organism, is a fiction. We, as individuals, are the organisms; institutional habits are really part of *us*. And like all our other habits and dispositions, they must be made to harmonize not merely with the pattern of society, but with *our own characters as individuals*. If such a personal integration is not accomplished, no matter how expertly the "societal" pattern is conceived, our institutional habits (that is, our "institutions") will lead eventually to our ruin. (pp. 471–472)

Voicing the same sort of fears as his brother, Floyd Allport characterized socially engaged beliefs, emotions, and behaviors as posing a threat to human individuality and to society itself. In fact, Allport's *Institutional Behavior*, published in 1933, was as much a polemical work of social and political philosophy as a work of theoretical and empirical psychology. Most of this work recognized the existence of socially engaged psychological states and behavior but railed against the powerful role played by illusory beliefs in the supraindividual reality of social groups (beliefs that no doubt do play a significant role in cementing socially engaged psychological states and behavior).

## IV

Another reason the social dimensions of human psychology and behavior were held to pose a threat to autonomy individualism was the tendency of some American social psychologists to follow European theorists such as Gustav Le Bon (1895/1896) and Gabriel Tarde (1890/1903) in equating social behavior with the irrational and emotional behavior of crowds or mobs.[21] Thus Floyd Allport (1924a), for example, characterized the conforming behavior represented by the J-curve distribution as a pathological form of "crowd-like subservience" (p. 396).[22]

Le Bon (1895/1896) maintained that "a crowd being anonymous, and in consequence irresponsible, the sentiment of responsibility which always controls individuals disappears entirely" (p. 29) and that the unconscious suggestibility of individuals in crowds is analogous to the behavior of individuals under the influence of hypnosis. Gabriel Tarde, who founded his conception of social behavior on the principle of "imitation," was also scathing in his denunciation of the irrational and irresponsible behavior of individuals in crowds. He likened social behavior to a form of sleepwalking: "Society is imitation and imitation is a form of somnambulism" (1890/1903, p. 87).

Conceptions of social influence as analogous to crowd influence posed a threat to personal autonomy by postulating factors that were held to determine human psychology and behavior independently of considerations of rationality. Significantly, Jackson and Pettit (1992), in their discussion of subversive social structural determinants of human psychology and behavior, liken a person insensitive to evidence, argument, and considerations of consistency to a person "under a hypnotically or neurally induced compulsion" (p. 129). This was precisely how theorists such as Le Bon and Tarde characterized the behavior of individuals in crowds: "The social like the hypnotic state is only a form of dream" (Tarde, 1890/1903, p. 77).

McDougall (1920), once again, was an exception, and he steadfastly resisted the general assimilation of crowds and social groups. As he noted, it was primarily crowd theorists such as Le Bon and Sighele who believed that the principles of a distinctively social or group psychology posed a threat to autonomy and rationality. Such crowd theorists claimed to have shown "how participation in the group life degrades the individual, how the group feels and thinks and acts on a much lower plane than the average plane of the individuals who compose it" (p. 20). In contrast, McDougall aligned himself with those who

> insisted on the fact that it is only by participation in the life of society that any man can realise his higher potentialities; that society has ideals and aims and traditions

[21] The consequences of this equation are explored in more detail in Chapter 7.

[22] It was precisely this assimilation of the social and the psychologically abnormal that led Morton Prince in 1918 to temporarily unite social psychology and psychopathology within the pages of the *Journal of Abnormal and Social Psychology*.

loftier than any principles of conduct the individual can form for himself unaided; and that only by the further evolution of organised society can mankind be raised to higher levels; just as in the past it has been only through the development of organized society that the life of man has ceased to deserve the epithets "nasty, brutish, and short" which Hobbes applied to it. (p. 20)

As McDougall recognized, these opposing positions appeared to generate a paradox:

Participation in group life degrades the individual, assimilating his mental processes to those of the crowd, whose brutality, inconstancy and unreasoning impulsiveness have been the theme of many writers; yet only by participation in group life does man become fully man, only so does he rise above the level of the savage. (p. 20)

McDougall conceived of the resolution of this paradox as "the essential theme" of *The Group Mind*. He purported to resolve the paradox precisely by distinguishing between crowds and social groups:

[*The Group Mind*] examines and fully recognises the mental and moral defects of the crowd and its degrading effects upon all who are caught up in it and carried away by the contagion of its reckless spirit. It then goes on to show how organization of the group may, and generally does in large measure, counteract these degrading tendencies; and how the better kinds of organisation render group life the great enabling influence by aid of which alone man rises a little above the animals and may even aspire to fellowship with the angels. (p. 20)

However, McDougall's position was, and has remained, a minority view.

## V

As in the case of the relation between the original conception of the social dimensions of human psychology and behavior and supraindividual theories of the social mind, there is no intrinsic connection between the original conception of the social dimensions of human psychology and behavior and the moral and political forms of communitarianism and collectivism that appeared so threatening to the Allports and their followers. However, their association may partly explain the reluctance of many social psychologists to wholeheartedly embrace a social psychology committed to the empirical investigation of socially engaged psychological states and behavior. It may also partly explain why social psychologists, like general psychologists, political scientists, and economists, came to label their discipline as a "behavioral" rather than a "social science"

(Manicas, 1987, p. 237), under pressure from grant-awarding agencies such as the Ford and Russell Sage Foundations. These foundations may have been perceived as unlikely to fund research that would undermine the psychological foundations of autonomy individualism and political liberalism.

It is certainly true that from the 1930s onward the social dimensions of human psychology and behavior were generally conceived negatively, in terms of threats to the autonomy of the individual. Krech, Crutchfield, and Ballachey (1962) lamented the "other-directed" nature of the behavior of many Americans (notably upper middle class urban Americans) highlighted by Reisman (1950),[23] but they assured their readers that "every age has had an effective number of persons who have held and acted on attitudes that are *counter* to the majority" (p. 192). Social psychologists expressed their deep concerns about "risky-shift" behavior (Dion, Barry, & Miller, 1970), "group-think" (Janis, 1968), "conformity" (Kiesler & Kiesler, 1968), and "obedience to authority" (Milgram, 1963, 1974), which were treated as threats to society as well as the autonomy of the individual.

[23] See Reisman (1950): "What is common to all the other-directed people is that their contemporaries are the source of direction for the individual" (p. 22).

# 7

# Crowds, Publics, and Experimental Social Psychology

A restrictive form of methodological individualism, in conjunction with a particular vision of moral individualism, played a significant role in the historical neglect of the social by later generations of American social psychologists, especially those who followed Floyd Allport in his commitment to an objective experimental science of social psychology. However, this was not because commitment to an experimental science was itself antithetical to the exploration of the social dimensions of human psychology and behavior but because of the impoverished conception of social groups that came to inform the experimental program of American social psychology.

European social theorists such as Durkheim, Weber, and Simmel and early American social psychologists such as Bernard, Bogardus, Dunlap, Katz, McDougall, Ross, Schanck, Thomas, and Wallis had a fairly sophisticated grasp of the distinction between social and merely common forms of cognition, emotion, and behavior and of the distinction between genuine social groups and aggregate groups. However, many American social theorists, including some early American social psychologists, inherited the impoverished conception of social phenomena advanced by European crowd theorists such as Gustav Le Bon (1895/1896), Gabriel Tarde (1890/1903, 1901/1967), and Scipio Sighele (1892).[1] They tended to assimilate, if not directly equate, social or collective behavior and crowd or mob behavior (i.e., the behavior of aggregations of physically proximate individuals irrespective of whether such individuals are members

[1] Le Bon was accused of plagiarism by Sighele and certainly appears to have "borrowed" significantly from Tarde (van Ginneken, 1985).

of social groups). This made it very easy for critics of a distinctive social psychology to deny or depreciate the social dimensions of psychological states and behavior.

As noted earlier, European social theorists who likened social behavior to crowd behavior stressed the irrational and emotional nature of crowd behavior. They also focused on the threats to civilization posed by democratic assemblies and jury trials, conceived as small crowds (R. A. Nye, 1975; van Ginneken, 1992). Le Bon declared the modern age the "era of crowds," a time in which "the substitution of the unconscious action of crowds for the conscious activity of individuals is one of the primary characteristics" (p. 5). Consequently, universal suffrage would result in a "sovereign crowd" determined "to utterly destroy society as it now exists" (p. 16).

Gabriel Tarde (1890/1903), who defined a crowd as "a collection of psychic connections produced essentially by physical contacts," also condemned the irrational and irresponsible behavior of individuals in crowds but maintained that the modern age was not an age of crowds but of "publics." Publics, the product of technological developments in travel and communication (the railroad, the newspaper, and the telephone), were defined as "a purely spiritual collectivity, a dispersion of individuals who are physically separated and whose cohesion is purely mental" (Tarde, 1901/1969, pp. 277–278).

## I

It must be acknowledged that there are some grounds for assimilating the psychology and behavior of individuals in crowds and in social groups. Individual members of crowds and individual members of social groups tend to instantiate forms of cognition, emotion, and behavior different from the forms they would instantiate "in isolation" from the crowd or social group of which they are members. Thus Le Bon (1895/1896) maintained that individuals in a crowd "feel, think, and act in a manner quite different from that in which each individual of them would feel, think and act were he in a state of isolation" (p. 2). And McDougall (1920) maintained that "the actions of the society are, or may be, very different from the mere sum of the actions with which its several members would react to the situation in the absence of the system of relations that render them a society" (p. 9).

However, despite this acknowledged similarity, there is a fundamental difference between the influence of the crowd and the social group.

Crowd behavior is a (postulated) function of the interpersonal influence of physically proximate individuals independent of their membership of social groups. Individuals in crowds behave (and think and feel) differently than they do in physical isolation from other individuals. In contrast, social behavior is a (postulated) function of the orientation of behavior (and thought and feeling) to the represented behavior (and thought and feeling) of members of social groups independent of the physical presence of other members of social groups (or any other persons).

Thus the two senses of behavior "in isolation" with which crowd and social behavior were frequently contrasted are quite different. In the case of crowd psychology, the contrast was with the psychology of individuals in physical isolation from other individuals; in the case of social psychology, the contrast was with the psychology of individuals independent of their social group orientation. Recalling Cooley's (1902) remark that "a separate individual is an abstraction unknown to experience" (p. 1), we might note that, while a socially separate individual, an individual with no actual or potential ties to any social group, is an abstraction (almost) unknown to experience,[2] a physically separate individual is certainly not. Indeed, many individuals engage in social forms of cognition, emotion, and behavior, such as solitary prayer or hunger strike, in physical isolation from other individuals.

It was precisely this issue that divided Durkheim and Weber on the one hand and Le Bon and Tarde on the other. Durkheim and Weber denied that forms of psychology and behavior that are simply a product of interpersonal imitation and influence are social forms of human psychology and behavior. As Weber (1922/1978) put it, behavior that is merely a product of interpersonal imitation, such as following the other members of a crowd in a certain direction, is not social action because one's "action is... causally determined by the action of others, but not meaningfully" (p. 114): that is, not by reference to the represented behavior of members of a social group.

The difference in these postulated forms of interpersonal and social causation is not accidental but relates directly to the fundamental difference between crowds or mobs on the one hand and social groups on the other. Crowds or mobs are constituted in a quite different fashion from

[2] Aside from feral children and Gordon Allport's congenital Robinson Crusoe (1935, p. 838), who (at least according to Allport) would retain some elements of personality (albeit highly impoverished elements).

social groups and only rarely and contingently correspond with them. Crowds or mobs are aggregate groups composed of physically proximate individuals. Social groups, by contrast, are composed of individuals whose psychology and behavior are oriented to the represented psychology and behavior of other members of the social group, independently of whether the other members of the social group happen to be physically assembled in any particular place at any particular time.

The members of a crowd need not conceive of themselves as constituting a social group (or any form of "unity"), and their common psychological states and behavior may be generated individually: that is, independently of whether members of any social group are represented as engaging in these psychological states and behavior. The behavior of individuals in a crowd attacking stallholders and stealing their bread may be a product of their individually engaged states of starvation or their individually engaged reaction to hearing the news that the price of bread has been raised. The common pride or anger of individuals in a crowd may be a product of their individually taking pride or individually being angered by the president's speech. The common pride or anger of individuals in a crowd may also be transmitted by some individuals in the crowd to others via behavioral cues and signals, in the fashion in which originally unperturbed individuals tend to become afraid when they enter the dentist's waiting room and perceive the fear of others (Wrightsman, 1960), or in which individuals in the Schachter-Singer experiment labeled their (artificially induced) arousal as anger or euphoria based on an inference from the observed "angry" or "euphoric" behavior of experimental stooges (Schachter & Singer, 1962).

This is not to deny that the psychology and behavior of individuals in a crowd are sometimes *also* social in nature: for example, when the individuals in a crowd also happen to be members of a social group, as when the trade union members agree to picket the factory on Monday morning, or when the crowd of Native Americans delivering a petition to the town hall become angered by the flag honoring the local "Redskins" football team. However, these psychological states and behavior are social because they are socially engaged by members of a social group, not because they are *common* psychological states and behavior engaged by physically proximate individuals stimulated by common objects or by the psychological states and behavior of others. Individuals in a crowd need not be and frequently are not members of any distinctive social group. Often enough they are strangers who are individually members of quite different social groups (lawyers, psychologists, Catholics, Muslims, and so forth).[3]

Unfortunately, too many early American social psychologists did follow Le Bon and Tarde in assimilating crowds and social groups. This made things very easy for critics of a distinctive social psychology, by enabling them to dismiss the notion of a distinctive social psychology via analyses of crowd emotion and behavior as individually engaged forms of emotion and behavior, which are easy to provide. As Floyd Allport (1924a) and others noted, crowd emotion and behavior are often generated via a common stimulus that individually engages common forms of emotion and behavior in a large number of individuals (many of the individuals in the socially heterogeneous crowd may be individually angered by the president's speech). By assimilating social forms of cognition, emotion, and behavior and common forms of cognition, emotion, and behavior in a crowd, it was possible to advocate and develop a scientific social psychology that methodologically acknowledged *only* individually engaged psychological states and behavior.

[3] That said, it should be noted that Durkheim himself did tend to assimilate crowds and social groups despite his own best efforts to distinguish between a genuine social group and a "mere sum of individuals." He noted that the external forms of coercion operative in assemblies of physically proximate individuals are similar to those that operate in social groups:

> Thus in a public gathering the great waves of enthusiasm, indignation and pity that are produced have their seat in no one individual consciousness. They come to each one of us from outside and can sweep us along in spite of ourselves. (1895/1982a, pp. 52–53)

Durkheim may have been thinking of crowds that also happened to contingently constitute social groups: the "public gathering" or "assembly" he had in mind may have been a public gathering or assembly of some social group, such as an assembly of political supporters, trade unionists, or religious converts. Alternatively, he may have been noting nothing more than the fact (or may have been simply misled by the fact) that both crowd influence and social influence involve forms of *external* determination of cognition, emotion, and behavior.

Durkheim may also have been concerned in this passage to defend his formal definition of social facts in terms of externality and constraint (their "thing-like" nature) and may have (mistakenly) thought that crowd emotion served as a good illustration of the supraindividuality of social facts. Durkheim in fact employed this example to make a legitimate independent point (independent of the adequacy of the illustrative example), that social forms of cognition, emotion, and behavior are not restricted to social groups with an established and "well-defined social organization" but are also to be found in more fluid "social currents," both "transitory" and "more lasting" (1895/1982a, pp. 52–53). Finally, whatever exactly Durkheim had in mind when he talked of the "waves of enthusiasm, indignation and pity" produced at a public assembly, he would have categorically denied that forms of enthusiasm, indignation, or pity count as "social facts" (social emotions) if they are merely the product of interpersonal imitation (1895/1982a, p. 59).

Many early American social psychologists, such as Floyd Allport (1920, 1924a), Baldwin (1897), Cooley (1909), Giddings (1896), Park (1902), and E. A. Ross (1908), followed Le Bon and Tarde[4] in assimilating crowds and social groups. Moreover, they sought to avoid the irrationalist and antidemocratic implications of European crowd theories by developing Tarde's distinction between physically proximate crowds and dispersed crowds or publics. They maintained that so long as aggregations of individuals are physically dispersed, the irrationalist influences of physically proximate crowds can be resisted (King, 1990).

Therefore, while Franklin Giddings (1896, pp. 150–151) agreed with Le Bon's characterization of the behavior of individuals in crowds as "subject to a swift contagion of feeling" and "devoid of a sense of responsibility," he denied that this necessarily applied to dispersed crowds or publics:

> In recognizing the deliberate action of the social mind I am of course by implication rejecting the conclusion of those who hold that the social mind never acts rationally, or that its action at the best must be less rational than is that of individuals. M. Le Bon argues that unconscious action, passion, and sentiment predominate in the crowd, because individuals differ less in feeling than in intelligence. His conclusion is beyond doubt true of crowds in the usual English meaning of the word, but M. Le Bon gives a wide extension to *foule*, and makes it cover not only a number of persons congregated in one place, but also any class of persons that communicate about their common interests. Of associations in this latter sense his conclusion will not always hold good. In the prolonged deliberations of a group of men that alternatively meet and separate, or that communicate without meeting, the highest thought of the most rational mind among them may prevail. (p. 137)

Indeed, Giddings was so enthusiastic about this conclusion that he maintained that the psychological processes characteristic of individual members of publics are often identical to those of individuals in isolation.[5]

[4] Many early American social psychologists acknowledged their intellectual debt to Tarde. Ross (1908) paid "heartfelt homage to the genius of Gabriel Tarde" (p. viii); Baldwin translated and provided an introduction to Tarde's (1899) *Les lois sociales* (Laws of Society); Giddings translated and provided an introduction to Tarde's (1890) *Les lois de l'imitation* (Laws of Imitation).

However, not all theorists who acknowledged a debt to Tarde accepted the specifics of his account. Ross admitted that his system "swung wide" of Tarde's, and Baldwin complained about being lumped together with Tarde. Giddings was actually much closer to Durkheim than Tarde, despite his ostensive criticism of Durkheim (Giddings, 1896, pp. 146–147) and support of Tarde: he distinguished between "impulsive" behavior based on interpersonal imitation and "traditional" behavior oriented to social group membership.

[5] Thus in 1896 Giddings answered in the affirmative the question cautiously raised as an open question by Durkheim in 1901, as to whether the laws governing social and

He also held that this optimistic analysis extended to those crowds that constitute parliamentary assemblies, about which Le Bon, Tarde, and Sighele had the darkest and most pessimistic of thoughts:

> Alternative meeting and separation is, in fact, the one essential condition of true social deliberation. For the social mind is far from being, as M. Le Bon attempts to prove, very unlike the individual mind in its operations. It is astonishingly like the individual mind, and in no respect more so than its rational processes. When the individual deliberates he permits new ideas to interpose themselves between suggestion and act, or between hypothesis and judgment. He diverts his attention, as he says, which simply means that he breaks the continuity of idea and impulse by opening the mind to new influences. Time and new associations are necessary to deliberation. If the social mind would deliberate it must follow a similar course. The spell that holds the crowd must be broken. The orientation of its thought must be disturbed; the catch-word fetishes must cease to hypnotize. To this end the crowd must disperse; the assembly must adjourn; the legislator must now and then go back to his constituents. When this is done the social mind may deliberate as rationally as the individual mind. (p. 151)

This was little more than wishful thinking unsupported by evidence.[6] But it was a common enough opinion among some early American social psychologists, who argued that properly structured and educated publics, along with properly regulated democratic assemblies, have the "luxury of extended deliberation" (Baldwin, 1897, p. 108) and the capacity for "cool discussion and leisurely reflection" (E. A. Ross, 1908, p. 47).

European theorists such as Le Bon, Tarde, and Sighele were extremely pessimistic in their views about democracies and democratic assemblies but somewhat grudgingly accepted them as facts of social life. They consoled themselves by maintaining that knowledge of the irrationalist and imitative tendencies of crowds and publics could be exploited and manipulated by elites for the greater good of society. Le Bon, for example,

individual representations are identical or distinct. He also answered it by reference to the laws of the association of ideas developed within individual psychology, which Durkheim (with good cause) doubted were capable of accommodating social forms of cognition.

6 Other than by an appeal to rather contentious (and selective) historical examples, such as the repeal of the corn laws in England in 1849 and the abolition of slavery in 1865 (both arguably examples of political and military expediency).

Nonetheless, Giddings (1896, p. 147) did recognize socially engaged beliefs and attitudes grounded in "consciousness of kind": a form of common "like-mindedness" of members of social groups based on their awareness of themselves as members of distinct social groups. He avoided the threat posed to autonomy and rationality by distinguishing between *traditional* social behavior, grounded in socially engaged beliefs and attitudes, and *rational* social behavior, based on individually engaged beliefs and attitudes (based on critical reasoning).

was a big hit with Mussolini and Hitler.[7] Americans appeared much less eager to approve the psychological exploitation of the masses by elites (King, 1990), although Floyd Allport was a possible exception (1924a, p. 396). What concerned them most was what continues to concern many today: the danger of the manipulation of public opinion, emotion, and behavior by the government and the press via the selective and suggestive presentation of information.[8] However, their concern was with what they perceived as an external threat to the intrinsic rationality of educated and responsible members of publics in mature democracies. It was not a concern with the fundamentally subversive social dimensions of everyday thought, emotion, and behavior.

## II

Although Americans managed to resolve their doubts about the irrationalist and antidemocratic implications of crowd theories by distinguishing between physically proximate crowds and dispersed publics, their assimilation of social groups and physically proximate crowds and dispersed publics had an important impact on the development of American social psychology. It effectively restricted experimental social psychology to the study of the interactions of small aggregations of strangers: to the study of the interpersonal behavior of small local "crowds." It also effectively restricted the study of social attitudes and opinions to surveys of the attitudes and opinions of dispersed aggregations of individuals: to surveys of the attitudes and opinions of dispersed "publics." Earlier studies of socially held attitudes came to be replaced by studies of "public opinion," of publicly expressed attitudes toward social objects (other persons and social groups), which were naturally analyzed as merely common attitudes and opinions (G. Allport, 1935) held independently of their orientation to social groups (Jaspars & Fraser, 1984). This impoverished conception of the social was supported and reinforced by the rhetorical emphasis on the study of individual attitudes and behaviors as opposed to reified and personified social minds.

[7] Although he was almost completely rejected by the French academic establishment, who treated him as a bit of a joke (R. Smith, 1997, p. 753).

[8] There is some irony in the fact that the experimental program of research on persuasion and communication initiated in the 1930s was eventually forced to recognize that the limits of public persuasion are generally the limits of individually engaged forms of cognition, emotion, and behavior (Cohen, 1964). Socially engaged (or "socially anchored") forms of cognition, emotion, and behavior proved to be highly resistant to change (see, e.g., E. Katz, 1957; Kelley & Volkart, 1952; Lewin, 1947a).

This orientation – or, strictly speaking, reorientation – of early American social psychology got a special twist in the work of Floyd Allport, the most effective proponent of an individualistically conceived scientific and experimental social psychology (Post, 1980; Parkovnick, 2000).[9] Allport (1924a) followed Baldwin, Cooley, Giddings, Park, and Ross in assimilating crowds and social groups by treating social groups as either crowds composed of physically proximate individuals or publics composed of physically dispersed individuals.

Two features of Allport's treatment deserve special attention. In the first place, Allport (1924a) resolved the threat of the external determination of cognition, emotion, and behavior by crowd or social group influence by *denying that it actually takes place in the crowd situation.* While he acknowledged that individuals in crowds behave in emotional, primitive, and irrational ways and that the behavior of individuals in crowds is influenced by the behavior of physically proximate others, he claimed that such interpersonal influences merely promote or enhance dispositions (which he called "prepotent individual reactions") that are activated individually by independent stimuli:

> The crowd is a collection of individuals who are all attending and reacting to some common object, their reactions being of a simple prepotent sort and accompanied by strong emotional responses. (p. 292)
>
> All of the fundamental, prepotent reactions are . . . operative in crowds of various sorts; and conversely, all spontaneous, mob-like crowds have their driving force in these basic individual responses. (p. 294)
>
> By the similarity of human nature the individuals of the crowd are all set to react to their common object in the same manner, quite apart from any social influence. *Stimulations from one another release and augment these responses; but they do not originate them.* (p. 299)
>
> It seems likely, therefore, that our preceding interpretation of crowd excitement holds true in general. The origin of responses is determined not by crowd stimuli but by the prepotent trends of the individual himself. The increase in the violence of emotion and action in crowds is due to the effect of behavior stimuli from others in releasing and reinforcing these prepared responses of individuals. (p. 300)

According to Allport, individuals in crowds *never* think, feel, or act because and on condition that other individuals in the crowd think, feel, or

[9] D. Katz (1991) claims that Floyd Allport effectively founded the scientific and experimental discipline of American social psychology with the publication of *Social Psychology* in 1924. Allport (1974) himself claimed that his work suggested "the possibility of a new experimental science of social psychology" (p. 9).

act in particular ways or because they are represented as thinking, feeling, or acting in particular ways: their individually engaged responses are merely *enhanced* by the responses of other individuals in the crowd. By attributing what is true of crowd excitement to social influence in general, Allport effectively denied the existence of socially engaged psychological states and behavior.

Allport (1924a) generalized his account of the crowd enhancement of individually engaged prepotent responses to the responses of dispersed publics by invoking the notion of an "impression of universality." Allport recognized that the enhanced responses of individuals in a crowd could not be explained simply as a direct function (a "geometrical relation") of the number of interpersonal stimulations among crowd members. This was because, although enhancement of response is greater in a large as opposed to a small crowd, any particular individual is only physically proximate to a limited number of other individuals:

> If one is surrounded by a throng, those near at hand shut out the view of those more distant. Barring volume of sound, therefore, a man in the center of a crowd of five hundred should receive as many contributory stimulations as the man in the midst of a crowd of five thousand. It will be agreed, however, that excitement runs higher in the vast throng than in the smaller body. We must therefore find some explanation, other than facilitation through social stimuli, to account for this dependence of crowd excitement upon numbers. (p. 305)

The explanation Allport advanced postulated an "impression of universality," according to which individuals in a crowd attribute identical responses to other individuals in a crowd to whom they are not physically proximate:

> A number of references have been made to the attitude assumed by the individual when he knows that he is in the presence of a large company. The situation is more complex than that of a small crowd with actual all-to-all contacts, the form of the response being largely determined by a central adjustment in the individual's nervous system, as well as by the external stimulations which call it forth. In terms of behavior we may say that the individual reacts to stimuli which he actually receives *as if* they were coming from an enormously greater number of individuals. In terms of consciousness he imagines that the entire vast assembly is stimulating him in this fashion. He has mental imagery – visual, auditory, and kinaesthetic – of a great throng of people whom he knows are there, although he does not see them. These people moreover are imagined as reacting to the common crowd object. There is vivid visual and motor imagery of their postures, expressions, and settings for action. We have already seen that there is an attitude to react as the other members of a crowd are reacting. There must of course be some *evidence* of how they are reacting in order to release this attitude. In default

of evidence through stimulation (as in the case of those concealed from view) mental imagery supplies the necessary cues.

It will be convenient to speak of the attitude of responding as if to a great number of social stimuli and the accompanying imaginal consciousness of the crowd's reaction as the *impression of universality*. (pp. 305–306)

According to Allport, the limited evidence we have from physically proximate others that physically distant others are responding to common stimuli in a similar fashion to ourselves is supplemented and complemented by the *social projection* of our responses to others:

A further imaginal factor is revealed in the behavior of individuals in a crowd. Whence comes this impression that the entire crowd is accepting and acting upon *the suggestions given by the speaker*? Why does the individual suppose that the attitude of those whom he cannot observe is favorable rather than hostile to the words uttered? The sight of compliance in one's immediate neighbors in part affords an impression which is extended to the entire crowd. The mere fact that the speaker is known to have prestige also counts. But a further explanation probably applies here. It may be stated as follows: As we catch a glimpse of the expressions of the others we "read into them" the setting which for the time is dominating us. The tendency is true of all perceptions under the influence of a special attitude. *We ourselves accept and respond to the words of the leader; and therefore we believe and act upon the assumption that others are doing so too.* The attitude and imagery involved in this reference of self-reaction to others we may call by a figurative term, *social projection*.

In crowds social projection and the impression of universality "work hand in hand." (pp. 306–307)

This account of the enhancement of common reactions to common stimuli (and denial of socially engaged forms of cognition, emotion, and behavior) was simply generalized by Allport to publics, conceived as dispersed and "imagined" crowds:

Psychologically speaking, the "public" means to an individual an imagined crowd in which (as he believes) certain opinions, feelings and overt reactions are universal. What the responses are imagined to be is determined by the press, by rumor, and by social projection. Impressed by some bit of public propaganda, the individual assumes that the impression created is universal and therefore of vital consequence. Thus the impression of universality is exploited and commercialized both on the rostrum and in the daily press. Newspaper columns abound in such statements as "it is in the consensus of opinion here," "telegrams [of remonstrance or petition] are pouring in from all sides," "widespread amazement was felt," and the like. (p. 308)

In generalizing his account of crowd reactions to public reactions, Allport was forced to modify his position and grant that a plurality of persons in a public could come to adopt certain attitudes (and emotions and behaviors) that are *not* a common response to common stimulation. According to Allport, some attitudes may come to be held because other individuals are merely *represented* as holding such attitudes:

Public opinion is merely the collection of individual opinions. It has no existence except in individual minds; and these minds can only conjecture what the general consensus is. Like the other unorganized forms of social control public opinion acquires its power through the attitude of the individual. The attitude is one of ascribing universality to certain convictions and then supporting them strongly in order to conform with the supposed universal view. (p. 396)

Yet while he acknowledged a psychological phenomenon closely akin (if not identical) to socially engaged attitudes (and emotion and behavior), he stressed that the adoption of attitudes in this fashion was generally based upon an error, upon a mere "illusion of universality." According to Allport (1924a), attitudes adopted in this fashion border on the pathological:

In certain types of insanity unconscious and dissociated thought reactions are projected to others, so that the patient does not recognize them as his own, but alleges that they are the ideas or accusations of others concerning him. This is the "projection" of psychoanalysis. (pp. 307–308)

Allport documented this "error" in tedious detail in *Institutional Behavior* (1933). In *Social Psychology* (1924a), he followed European theorists in recognizing the threats posed to autonomy and rationality by the media, while noting that such psychological phenomena could be usefully exploited by enlightened elites:

Newspapers and journals are self-constituted exponents of that which they assert to be the voice of the public. Their assertions are often hasty generalizations and sometimes deliberate propaganda. By pretending to express public opinion they in reality create and control it (p. 309). The illusion of universality may of course be used to establish a popular acceptance of *enlightened* views. The press thus has great possibilities, and indeed responsibilities, for creating solidarity in constructive citizenship.

One of the serious evils of American democracy is the exaggerated susceptibility to crowd-like control of public opinion. Impression of universality and the conformity attitude are so powerful that liberty of thought is scarcely tolerated. This fettering of free expression continues as an after-effect of the censorship necessary in the World War. Crowds and crowd-like publics dominate the thinking of the individual and tend to stifle independence of judgment. (p. 396)

In assimilating social groups and crowds and publics, Allport denied the existence of social forms of cognition, emotion, and behavior (in crowds) and dismissed acknowledged approximations to them (in publics) as pathological forms of cognition, emotion, and behavior based on error and projection. Allport never acknowledged that social forms of cognition, emotion, and behavior are sometimes voluntarily engaged by individuals because and on condition that they are accurately represented as engaged by members of social groups.[10] Thus, while he recognized the residual threats to autonomy and rationality posed by the press and demagogues in the case of publics, he could maintain both that they are much less of a threat than the threat posed by crowds (since they are cognitively rather than behaviorally based) and that they do not really involve any independent determination (but only the facilitation) of individually engaged drives and desires. Thus, he effectively denied and dismissed the threat posed to his moral and political individualism by the social dimensions of human psychology and behavior.

## III

The second feature of Floyd Allport's assimilation of crowds and social groups that deserves special attention is the role that it played in shaping his individualistic program for the experimental analysis of social groups. Allport (1924a) defined social groups in experimental settings as small crowds, as physically proximate aggregates of individuals, irrespective of their social group orientation:

> We may define a group as any aggregate consisting of two or more persons who are assembled to perform some task, to deliberate upon some proposal or topic of interest, or to share some affective experience of common appeal. (p. 260)[11]

[10] Despite his acknowledgment of fairly distinctive statistical facts about social groups, according to the J-curve hypothesis (1934). Yet this is perhaps not that surprising, since deviations from the normal distribution were treated by Allport as *pathological* as well as statistically abnormal, representing instances of crowd-like social control:

> The term public opinion usually signifies some conviction, belief, or sentiment common to all or to the great majority. The distribution of opinion on a question, excluding the bias of factions or parties, probably follows the general form of the probability curve. The opposite views on any issue are represented by fewer and fewer individuals as we approach the extreme forms of these views. The moderate position expresses the opinion of the majority. This high peak of the curve is the consideration which guides political leaders in their quest for public favor. It is also exploited by the press. Revolutionary mobs, crowd-like subservience to party principles, and like phenomena destroy this sober balance between opposing extremes. (pp. 395–396)

[11] In a footnote to his definition of groups, Allport acknowledged that social groups are often defined without the restrictive "physical presence condition":

> The word "group" is sometimes used in a sociological sense to denote a collection of individuals, not assembled in another's presence, but joined by some common bond of

In defining groups in this way, Allport acknowledged that they are not fundamentally different from crowds. Crowds are just those types of groups distinguished by their heightened emotionality and the operation of more primitive drives:

> The crowd we shall distinguish from such formations by the presence of emotional excitement and the replacing of the deliberate group activities by drives of the more primitive and prepotent level. (p. 260)[12]

Allport distinguished between "co-acting" and "face-to-face" groups. Co-acting groups were defined as groups in which "the individuals are primarily occupied with some stimulus rather than one another. The social stimuli in operation [from other persons] are therefore merely *contributory*" (p. 260). Face-to-face groups were defined as groups in which "the individuals react mainly or entirely to one another.... The social stimulations in effect are of the *direct* order" (p. 261).[13]

What is striking about these definitions is that they make reference to physically proximate individuals only, not to members of social groups. Both types of groups are defined without reference to any social group, and social influence is conceived entirely in terms of (direct or indirect) interpersonal influence, without reference to the social group membership of the individuals engaged in interpersonal interaction. Allport's own experimental studies of "social facilitation" in co-acting groups were studies of interpersonal influences on response rates and types of response to

> interest or sympathy. In so far as the behavior of individuals in such groups may be termed social it has its original basis in the actual contacts described in this and the following chapters. (1924a, p. 260, n. 1)

However, a starker example of the genetic fallacy would be difficult to find. It may be the case (and doubtless is most of the time) that social groups and social forms of cognition, emotion, and behavior come into being as a product of physically proximate interpersonal interactions. It does not of course follow that the dynamics of social forms of cognition, emotion, and behavior can be explained in terms of physically proximate interpersonal dynamics. This would be like assuming that the radioactive properties of uranium isotopes can be explained in terms of physical mechanics just because they are the product of neutrons colliding with uranium atoms. Moreover, even it were true that some form of physically proximate interaction between individuals is a causally necessary condition for the creation of social groups, it is clearly not a sufficient condition: the victims and perpetrators of random aggression or rape do not form a social group just by virtue of their physically proximate interaction.

[12] As Allport (1924a) noted, the distinction between groups and crowds is "not sharply drawn, and one form is capable of passing into the other" (p. 260).

[13] Allport (1924a) also noted that in real life "many groups, of course, combine the direct and contributory social influences, and are thus neither exclusively co-acting nor face-to-face" (p. 261).

certain tasks. The rates of response when individuals performed certain types of tasks alone were compared with rates of response when they performed these tasks in the company of others. On some tasks, their response rates increased and quality of performance improved while working with others; on other tasks, their response rates decreased and quality of performance deteriorated. Individual judgments of the hedonic quality of smells and estimates of relative weights were compared when made alone and in the company of others making the same types of judgments: judgments made in the company of others tended to be less polarized than judgments made in their absence. As in the case of the responses of individuals in a crowd, such individually engaged responses were enhanced or facilitated by the physical presence of others but were not (socially) engaged because and on condition that other members of a social group were represented as responding in this fashion.

This is important to stress. There is no reference to socially engaged forms of cognition, emotion, and behavior in Allport's *Social Psychology* (1924a). Social facilitation is simply interpersonal facilitation, and forms of social influence beyond interpersonal influence are pathologically grounded in illusions of universality and social projection.[14] The tradition of scientific and experimental psychology initiated by Allport played a major role in the progressive elimination of the social dimensions of cognition, emotion, and behavior as objects of social psychological investigation. This increasingly restrictive experimental tradition focused almost exclusively on the interpersonally generated cognition, emotion, and behavior of small aggregates of individuals (usually strangers) independently of their social group orientation.

The tradition was already well represented in the 1930s by Murphy and Murphy's book-length survey entitled *Experimental Social Psychology* (1931; Murphy et al., 1937), Dashiell's chapter on experimental social psychology in the 1935 *Handbook of Social Psychology*, and Gurnee's (1936) *Elements of Social Psychology*. This is the tradition in which Norman Triplett's (1898) study of pacemaking and competition came to be

[14] D. Katz (1991) has suggested that Allport's "impression of universality" anticipated the "near contemporary concept of reference group – any of the groups with which an individual identifies" (p. 131). However, this is a rationalizing reconstruction. Allport's "impression of universality" did not anticipate the concept of a reference group because it did not involve any reference to a represented social group. If anything, it anticipated the contemporary asocial notion of "false-consensus" (Marks & Miller, 1987; L. Ross, Greene, & House, 1977) rather than the social notion of a reference group (Hyman, 1942; Merton & Kitt, 1952).

seen as the first experiment in social psychology (Haines & Vaughan, 1979), even though there is nothing social about this experiment on interpersonal influence. As Murphy et al. (1937) noted, "From the experiments of Triplett (1900) to the appearance of Allport's (1920) studies it was becoming steadily clearer that social psychology could advance by the experimental method" (p. 13).

The origin of this form of experimental social psychology in European crowd psychology was clearly acknowledged by its advocates. As early as 1919, Floyd Allport identified crowd theorists such as Moede (1914, 1920)[15] as the anticipators of experimental social psychology.[16] Murphy et al. (1937) cited Allport and Moede as having "defined a social psychology which should have the same solid experimental foundations as had already been achieved for general psychology" (p. 8), and they said of Le Bon and Tarde that "it is generally agreed that modern social psychology was founded by these men and their followers" (p. 4).

Studies of human performance in front of an audience or in conjunction or cooperation with other physically proximate strangers (Gates, 1923; Shaw, 1932; Travis, 1925; Wheeler & Jordan, 1929) came to be treated as "paradigms," in Kuhn's (1970) sense of "exemplar," for the tradition of experimental research that began in earnest in the 1930s. Although interest in Allport's research on social facilitation waned in the 1930s (until revived by Zajonc's 1965 review), the basic paradigm for analyzing the actions and interactions of individuals[17] in small aggregate groups or small crowds, irrespective of their social group orientation, was well

[15] Wolfgang Moede wrote an introduction to Le Bon's *The Crowd* when it was republished in Germany in 1932 (Danziger, 2000), and he called his own form of experimental social psychology "experimental crowd psychology" (*experimentelle Massenpsychologie*) (Moede, 1914, 1920).

[16] Allport also cited German educational psychologists such as Alfred Mayer and Ernst Meumann as anticipators of his "social facilitation" research. Allport was introduced to their work at Harvard by Hugo Münsterberg, who suggested "social facilitation," as it came to be called, as a dissertation topic (Allport, 1974, cited in Danziger, 2000). Meumann, like Müsterberg, was a student of Wundt: he directed work at the Leipzig laboratory on educational psychology and edited a number of journals on genetic psychology and "experimental pedagogy." Although Wundt withdrew his support of Meumann's program (because it was not sufficiently experimental), it was enormously influential in Germany and beyond. His *Introductory Lectures on Pedagogy and Its Psychological Basis* (1907) was required reading for generations of German educators and was used in the Soviet Union and South America.

[17] Kay Karpf (1932) described what she took to be the distinctively American approach to social psychology that developed in the 1930s as follows: "It has resulted in the formulation of a point of view which might, for the want of a better designation, be termed 'interaction' psychology, or, rather, 'interaction' *social* psychology" (p. 419–420).

established by this time,[18] and it has remained more or less unchallenged (*pace* the "crises") up to the present day.

Of the 138 studies of aggression, competition, and social behavior surveyed by Murphy et al. (1937), for example, only two employed experimental groups of subjects preselected from social groups: Radina (1930) employed groups comprised of the children of laborers and domestic servants; Vetter and Green (1932) employed members of the American Association for the Advancement of Atheism. All the other studies employed aggregate groups of subjects defined by age, gender, intelligence, socioeconomic status, and the like, or "arbitrary groups" (Maller, 1929) assembled for the purposes of the experiment. Of the 71 studies of social attitudes and social attitude change surveyed, only a handful employed putative social groups or explored relations to represented social groups: for example, Kulp (1934) studied changes of attitude based on "feedback" from various "prestige" groups, and Kolstad (1933) compared the attitudes of history, household arts, and nursing majors. Most of the studies of attitude change anticipated the later "communication and persuasion" literature (Hovland, Janis, & Kelley, 1953) by focusing almost exclusively on variables such as the properties of the source and mode of communication. Virtually all of the studies defined social attitudes by reference to their social objects (races, criminals, public employment, prohibition, religion, unemployment insurance, and so forth) rather than by reference to their orientation to social groups. Murphy and Murphy referred their readers to Dashiell (1935) for the study of social influences on adult

[18] Murphy and Murphy complained that too many social psychologists seemed to think that a narrowly defined experimental social psychology constituted the whole of scientific social psychology:

> The assumption... spread that the *one* inevitable method of social psychology was the laboratory control of discrete variables, the isolation and measurement of each variable in terms of its effects. Despite the vigorous protests in Chapter 1 of our first edition regarding experimental social psychology, the publication of our book was unfortunately assumed to be further evidence that the experimental method must always come first and that all problems must fall willy-nilly into a form recognized by the laboratory worker. (Murphy et al., 1937, p. 13)

However, their complaint was somewhat disingenuous, since they went on to bemoan the fact that earlier social psychologists had left "pressing and vital problems to those who were ill-equipped for serious scientific work" (p. 13), and they maintained that "the best research in social psychology seems to come from those who have the broader perspective, who know how to see the big problems and, at the same time, have some familiarity with the possibilities of the laboratory" (pp. 13–14). Those they had in mind were Piaget, Lewin, and Sherif.

behavior, but none of the experimental studies surveyed by Dashiell were studies of socially engaged (or learned) forms of cognition, emotion, or behavior.

An essential feature of the metatheoretical and methodological position developed by Floyd Allport is that there is no discipline of social psychology distinct from individual psychology (Allport, 1924a): that the principles of individual psychology governing individually engaged psychological states and behavior also govern the psychological and behavioral products of social – that is, interpersonal – interaction. This individualistic presumption was clearly if somewhat quaintly expressed by J. P. Dashiell (1935) in his chapter on experimental social psychology in the 1935 *Handbook of Social Psychology*:

> Particularly is it to be borne in mind that in this objective stimulus-response relationship of an individual to his fellows we have to deal with no radically new concepts, no principles essentially additional to those applying to non-social situations. (p. 1097)

It was also manifest in the theoretical orientation and publications of the Institute of Human Relations at Yale University, one of the few interdisciplinary programs in the social sciences created in the prewar years. Funded by the Rockefeller Foundation and directed by Mark May, the institute was devoted to the integrative development of the various sciences of man (biology, psychology, psychiatry, sociology, and anthropology), including the study of both group and individual behavior. The work of the institute came to be dominated by the theoretical orientation of the neo-behaviorist Clark L. Hull (1952), whose goal was to identify the "primary laws of human behavior" based on principles of conditioning, which supposedly would yield "unambiguous deductions of major behavioral phenomena, both individual and social" (p. 162; cf. Hull, 1943, p. v). Perhaps the best known "social psychological" publication that grew out of the work of the Institute of Human Relations was N. E. Miller and Dollard's *Social Learning and Imitation* (1941), which extended Hull's principles of conditioned learning to interpersonal behavior, or, as Mark May (1950) put it, "illustrated how the principles of learning operating under the conditions of social life produce social habits" (pp. 165–166). N. E. Miller and Dollard followed Tarde and Allport in equating social and interpersonal learning.

The same basic presumption is to be found in Murphy and Murphy (1931; Murphy et al., 1937), despite the fact that they explicitly restricted their findings (tentatively at least) to American culture in the early twentieth century.[19] Murphy and Murphy (Murphy et al., 1937) defined social psychology as the "study of the way in which the individual becomes a member of, and functions in, a social group" (p. 16), but they clearly followed Allport in treating social forms of cognition, emotion, and behavior as merely common forms of cognition, emotion, and behavior. For the Murphys as much as for Allport, human social behavior is merely interpersonal behavior, behavior stimulated by members of one's own species. Thus, like Allport (1924a, pp. 154 ff.), the Murphys maintained that social behavior can be traced all the way down the phylogenetic scale:

> Even at the level of unicellular organisms a social factor is present. The interaction between organisms is one of the most fundamental of biological facts. If chasing and pursuing among human beings is a social fact, why is it not when it occurs in the amoeba? If the formation of human groups with mutual interstimulation is a social fact, why is not the formation of groups in the protozoa a social fact? The social is literally an *aspect of the biological*. (p. 19)
>
> The learning of an activity from another human being is just as biological a fact as any to be found in nature; and if social psychology is to contribute anything real to the social sciences, it must remember that there is no event in its entire subject-matter which is not in a sense a biological event, that is, an activity of a living organism or a group of living organisms stimulating one another in ways which can, if one wishes, be described through the use of biological concepts. (p. 20)

They also maintained that the distinction between social and individual psychological states and behavior, and thus between social and individual psychology, is entirely artificial, being based entirely on differences

[19] See Murphy et al., (1937):

> It must be recognized that nearly all the experimental work in social psychology, such as makes up the subject matter of this book, has value and is definitely meaningful only in relation to the particular culture in which the investigation was carried on. Such psychological laws as we can discover are for the most part statements of relations between stimuli and responses in civilized man, and perhaps many of them hold good only in specific groups or under specific social conditions. The social psychologist is, of course, not content with such generalizations as these; he wishes to find laws which are universal for the entire human family and for all existing or historically known cultures. It may reasonably be conjectured that a few of the laws already discovered – for example, some of the laws relating to suggestion and social facilitation – hold good among oriental as well as among occidental peoples, and among primitive peoples as well as among the more advanced. But it would be going outside the domain of experimental social psychology to insist even upon such a cautious statement as this. Whether any of our laws are really fundamental and necessary laws, deriving inevitably from human nature wherever it exists, can be determined only by experiment itself. (p. 7)

in the types of objects to which psychological states and behavior are directed:

We are forced to make a rough distinction between social psychology and "general" or "individual" psychology. But the concept of a "social" psychology as contrasted with an "individual" psychology involves the assumption that some of our behavior is stimulated exclusively by persons, the rest of it exclusively by things. If it is put thus baldly, it is necessary to concede that there are all sorts of responses for which the stimulating situation is a *combination* of personal and interpersonal stimuli.... For our purposes, individual psychology will be simply that psychology in which social factors (past or present) play a *relatively small* part, and social psychology will be that psychology in which social factors play a *relatively large* part. (pp. 22–23)

## IV

To maintain that the asocial theoretical and experimental paradigm developed by Floyd Allport was well established by the 1930s is not to deny that some social psychologists continued to study social forms of cognition, emotion, and behavior. Many in fact did, at least until the 1960s. Specific examples include Asch (1952); Cantril (1941); Converse and Campbell (1953) Edwards (1941); Festinger, Riecken, and Schachter (1956); Festinger, Schachter, and Back (1950); Hyman (1942); Lewin (1947a); Newcomb (1943); Schachter (1959); Schanck (1932); Sherif (1935, 1936, 1948); Sherif and Cantril (1947); Stouffer, Lumsdane, et al. (1949), Stouffer, Suchman, et al. (1949), and W. S. Watson and Hartmann (1939). Moreover, some continued to *experimentally* investigate social forms of cognition, emotion, and behavior: specific examples would include Asch (1951); Asch, Block, and Hertzman (1938); Charters and Newcomb (1952); Festinger (1947); French (1944); Kelley (1955); Kelley and Volkart (1952); Kelley and Woodruff (1956); Lewin, Lippitt, and White (1939); Sherif (1935); and Siegel and Siegel (1957). Yet these studies appear to have represented only the residue of the older social psychological tradition, not the increasingly dominant asocial and experimental tradition initiated by Allport in the 1920s and developed in the 1930s. What is striking about the experimental studies of social forms of cognition, emotion, and behavior cited above is their relative paucity amid an ocean of asocial experimental studies and the increasingly unrepresentative nature of such studies as the century advanced.

The work of Muzafer Sherif and Solomon Asch perhaps deserves special attention, since their classic experimental studies of the formation of group norms (Sherif, 1935) and conformity (Asch, 1951) have become classics in the experimental social psychological literature. Both Sherif and Asch clearly recognized the social dimensions of human psychology and behavior, although this aspect of their theoretical and experimental contribution came to be neglected by the increasingly asocial experimental tradition that appropriated their work.

Sherif's (1935) studies of individually and socially engaged "frames of reference" in relation to the "autokinetic effect" established the potency of social forms of perception: that is, perceptual judgments oriented toward previously established social norms. It is worth noting that in the original Sherif experiment the form of social perception investigated was defined as social by virtue of its orientation to the social group norm, not by virtue of its reference to a social object (since the object of perception was in fact nonsocial, namely, the position of the light source).[20] Moreover, socially engaged perceptual judgments oriented to prior social group norms were later maintained by experimental subjects in *physical isolation* from the members of the original social group:[21]

> When a member of a group faces the same situation subsequently *alone*, after once the range and norm of his group have been established, he perceives the situation in terms of the range and norm that he brings from the group situation. (Sherif, 1936, p. 105)

Although Sherif maintained his interest in social forms of cognition, emotion, and behavior throughout his academic career, three factors tended to dilute the specifically social aspect of his original study. First, Sherif (1936) tended to rhetorically present his conception of social norms as supportive of the notion of an emergent social mind:

> The fact that the norm thus established is peculiar to the group suggests that there is a factual psychological basis in the contentions of social psychologists and sociologists who maintain that new and supra-individual qualities arise in the group situations. This is in harmony with the facts developed in the psychology of perception. (p. 105)

[20] Sherif (1948) also made it very clear (following Thomas & Znaniecki, 1918, pp. 30–31) that such socially engaged frames of reference relate to both social objects (persons and groups) and nonsocial objects (tables, trees, and tarantulas):

> Once such frames of reference are established and incorporated in the individual, they enter as important factors to determine or modify his reactions to the situations that he will face later – social, and even non-social at times, especially if the stimulus field is not structured. (p. 174)

[21] Compare the fashion in which green recruits who joined combat units in World War II adopted the attitudes of the veterans in these units toward combat and maintained these attitudes in physical isolation from their comrades (e.g., when on home leave). See Stouffer, Lumsdane, et al. (1949, pp. 242 ff.).

This supraindividual characterization made it unlikely that others would rush to embrace the distinctively social dimensions of his work.

Second, Sherif (1936) tended to see the formation of social norms as an illustration of more generally holistic or relational psychological processes of the sort made familiar by Gestalt psychology:

> The experiments, then, constitute the study of the formation of a norm in a simple laboratory situation. They show in a simple way the basic psychological process involved in the establishment of social norms. They are an extension into the social field of a general psychological phenomenon that we found in perception and in many other psychological fields, namely, that our experience is organized around or modified by frames of reference participating as factors in any given stimulus situation. (pp. 105–106)

Thus it was common for Sherif (1936) to present his work in terms of merely common cognitive (or emotive or behavioral) frames of reference:

> The psychological basis of the established social norms, such as stereotypes, fashions, conventions, customs and values, is the formation of common frames of reference as a product of the contact of individuals. (p. 106)

As in the case of Kurt Lewin,[22] later generations of social psychologists tended to focus on the cognitive components of common frames of reference rather than their social nature.

Third, many of Sherif's studies, including the original 1935 study, were concerned with the *origin* and *development* of norms. Theoretically assuming that some psychological states and behavior are socially engaged, Sherif explored (via both laboratory and field experiments) the development of social norms and in consequence the formation of social groups themselves (much of his later work was explicitly devoted to this: Sherif, 1951; Sherif, Harvey, White, Hood, & Sherif, 1954; Sherif & Sherif, 1953). Since he experimentally explored the development of such social norms among strangers interacting in small experimental groups, his work was easily assimilated to the already dominant experimental paradigm – and easily misrepresented as concerned with common and individually engaged psychological products of interaction.[23]

Asch's (1951) study of "group pressure upon the modification and distortion of judgments" was also concerned with the social dimensions of

[22] Lewin's contribution is discussed in Chapter 8.

[23] In fact, it may reasonably be doubted if subjects in the original Sherif (1935) study did develop social forms of perception. Given that the stimuli presented were impoverished and ambiguous, subjects may have simply revised their perceptions individually by using the feedback of other subjects as an information source rather than being socially influenced by their judgments (Deutsch & Gerard, 1955).

cognition, emotion, and behavior. However, it has come to be represented as one of the original sources of the experimental study of conformity, defined in terms of individually engaged responses to interpersonal pressure irrespective of whether it is socially influenced.[24] Three points are worth noting about the Asch study.

In the first place, the length-of-line estimation study, in which subjects submitted to "group pressures" by repeating the erroneous judgments of experimental confederates, was ostensibly a study of social perception or judgment. However, since few of the subjects were persuaded to change their minds about their original estimates of the length of the presented lines, the study was really about social influences on *public avowals* of perceptual judgments.

In the second place, while it may be doubted that the pressure to which experimental subjects were exposed was genuinely social (as opposed to merely interpersonal), there is some reason to suppose that it in fact was. Unlike subjects in later developments of this experimental paradigm for the study of conformity, the subjects in the original Asch study were not randomly selected and were not strangers to each other:[25]

> The critical subjects were recruited by members of the cooperating group from among their acquaintances. They were told that an experiment in psychology was being performed for which additional subjects were required. (Asch, 1952, p. 454)

Finally, most commentators ignore the fact that some of the experiments reported in the Asch study were explorations of social rather than merely interpersonal influence. In addition to studies that explored the influence of copresent subjects on the judgments of individuals, Asch also conducted studies of the influence of information relating to the judgments of different social groups. Thus individual subjects in physical isolation from other subjects oriented their judgments to the reported judgments of

[24] For example, it is cited in Proshansky and Seidenberg's (1965) *Basic Studies in Social Psychology* under "Interpersonal Influence."

[25] This may or may not be significant in light of the fact that later studies, employing randomly selected subjects, failed to replicate the results of the original Asch experiment. See Perrin and Spencer (1980) and the consequent correspondence relating to such failures. In this connection it is also worth noting that some cross-cultural studies of conformity are similarly suggestive. Japanese subjects manifest considerably less conformity than Asch's original subjects, and in fact they manifest significant "anti-conformity" responses (Frager, 1970), at least in experiments in which subjects are randomly selected. They manifest greater conformity when the experimental confederates are members of their own social groups (Kinoshita, 1964, quoted in Frager, 1970; see also Williams & Sogon, 1984).

fellow students but not to those of Nazi storm troopers (Asch, Block, & Hertzman, 1938). Certainly this form of experimental investigation of social attitudes and judgment was not well represented in the subsequent experimental literature, unlike the study employing copresent subjects, which became a paradigm for future conformity research. Conformity research came to be experimentally focused (with only few exceptions) on external interpersonal or "situational" determinants of conformity rather than on the social dimensions of cognition, emotion, and behavior (Moscovici, 1985). Conformity came to be operationally defined in terms of interpersonal pressures that cause individuals to act differently from how they would act if alone (Kiesler & Kiesler, 1969, cited in Cialdini & Trost, 1998, p. 162).

## V

Of course, American social psychology did not become a predominately experimental discipline overnight. The experimental program initiated by Allport represented only a relatively small proportion of the social psychological research produced in the 1930s and 1940s. For example, experimental social psychology was represented by only one chapter in the 1935 *Handbook of Social Psychology*, and many of the studies included in Murphy and Murphy's *Experimental Social Psychology* (1931; Murphy et al., 1937) were correlational or developmental studies. It was only after World War II that experimentation came to displace all other methods (field studies, interviews, and so forth) as *the* research paradigm in psychological social psychology. What is important to recognize, however, is that the basic asocial experimental program that eventually came to dominate in the postwar years was already firmly in place by the 1930s.

When American social psychology expanded and developed rapidly in the postwar years, its basic asocial theoretical and experimental trajectory was already fixed. However, the original conception of the social dimensions of human psychology and behavior and of a distinctive social psychology did not simply fade away from the 1930s onward. On the contrary, it reached its high-water mark in the immediate postwar years, partly as a result of the interdisciplinary links forged and interdisciplinary enthusiasm generated during the war. Unfortunately, the development of postwar social psychology did not live up to its original social promise.

Nonetheless, it is important to stress that before and after the war some American social psychologists unambiguously acknowledged the social dimensions of human psychology and behavior and did conduct empirical studies, including experimental studies (both laboratory and field), directed toward the exploration of the social dimensions of cognition, emotion, and behavior. This is because, in the first place, it illustrates that although the asocial metatheoretical program initiated by Floyd Allport was "tailor made for an experimental social science" (Danziger, 2000, p. 333), it was not itself a product of the commitment to an experimental social psychology.[26] Allport's reasons for rejecting the social dimensions of cognition, emotion, and behavior were largely independent of his commitment to an experimental program,[27] and a fair number of social psychologists did experimentally explore the social dimensions of cognition, emotion, and behavior from the 1930s to the 1950s.

In the second place, it demonstrates the *possibility* of a social psychological science, including an experimental social psychological science, devoted to the exploration of the social dimensions of human psychology and behavior, even if this was not what was actually developed in the postwar years. What came to be neglected was not any mysterious supraindividual entity distinct from the psychological states of individuals but a set of social psychological phenomena oriented to the represented psychology and behavior of members of social groups. It continues to be neglected at the beginning of the new millennium.

[26] See Chapter 9 for a detailed defense of this claim. A similar point can be made with respect to the abandonment of the original conception of social attitudes as socially engaged attitudes (attitudes oriented to the represented attitudes of members of a social group) (Faris, 1925; Schanck, 1932; Thomas & Znaniecki, 1918) in favor of the later conception as common attitudes (G. W. Allport, 1935) directed toward social objects, namely, other persons (F. H. Allport, 1924a). The development of research on social attitudes was stimulated in large part by the development of instruments for measuring attitudes (Likert, 1932; Thurstone, 1928; Thurstone & Chave, 1929) and by the huge social and political interest in the determination of attitudes toward immigrants, minority groups, prohibition, labor unions, and the like during the interwar years (Danziger, 1997). However, the development of such instruments does not itself explain the abandonment of the earlier conception of social attitudes. Floyd Allport (1932) maintained that such instruments could be employed in the identification of socially engaged attitudes, that is, attitudes related to "fictions" such as social groups and institutions, and in fact he employed them in his analysis of conforming attitudes represented by J-curve distributions (F. H. Allport, 1934). Later studies of socially engaged attitudes (Edwards, 1941; W. S. Watson & Hartmann, 1939) continued to employ these measures, as did probably the last major exploration of socially engaged attitudes in American social psychology, the studies that made up *The Authoritarian Personality* (Adorno et al., 1950).

[27] Although it no doubt played some role. One of Allport's (many) reasons for rejecting the group mind was that it "impeded experiment in social science" (1919, p. 333, quoted in Danziger, 2000).

# 8

# Crossroads

Dorwin Cartwright, in his 1979 paper "Contemporary Social Psychology in Historical Perspective," claimed that the person who had the greatest impact on the development of American social psychology was Adolf Hitler. Certainly World War II functioned as a great catalyst for intellectual cooperation between psychologists, sociologists, anthropologists, and psychiatrists, who worked within various wartime government agencies, such as the Army Information and Education Division, the OSS Assessment Staff, the Food Habits Committee of the National Research Council, and the Bureau of Program Surveys of the Department of Agriculture. These groups conducted studies on the attitudes, morale, and adjustment of combat troops (Stouffer, Lumsdane, et al., 1949; Stouffer, Suchman, et al., 1948b), French civilian reaction to the D-Day landings (Riley, 1947), the relationship between enemy morale and saturation bombing (U.S. Strategic Bombing Survey, 1946; Janis, 1951), and civilian morale and propaganda (Berelson, 1954; Watson, 1942). In a variety of quasi-experimental and field studies, Kurt Lewin (1947a) and Dorwin Cartwright (1949) explored "group dynamics" via programs designed to persuade housewives to change their food habits and to promote the sale of U.S. war bonds, initiating what came to be known as the Lewinian tradition of "action research."

Another major impact on American social psychology of the actions of Adolf Hitler was the migration of academic refugees from Western Europe, which provided a massive infusion of talent into American science, culture, and psychology, including social psychology. Scholars such as Egon Brunswik, Else Frenkel-Brunswik, Fritz Heider, Wolfgang Köhler, Paul Lazerfield, Kurt Lewin, and Max Wertheimer not only introduced novel theoretical perspectives but also played a significant role in shaping the theoretical perspectives of many of the major figures in postwar American social psychology, such as Solomon Asch, Dorwin Cartwright, Leon Festinger, Harold Kelley, David Krech, Stanley Schachter, and Alvin Zander. The downside of this migration was the prewar emaciation of social psychology in Europe, so that postwar social psychology became a decidedly American achievement (G. W. Allport, 1968b; Cartwright, 1979; R. V. Levine & Rodrigues, 1999), shaped by American interests and American prewar metatheoretical and methodological orientations (Israel & Tajfel, 1972). Given its immediate postwar dominance of the field, American social psychology effectively colonized the international academic social psychological community over the ensuing decades (Farr, 1996; van Strien, 1997).

The collaborative interdisciplinary efforts of the war promoted a genuine optimism about the possibilities of developing social psychology as a mature and genuinely interdisciplinary science. New research facilities were established, such as the Survey Research Center in Washington, DC, and the Research Center for Group Dynamics at MIT (which jointly became the Institute for Social Research at the University of Michigan), the Laboratory for Social Relations, and the National Training Laboratories for Group Development at Bethel, Maine. These were often associated with graduate training programs at major universities and provided research facilities for faculty and students from a variety of social science disciplines. New interdisciplinary programs in social psychology were set up at major universities, such as Harvard's Department of Social Relations (headed by Talcott Parsons), the Communication Research Center at Yale[1] (headed by Carl I. Hovland), Columbia's Bureau of Applied Research, and the doctoral program in social psychology at the University of Michigan. The Research Center for Group Dynamics at MIT, which moved to the University of Michigan in 1948 (after Lewin's death in 1947), included on its staff such eminent figures as Dorwin Cartwright, Leon Festinger, Jack French, Ronald Lippitt, Marion Radke, and Alvin Zander, whose graduate students included later luminaries such as Kurt Back, Morton Deutsch, Murray Horwitz, Harold Kelley, Albert Pepitone, Stanley Schachter, and John Thibaut (whose own associates and graduate students included Elliot Aronson, John Darley, Edward E. Jones, Lee Ross, and Philip Zimbardo).

[1] In addition to the Institute of Human Relations, the interdisciplinary program founded in 1929.

There was also pressure for integration from the funding agencies, as the newly designated "behavioral sciences" scrambled for financial support to apply the theoretical and practical fruits of their wartime experience to the resolution of postwar problems.[2] Interdisciplinary conferences, many generously funded by government agencies, were a common feature of the 1950s. At these conferences, there were regular and explicit calls for a more integrated and interdisciplinary social psychology that would exploit the talents and traditions of sociologists and psychologists, and many identified the main threat to the scientific integrity of social psychology as the historical division between psychological and sociological social psychology. At one of these conferences, presciently titled *Social Psychology at the Crossroads*, Theodore Newcomb (then head of the interdisciplinary doctoral program in social psychology at the University of Michigan) complained of "the unfortunate circumstance that there are two social psychologies thriving in the land" (1951, p. 31) and made a plea for the integration of individual (psychological) and social (sociological) approaches – a plea echoed by many other psychologists and sociologists in the postwar era.

There were real grounds for optimism. While prewar social psychologists had increasingly come to adopt the asocial metatheoretical and methodological paradigm promoted by Floyd Allport and his followers, the earlier conception of socially engaged psychological states and behavior (as psychological states and behavior oriented to the represented psychology and behavior of members of social groups) and of a distinctive social psychology devoted to their study remained theoretically vibrant. Indeed, some of the best theoretical descriptions of the social dimensions of human psychology and behavior were articulated during the late 1940s and 1950s.

## I

Theoretical descriptions of the social dimensions of cognition, emotion, and behavior and dedicated attempts to integrate psychological and sociological traditions of research can be found in a variety of popular texts of this period, such as Newcomb and Hartley's *Readings in Social Psychology* (1947), Krech and Crutchfield's *Theory and Problems*

[2] One example of this was the establishment of the Commission of Community Interrelations (by Kurt Lewin in New York City), supported by the American Jewish Committee. The commission's membership included Dorwin Cartwright, Kenneth B. Clark, Morton Deutsch, Leon Festinger, Marie Jahoda, Ronald Lippett, and Goodwin Watson.

of *Social Psychology* (1948), Sherif's *An Outline of Social Psychology* (1948), and Asch's *Social Psychology* (1952). These works tried to integrate past psychological and sociological work and set a future program of integrative research that would shape the development of postwar social psychology.

Newcomb and Hartley's *Readings in Social Psychology* (1947), sponsored by the Society for the Study of Social Issues and backed by a distinguished interdisciplinary committee, explicitly aimed to define the field of social psychology. The preface began with a clear commitment to the social dimensions of human psychology and behavior:

> It is the peculiar province of the social psychologist to bring to bear upon his study of the behaving organism all relevant factors, from whatever sources and by whatever methods ascertained, which inhere in the fact of association with other members of the species. Most of these factors in the case of human beings have to do in some way with membership of groups. (p. vii)

Krech and Crutchfield, in *Theory and Problems of Social Psychology* (1948), began with a refreshingly open-minded statement of the objectives of science, including social psychological science. In place of the standard positivist and empiricist equation of explanation with prediction and control, they insisted that "the major objective of science is not primarily to control and predict, but to *understand*. Effective control is a reward of understanding, and accuracy in prediction is a check on understanding" (p. 3). Despite the fairly amorphous nature of their definition of social psychology as "the science of the behavior of the individual in society" (p. 7), Krech and Crutchfield (1948) clearly acknowledged the social dimensions of human psychology and behavior, citing a variety of studies of socially engaged attitudes and beliefs, such as Hirschberg and Gilliland (1942) and Newcomb and Svehla (1938) on the influence of family, Winslow (1937) on the influence of friendship groups, and Carlson (1934) and A. J. Harris, Remmers, and Ellison (1932) on the influence of religious groups. They noted that

> most beliefs and attitudes receive social support. This not only tends to make beliefs and attitudes resistant to change... but can also help to induce change. Effective measures designed to control beliefs and attitudes must seek, wherever possible, to create new group identifications for the people it would change to the end that social support for the new beliefs and attitudes will be forthcoming. Social support for beliefs and attitudes is effective only in so far as the individual is or wants to be a member of the group that has those beliefs and attitudes. (pp. 200–201)

Moreover, as the above quote suggests, Krech and Crutchfield followed other theorists (notably Sherif and Asch) in maintaining that socially engaged forms of cognition, emotion, and behavior are intimately bound up with representations of personal identity.[3] Thus Sherif, for example, in *An Outline of Social Psychology* (1948), detailed the social dimensions of attitudes and their intimate link to an individual's sense of identity as an individual:

> It has become clear by this time that groups play a major role in shaping attitudes in man. In fact, it may be safe to assert that the formation and effectiveness of attitudes cannot be properly accounted for without relating them to the group matrix. (p. 138)

> We can state that the individual in any human grouping develops an ego which more or less defines in a major way the very anchorages of his identity in relation to other persons, groups, institutions, etc.... disruption of these anchorages, or their loss, implies psychologically the breakdown of his identity.
>
> In a good many cases the individual forms his attitudes on the basis of the values and the norms of the groups he joins. He becomes a good member to the extent to which he assimilates these norms, conforms to them, and serves the aims demanded by them. (p. 105)

Some of the clearest theoretical descriptions of the social dimensions of human psychology and behavior are to be found among the various contributions on "reference groups" in Newcomb and Hartley's (1947) *Readings in Social Psychology* (and in the revised editions, Swanson, Newcomb, & Hartley, 1952, and Maccoby, Newcomb, & Hartley, 1958). The notion of a "reference group" was introduced by Hyman (1942) and was quickly adopted by social psychologists to accommodate the fact that although socially engaged psychological states and behaviors are regularly oriented to social groups of which individuals are members (or anticipated or apprentice members), they can also be oriented to social groups of which individuals are not members.

Thus Harold Kelley (1952), fresh from a 1951 summer conference on reference groups at Yale University, acknowledged the social dimensions of attitudes: "A considerable number of every person's attitudes are related to or anchored in one or more social groups" (p. 410). He also recognized the theoretical utility of the concept of a reference group, which

[3] I talk here of "personal identity" rather than "social identity" because I want to stress that although a person's sense of identity is undoubtedly socially grounded (Greenwood, 1994), it is a sense of one's identity *as an individual* within a social matrix (a sense of one's achievements and failures in relation to the norms of social groups).

enabled social psychologists to accommodate socially engaged cognition, emotion, and behavior oriented to both membership and nonmembership social groups, or to accommodate, as he put it, their "anchorage in both membership and nonmembership groups":

> Although this theory is still in the initial stages of development, because of the problems it formulates it promises to be of central importance to social psychology. In particular it is important to those social scientists who desire to interpret the development of attitudes, to predict their expression under different social conditions, to understand the social basis of their stability or resistance to change, or to devise means of increasing or overcoming this resistance. (p. 10)

Kelley enthusiastically anticipated that this concept would play a major role in the future theoretical and empirical development of social psychology: "Through ... research and conceptual development ... we may expect great advances in our understanding of the social basis of attitudes" (p. 414).

Newcomb himself (one of the editors of *Readings in Social Psychology*) was in no doubt that certain attitudes are socially held. His own longitudinal study of student attitudes at Bennington College (Newcomb, 1943) demonstrated their orientation to the represented attitudes of members of approved social groups:

> In a membership group in which certain attitudes are approved (i.e. held by majorities, and conspicuously so by leaders), individuals acquire the approved attitudes to the extent that the membership group (particularly as symbolized by leaders and dominant subgroups) serves as a positive point of reference. The findings of the Bennington study seem to be better understood in terms of this thesis than any other. (Newcomb, 1952, pp. 410–11)

*The American Soldier*, a four-volume collaborative study by psychologists, sociologists, psychiatrists, and anthropologists published at the end of the war, provided a rich source of data and hypotheses concerning socially engaged cognition, emotion, and behavior. For example, many green recruits appear to have changed their attitudes toward combat after joining veteran units as replacements (and prior to actual combat). Many appear to have abandoned their previous "gung-ho" attitude and to have adopted the more measured attitude of the combat veterans (which they maintained when physically absent from the combat group, such as when they were home on leave). Thus among sampled green recruits, 45 percent expressed willingness for combat; among sampled veterans, only 15 percent expressed willingness for combat. Many green recruits who joined veteran groups changed their attitude in line with those of the veterans;

among sampled replacements, only 28 percent expressed willingness for combat. As Stouffer, Lumsdane, et al. (1949) noted,

To some extent the replacements took over the attitudes of the combat veterans around them, whose views on combat would have for them high prestige. (p. 250)

...probably the strongest group code [among combat veterans]...was the taboo against any talk of a flag-waving variety.... The core of the attitude among combat men seemed to be that any talk that did not subordinate idealistic values and patriotism to the harsher realities of the combat situation was hypocritical, and a person who expressed such ideas a hypocrite. (p. 150)

The significance of such studies was recognized by those who hoped to employ the concept of a reference group in the social psychological exploration of the social dimensions of human psychology and behavior. Moreover, the studies documented in *The American Soldier* reminded theorists that individuals tend to socially orient their psychology and behavior to a *variety* of different social groups, a point emphasized earlier by Cooley (1902, p. 114), Dewey (1927, p. 129), Faris (1925, p. 405), James (1890, p. 29), and La Piere (1938, p. 15):

For, as the analysis of cases drawn from *The American Soldier* plainly suggests, the individual may be oriented toward any one *or more* of the various kinds of groups and statuses – membership groups and nonmembership groups, statuses like his own or if different, either higher, lower, or not socially ranked with respect to his own. This, then, locates a further problem: If *multiple* groups or statuses, with their possibly divergent or even contradictory norms and standards, are taken as a frame of reference by the individual, how are these discrepancies resolved? (Merton & Kitt, 1952, p. 432)

The issue of multiple group membership was focused on by a number of theorists in the 1950s, including Hartley (1951), Newcomb (1950), Rosen (1955), and Sherif (1948).[4] Krech and Crutchfield (1948) noted that:

The same individual is perforce a member of many different groups. His pattern of group memberships will not be identical with other people's. Sometimes he finds himself grouped together with one set of people, such as his coreligionists; at other times he finds himself grouped with other people, such as his fellow members of a political organization and in opposition to his coreligionists. In that sense we may

[4] Sherif (1949) exploited it creatively to locate possible threats to one's identity posed by the "cross-pressures" of multiple social group orientations (attempts to orient one's psychology and behavior toward different social groups with different psychological and behavioral norms) and to explain cross-situational inconsistencies in attitudes and behavior, which constituted a major problem for later attitude research (Mischel, 1968).

say that the membership of any specific group consists of "part personalities." The single member's loyalties to his various groups and to their members will be divided and often conflicting. (pp. 384–385)

Merton and Kitt (1952) not unreasonably saw the issue of multiple group membership as a fertile source of future theoretical and empirical developments in social psychology:

Theory and research must move on to consider the *dynamics of selection* of reference groups among the individual's several membership groups: When do individuals orient themselves to others in their occupational group, in their congeniality groups, or in their religious group? How can we characterize the *structure of the social situation* which leads to one rather than another of these several group affiliations being taken as the significant context? (p. 435)

One possibility, creatively and experimentally explored by Charters and Newcomb (1952),[5] is that potential conflicts or inconsistencies are "resolved" in line with the "relative potencies" of relevant social groups:

Many of the attitudes of an individual are greatly influenced by the norms of groups to which he belongs. Most individuals, however, are members of more than one group, and consequently may face a particular problem when these different groups prescribe opposing attitudes toward the same object. It seems reasonable to hypothesize that an individual's resolution of this problem will be a function of the relative potencies of his various group memberships. (p. 415)

## II

Perhaps the most articulate theoretical descriptions of the social dimensions of cognition, emotion, and behavior were offered by Solomon Asch in his *Social Psychology* (1952). Asch used the term "*intrinsically social attitudes*" to refer to those attitudes that are oriented to social groups that function as primary reference groups for individuals and thus provide the grounding of their individual identity. Intrinsically social attitudes were defined by Asch as "sentiments that many or all members of a group

[5] One distinguishing feature of this rare experimental study of socially engaged attitudes (among Catholics, Jews, and evangelical Protestants) was that, unlike most experimental studies at the time and thereafter, subjects were not randomly assigned to experimental conditions but were assigned on the basis of preselected social groups: "From a large class of introductory psychology students we selected those who we knew were also members of one of three strong religious organizations" (Charters & Newcomb, 1952, p. 415). Compare Seigel and Seigel (1957). The significance of this feature of the study is discussed in Chapter 9.

share. They are cognitively and emotionally crucial for its members and at the same time they control social action directly" (p. 575).

Asch also clearly recognized the conditionality or "mutual dependence" of socially held attitudes: social attitudes are social by virtue of being held because and on condition that other members of a social group are represented as holding them:

> Attitudes are the most concentrated expression of this relation of mutual dependence. They are social not merely because their objects are social or because others have similar attitudes. They are social principally in that they arise in view of and in response to perceived conditions of mutual dependence. (p. 576)

> Attitudes are not only causally connected with group conditions; they are also part of the mutually shared field. Therefore the investigation of attitudes brings us to the center of the person's social relations and to the heart of the dynamics of group processes. . . .
>
> . . . The racial sentiment of Southerners is only in part directed to Negroes; it is also a function of their most significant ties to family, neighborhood, and group. (p. 577)

Asch additionally diagnosed the inadequacy of the Allports' treatment of social attitudes. Following Durkheim (1895/1982a) and Weber (1922/1978), Asch denied that social attitudes could simply be equated with *common* attitudes (directed toward social objects):

> The simultaneous occurrence of the same (or similar) psychological process in a number of individuals in response to the same external conditions is not sufficient to turn these conditions into a social fact. (p. 128)

Although Asch was as clearly committed to the social dimensions of human psychology and behavior as McDougall (1920) and Wallis (1925), he carefully disassociated himself from their advocacy of supraindividual theories of the social mind:

> The order and system of a social field depends upon processes occurring in separate individual centers – in the brains and actions of individuals. (p. 251)

> There are no purposes or values of groups that are not the purposes and values of some individuals. Group goals must be held and cherished by individuals; the aims and needs of individuals are the only valid goals of groups. (p. 258)

However, he maintained, with McDougall and Wallis (and other early American social psychologists), that there is a real distinction between socially and individually engaged forms of cognition, emotion, and behavior. The orientation of psychological states and behavior to the represented psychology and behavior of members of social groups introduces new psychological forces distinct from those governing individually engaged psychological states and behavior. According to Asch, under such social conditions individuals "transform their own nature and bring into existence new psychological forces" (p. 136).

There is, according to Asch, nothing especially mysterious or miraculous about such postulated social forces. To acknowledge them theoretically is just to acknowledge socially (as opposed to individually) engaged forms of cognition, emotion, and behavior. As he rightly insisted, it is not enough to simply maintain, as the Allports and their followers maintained, that social psychological states and behaviors are the psychological states and behaviors *of individuals*. One has to recognize the distinction between the psychological states and behaviors of individuals that are *socially* engaged and those that are *individually* engaged. Yet this is precisely what was ignored or denied by individualists such as the Allports:

> The individualist doctrine denies the presence of a *socially structured field within the individual.* It fails to acknowledge that group facts may be represented within the individual in an ordered way and call upon him for action oriented to group realities. (p. 253)

Asch's (1952) statement of the relation between social groups and social forms of cognition, emotion, and behavior is perhaps the most sophisticated in the social psychological literature. Like Simmel (1908/1959), Asch was acutely aware of the joint psychological constitution of social groups and social psychological states and behavior. According to Asch, the jointly constitutive nature of this relation distinguishes this relation from all other known part-whole relations:[6]

[6] One way of expressing this point about the joint constitution of social groups and social forms of cognition, emotion, and behavior is as follows. In Chapter 4 it was argued that intrinsically social groups deserve to be regarded as the elemental components or "building blocks" of the social world, since without such social groups there would no society or social world. It may also be argued that the elemental units of social groups are not individuals per se but *social individuals*: that is, individuals whose psychology and behavior are oriented to the represented psychology and behavior of members of social groups. As Asch (1952) put it,

> Our task is to understand both the distinctness and inseparability of group and individual. Group conditions can act on individuals only because individuals have very definite properties. The individual possibilities of conversation must precede the actuality of conversation; the individual possibilities of a self must precede the actuality of a self that is socially related. We must understand also how group conditions penetrate to the very center of individuals and transform their character. In particular, we must understand that once a group is functioning, *the unit is not an individual but a social individual, one who has a place in the social order as a child, a husband, or a worker.* (p. 257 emphasis added)

> To understand the intimacy and separateness between individual and group we must grasp the unusual process that gives rise to groups at the human level. It is a process in which individuals play an extraordinary role, confronting us with a type of part-whole relation unprecedented in nature. It is the only part-whole relation *that depends on the recapitulation of the structure of the whole in the part*. Only because individuals are capable of encompassing group relations and possibilities can they create a society that eventually faces them as an independent, or even hostile, set of conditions. (p. 254, emphasis added)

Asch's theoretical sophistication was matched by an explosion of experimental studies on groups throughout the 1950s,[7] largely inspired by Kurt Lewin's theoretical and methodological contributions before and during the war (Lewin, 1936, 1939, 1947a, 1947b, 1947c, 1948, 1951; Lewin, Dembo, Festinger, & Sears, 1944; Lewin et al., 1939) and developed by Lewin's colleagues and students after his death in 1947 (Cartwright & Zander, 1953).[8] Lewin's (1947a) experimental studies during the war, which were designed to promote changes in food habits, were based on the assumption that many attitudes and behaviors are socially engaged, although (like Asch) Lewin harbored no illusions that such attitudes and behaviors are properties of supraindividual groups:

> One of the reasons why "group carried changes" are more readily brought about seems to be the unwillingness of the individual to depart too far from group standards; he is likely to change only if the group changes.... It should be stressed that in our case the decision which follows the group discussion does not have the character of a decision in regard to a group goal; it is rather a decision about individual goals in a group setting. (p. 337)

Lewin 1946/1997d clearly distinguished between socially and individually engaged psychological states of individuals, which he depicted as competing "forces" within the "life space" of the individual:

> The effect of group belongingness on the behavior of an individual can be viewed as the result of an overlapping situation; one corresponds to the person's own needs and goals; the other to the goals, rules and values which exist in him as a group member. Adaptation of the individual to the group depends upon the avoidance of too great a conflict between the two sets of forces. (p. 360)

Lewin also recognized (following Cooley, 1902; Dewey, 1927; Faris, 1925; James, 1890; and La Piere, 1938) that most individuals are members

[7] According to McGrath (1978), studies of group dynamics increased about tenfold during this period.

[8] Many of these studies were funded by the Defense Department, both during and after the war.

of a multiplicity of social groups and consequently subject to potentially conflicting forces:

> The individual is usually a member of many more or less overlapping groups. He may be a member of a professional group, a political party, a luncheon club, etc. The potency of any of these groups, that is, the degree to which a person's behavior is influenced by his membership in them, may be different for the different groups he belongs to. For one person, business may be more important than politics, for another, the political party may have the higher potency. The potency of the different groups to which the person belongs varies with the momentary situation. When the person is at home the potency of the family is generally greater than when he is in the office. (1940/1997b, p. 68)

Lewin's own Research Center for Group Dynamics at MIT was the first to offer a doctoral program in what he explicitly called *group psychology*, designed to "educate research workers in theoretical and applied fields of group life" (Patnoe, 1988, p. 8).

## III

Unfortunately, the promise of the immediate postwar years was not fulfilled. The postwar interdisciplinary enthusiasm was relatively short-lived. The scramble for funds by psychologists and sociologists encouraged as much competition as cooperation, and in practice one professional group or the other (usually the psychologists) tended to dominate at each interdisciplinary institution and set its research agenda. Psychologists continued to dominate the institutional academic delivery of social psychology, and the journals and textbooks, as they had done before the war. Sociologists became deeply resentful, and whether their resentments were justified, the effects of their resentment were real enough.[9] Given these tensions, many of the interdisciplinary programs in social psychology created after the war fractured along disciplinary lines (such as the programs at Michigan and Harvard). Beyond the interdisciplinary rhetoric, graduate students still had to specialize in psychology or sociology (or anthropology or psychiatry) and prepare themselves for an institutional academic world still rather rigidly divided along traditional disciplinary lines (Collier, Minton, & Reynolds, 1991). Most chose to specialize in psychology, and even within sociology psychological forms of social psychology eventually prevailed (Burgess, 1977; Liska, 1977).

[9] As W. I. Thomas and D. S. Thomas (1928) said, "If men define situations as real, they are real in their consequences" (p. 567).

However, the very idea of developing a mature scientific social psychology by integrating psychological and sociological social psychology was misconceived in the first place. The notion that there could be substantive psychological as opposed to sociological forms of social psychology[10] was part of the problem. The subject matter of social psychology, as recognized by its early proponents, is defined by what it attempts to explain and by its social psychological mode of explanation. In particular, social psychology is concerned with the explanation of the origin, maintenance, development, and dissolution of social forms of human psychology and behavior (oriented to the represented psychology and behavior of members of social groups), and its explanations are in terms of social forms of cognition, emotion, and behavior. For example, it offers explanations of aggression, altruism, attitude change, resistance to propaganda, and identity formation in terms of socially engaged aggression, altruism, attitude change, resistance to propaganda, and identity formation.[11] Such research can just as easily be conducted by psychologists or sociologists in departments of psychology or sociology, and quite properly so, since social psychological states and behavior represent the ontological realm in which the social and the psychological are jointly constituted.

As Charles Ellwood put it in 1925,

> We cannot understand the individual apart from his group, any more than we can understand the group apart from the nature of the individuals who compose it. Thus the dependence between sociology and psychology is reciprocal. Individual psychology must accordingly look to the study of group life for the explanation of much in individual behavior. It depends as much upon the psychology of society as the psychology of society depends upon it. (p. 22)

This last point is important to stress. On the level of the social dimensions of human psychology and behavior, the social and the psychological

[10] Social psychology conducted by psychologists as opposed to sociologists or conducted within departments of psychology as opposed to departments of sociology. To deny any substantive difference between psychological and sociological forms of social psychology is not, of course, to deny the substantive difference between social psychology and individual psychology.

[11] See, for example, McDougall (1908):

> Social psychology has to show how, given the native propensities and capacities of the individual human mind, all the complex mental life of societies is shaped by them and in turn reacts upon the course of their development and operation in the individual. (p. 18)

Compare Dunlap (1925), who maintained that social psychology was concerned with "those factors in individual minds which make social association possible or which result from social association" (p. 12).

constitute a *singularity*, not a duality. While no doubt influenced by individually engaged psychological phenomena (such as individually engaged motives, feelings, and perceptions) and external social phenomena (the hierarchical structures of established social groups and the relations between social groups), the social and psychological "components" of socially engaged psychological states and behavior do not operate as independent variables in interaction but jointly constitute certain psychological states and behavior of individuals *as* social psychological states and behavior (namely, as psychological states that are socially rather than individually engaged). To consider the relation between the individual and the social, or the psychological and the social, as a central problem of social psychology is already to misconceive the nature of social psychological phenomena.[12]

Again, Asch (1952) saw this more clearly than any of his contemporaries. Such a conception fails to acknowledge the internal or intrinsic relation between socially engaged forms of cognition, emotion, and behavior and social groups:

> We need a way of understanding group processes that retains the prime reality of individual *and* group, the two permanent poles of all social processes. We need to see group forces as arising out of the actions of individuals and individuals whose actions are a function of the group forces that they themselves (or others) have brought into existence. We must see group phenomena as both the *product and condition* of actions of individuals. We cannot resolve the difficulty by merging the two extreme views in some judicious way. (pp. 250–251)

Yet it was Floyd Allport's (1962) reductive resolution of the "master problem of social psychology" that prevailed, with the result that the study of common, interpersonally directed, and individually engaged psychological states and behavior came to dominate the field. As Asch himself put it in the Preface to Oxford University Press's 1986 reissue of *Social Psychology* (originally published in 1952), "Clearly I was swimming, often without realizing it, against the current. As one kindly and perceptive reviewer put it shortly after the book appeared: 'There is no doubt that Asch is a deviant'" (p. x).

Even the most hopeful proposals for an integrative and more social psychology were understated and easily interpreted – or reinterpreted – in terms of merely interpersonal and individually engaged psychological

[12] As Cooley (1902) stressed in the second clause of his much quoted claim that "a separate individual is an abstraction unknown to experience" (p. 1). Cooley continued, "and so likewise is society when regarded as something apart from individuals."

states and behavior. Thus, when Newcomb (1951) for example, in his introductory contribution to *Social Psychology at the Crossroads*, called for the integration of psychological and social forms of social psychology, he also stressed his commitment to the core position of psychological social psychology, namely, that the same basic explanatory principles should be employed in individual and social psychology:

> Psychological social psychologists... quite rightly insist that the same *basic* principles of behavior must be applied to human beings in social or in non-social situations. They decry the all-too-prevalent tendency to devise special principles to account for special forms of social behavior; much of the mythology of crowd behavior is of exactly this nature. All of this I can only applaud, but within the limits where the basic principles apply there is still room for much better refinement of the specifically human conditions in the environment which serve to determine the nature of human behavior. (p. 32)

Like Floyd Allport (1924a), Newcomb dismissed the idea of distinctively social forms of human psychology and behavior as part of the "mythology of crowd behavior." Like Dashiell (1935), he maintained that social psychology does not need to appeal to psychological principles distinct from those of individual psychology.

Nevertheless, Newcomb's (1951) own recipe for the postwar integrative development of social psychology was on the right track. A properly developed and integrative social psychology "would insist upon knowing how psychological processes function under the field conditions of group life" (p. 33). Newcomb himself had a fairly good grasp of the social dimensions of human psychology and behavior: his famous Bennington study (1943) was an unambiguous study of socially engaged attitudes. He also had a good sense of the distinction (insisted upon by Durkheim and Weber) between genuine social groups (those "amenable to social psychological study"), whose "members share norms about something," and mere aggregations of individuals who are engaged in interpersonal interaction, such as "persons at a street intersection," or who merely possess common properties, such as "all males in the State of Oklahoma between the ages of 21 and 25."[13] Unfortunately, it was all too easy for other social psychologists to reinterpret Newcomb's "field conditions of social

[13] The full quotation is provided in Chapter 3, in conjunction with the account of Durkheim's distinction between a genuine social group and a "mere sum of individuals" (1895/1982a, p. 129). Compare Asch's (1952) distinction between a genuine social group and a mere class or collection of individuals with common properties, such as "persons who are five years old or the class of divorced persons" (p. 260).

life" in terms of interpersonal objects and processes, especially given his antiholistic rhetoric.

Analogously, Krech and Crutchfield (1948), like Newcomb, felt obliged to maintain the identity of the principles of social and individual psychology upon which Allport (1924a) had insisted:

> As a basic science, social psychology does not differ in any fundamental way from psychology in general.... the principles of "social motivation," "social perception," and "social learning" are identical with the principles of motivation, perception, and learning found everywhere in the field of psychology. (p. 7)

They also appeared to maintain this identity for reasons analogous to Allport's own, namely fear of association with discredited theories of supraindividual social minds:

> ...there is nothing in social psychology that is not logically explainable at the level of the psychology of the individual. The study of groups as a whole can reveal nothing new beyond what is given by a synthesis of all the data pertaining to each of the group members. There is nothing superordinate to the individual, no "group mind." (pp. 366–367)

The subject matter of social psychology was characterized as the study of individually engaged psychological processes in the "social field," itself conceived on analogy with the "psychological field" of Gestalt psychology, interpreted in individual psychological terms. Thus, for Krech and Crutchfield (1948), there was no fundamental distinction between the objects and contents of nonsocial and social fields: "no theoretical distinction... can be drawn between a 'social field' and a 'nonsocial field'" (p. 8). The difference between the objects and contents of nonsocial and social fields is only a continuum representing different degrees of "capriciousness, mobility, loci of causation, power qualities, reciprocal reactivity" (p. 9).[14]

Although Krech and Crutchfield did document socially engaged forms of human psychology and behavior and endorsed the distinction between

[14] In fact, Krech and Crutchfield (1948) maintained the following:

> Theoretically it is not even necessary that the other objects in the perceiver's environment be animal; they might be plants or even inanimate objects. For example, we know that many natural, nonliving phenomena contain some or all of the essential properties we have been discussing. Clouds and storms and winds are excellent examples of objects in the psychological field that carry the perceived properties of mobility, capriciousness, causation, power of threat and reward.... Thus a man living in an environment completely without animate objects might still exhibit some types of behavior that we ordinarily think of as social. He might punish a "malicious" stone that trips him, try to appease thunder and lightning, imitate the noises of the waves, etc. (p. 10)

a social group and a mere aggregate group, they offered an interpersonal definition of a social group that was fundamentally no different from Floyd Allport's:

A group is different from a *class* of people or an aggregate of people who may be seen as grouped simply because the individuals are in close proximity. It should also be clear that a group is not the same as a "perceived group" that may exist for an individual. A person may think of himself and "see" himself as connected with other persons; yet for them he may not exist psychologically, and hence we cannot speak of these people as a group. On the positive side ... the criteria for establishing whether or not a given set of individuals constitutes a psychological group are mainly two: (1) All the members must exist as a group in the psychological field of each individual, *i.e.*, be perceived and reacted to as a group; (2) the various members must be in dynamic interaction with one another. (p. 368)[15]

Furthermore, like Floyd Allport, they defined social behavior as interpersonal behavior:[16]

... social behavior would be said to be that behavior which takes place in direct reference to other people ... (p. 8).

The Krech and Crutchfield (1948) text is often represented as an exemplar of the Gestalt psychological approach to social psychology, which is sometimes treated as having played a significant role in shaping the asocial forms of cognitive social psychology that developed in the 1960s

[15] The second "dynamic interaction" restriction appears to be unnecessarily strong, ruling out social psychological states and behavior oriented to the represented psychology and behavior of members of social groups with whom one does not dynamically interact:

A given Republican in California may have no psychological existence for another Republican in New York. Neither of them know each other or are ever aware of each other. These two Republicans, then, do not form a group. (Krech & Crutchfield, 1948, p. 18)

Earlier social psychologists who studied the social dimensions of political attitudes (e.g., Edwards, 1941) did not feel bound by this restriction. As Weber (1922/1978) stressed, the relevant others need not be known to us for our social attitudes and behaviors to be oriented toward them:

Social action ... may be oriented to the past, present, or expected future behavior of others.... The "others" may be individual persons, and may be known to the actor as such, or may constitute an indefinite plurality and may be entirely unknown as individuals. (Thus, money is means of exchange which the actor accepts in payment because he orients his action to the expectation that a large but unknown number of individuals he is personally unacquainted with will be ready to accept it in exchange on some future occasion). (p. 22)

[16] Compare Krech, Crutchfield and Ballachey's (1962) definition of social psychology:

In so far as any science can be specifically defined in terms of its basic unit or object of inquiry, social psychology can be defined as the science of interpersonal behavior. (pp. 4–5)

and 1970s (Farr, 1996). However, although the notion of a "social field" could be interpreted in terms of common, interpersonally directed, and individually engaged psychological states and behavior, as it was by Krech and Crutchfield, it could also be interpreted in terms of socially engaged psychological states and behavior, as it was by Asch (1952).

Although Gestalt psychologists such as Max Wertheimer, Wolfgang Köhler, and Kurt Koffka had little to say directly about social psychological phenomena, in their analysis of perceptual fields they constantly stressed the internal or intrinsic relation between the "elements" of perception and the perceptual fields in which they are configured: the identity of the elements of perception was held to be determined by their relational location in the perceptual fields within which they are configured (Wertheimer, 1987).[17] This original Gestalt emphasis on the internal or intrinsic connection between perceptual elements and perceptual wholes was carried over into the social psychological realm by Solomon Asch, who stressed the internal or intrinsic connection between socially engaged psychological states and behavior and represented social groups. Asch claimed that his social psychological theories were inspired by Wertheimer's lectures at the New School for Social Research in New York.[18]

Thus it does a great injustice to Asch's commitment to a distinctive social psychology to represent his position as basically equivalent to Floyd Allport's, just because both held – in their radically different ways – that social psychology ought to be concerned with the psychological states of individuals:

> While there is a strong theoretical contrast between Asch (1952) and Allport (1924a), which corresponds to the difference in perspective between Gestalt psychology and behaviorism, respectively, they share a common representation of the individual. This common element of individualism is part of American culture. We have seen earlier . . . that the same was true of the two Allport brothers, despite their quite different theoretical perspectives – cognitive in the case of G. W. and behavioral in the case of F. H. (Farr, 1996, p. 117).

Asch was one of the last social psychologists to retain the original conception of the social dimensions of human psychology and behavior and

[17] This was the feature that distinguished the so-called Frankfurt-Berlin school (Sprung & Sprung, 1997) of Gestalt psychology, represented by Wertheimer, Köhler, and Koffka, from the so-called Graz school (Fabian, 1997) of Gestalt psychology, represented by psychologists such as Stephan Witasek (1870–1915) and Vittorio Benussi (1978–1927).

[18] According to Kurt Danziger (personal communication).

to treat individual persons in the relational fashion of Gestalt psychology, as social individuals intrinsically related to other social individuals within represented social groups (Asch, 1952, p. 257).

## IV

The Lewinian program devoted to the study of "group psychology" and "group processes" also did not live up to its original promise. After Lewin's death in 1947, the Research Center for Group Dynamics moved from MIT to the University of Michigan (where it was incorporated with Lippett's Survey Research Center to form the Institute for Social Research). Faculty and students began to scatter and pursue their own careers and generally individualistic programs of research. The explosive growth of studies devoted to group processes in the 1950s initiated by Lewin's associates and students (Cartwright & Zandler, 1953) was followed by a precipitous decline in the 1960s (Steiner, 1974). Dorwin Cartwright, who headed the Research Center for Group Dynamics at MIT and Michigan after Lewin's death and taught the course in group dynamics at Michigan, noted that in later years most of the students taking the course were not psychologists (Cartwright, in Patnoe, 1988, p. 37).

The failure of the Lewinian program deserves special attention, both because it was so promising and because so many of the leading social psychologists of the late twentieth century, such as Leon Festinger, Dorwin Cartwright, Ronald Lippett, Morton Deutsch, Harold Kelley, Albert Pepitone, Stanley Schachter, and John Thibeau were students or associates of Lewin. The faculty and students of the MIT program "went on to shape the field of social psychology for nearly four decades after Lewin's death in 1947" and "trained a substantial proportion of the next generation of its most influential practioners" (Patnoe, 1988, pp. 10–11). These included social psychologists such as Edward Jones, Phillip Zimbardo, Jerome Singer, Lee Ross, John Darley, and Elliot Aronson.

There were a variety of reasons for the demise of the Lewinian program of group psychology. Lewin's original work was cognitive and motivational and largely inspired by Gestalt psychology (Lewin had been a colleague of Köhler and Wertheimer at the Berlin Psychological Institute in the 1920s). Although Lewin later extended the theoretical constructs of "field theory" (such as "tension system," "vector," "valency," and "quasi-stationary equilibrium") to cover group processes, these constructs were originally employed in the analysis of cognitive and motivational processes, such as satiation (Karsten, 1928), anger (Lissner, 1935;

Mahler, 1933), level of aspiration (Hoppe, 1930), and memory for interrupted tasks (Ovsiankina, 1928; Zeigarnik, 1927). Many of the students who joined Lewin at the University of Iowa Child Welfare Research Station (from 1935 to 1944) and later at the Research Center of Group Dynamics at MIT (from 1945 to 1947) knew only of this work. They were attracted by his commitment to bold exploratory theory and experimentation and his explicitly cognitive alternative to the prevailing behaviorist approaches.[19] Lewin's early cognitive and motivational studies had attracted American students to Berlin in the 1920s, long before he began work on group dynamics, and Lewin only became seriously interested in the subject of group dynamics when he came to America. He was introduced to group dynamics by his Iowa student Ronald Lippet (later one of the faculty at MIT), who had majored in group studies at Springfield College in Massachusetts (Cartwright, in Patnoe, 1988, pp. 30–31).[20]

Leon Festinger, one of Lewin's most influential students, frankly confessed that he had no real interest in social psychology and came to study with Lewin at Iowa because he was interested in Lewin's cognitive and motivational work. Festinger's doctoral dissertation was on level of aspiration, and he became a social psychologist merely "by fiat" when Lewin invited him to join the faculty of the Research Center for Group Dynamics at MIT (Festinger, in Patnoe, 1988, pp. 252–254). Analogously, Harold Kelley admitted that when he joined the MIT program (on the recommendation of Stuart Cook, with whom he worked during the war in the Aviation Psychology Unit), he "didn't really know what [he] was getting into" (Kelley, in Patnoe, 1988, p. 61). Kelley's (1950) own dissertation was on social perception, which by his own admission was "still very much individual psychology" (Kelley, in Patnoe, 1988, p. 66). Although Festinger, Kelley, and other students did produce notable theoretical and empirical work on group processes, it was perhaps not surprising

[19] A good number of the MIT students admitted to being attracted away from Yale for this reason. At the time, Yale was the only major university with an established graduate program (of sorts) in social psychology (the other was at Syracuse), but the Institute of Human Relations with which it was associated was heavily committed to behaviorist approaches, dominated by the research program of Clark L. Hull in the 1930s and 1940s. Although the MIT students took a bit of a risk opting for the new program at MIT, the risk was tempered by an arrangement through which students could take courses at Harvard University (Patnoe, 1988).

[20] He was also influenced by another student at Iowa, Alex Bavales, also a graduate of Springfield College. Bavales's genius for group work was probably the inspiration behind Lewin's vision for the MIT program (Cartwright, in Patnoe, 1988, pp. 31–32).

that after Lewin's death most of them shifted "to a more individualistic perspective" (Pepitone, in Patnoe, 1988, p. 88).

Most students at MIT had little contact with Lewin, who was frequently away fund-raising or serving as a consultant on various projects (including the Commission for Community Relations in New York). Lewin presided over the informal "Qasselstrippe" meetings, the "chatter line" (Ash, 1992) or "brainstorming" meetings in which evolving ideas were explored and discussed, but Lewinian theory at MIT was mainly taught by Marian Radke, who focused on Lewin's early Berlin work on cognition and motivation. Lewin did not generally supervise dissertations, and most MIT students had the same dissertation committee, composed of Festinger, Cartwright, Ranke, and Lippet.

Lewin's genius lay in the creation of a special group of talented theoreticians and experimentalists who saw themselves as an elite and pioneering team (Deutsch, in Patnoe, 1988, p. 95). He established a productive research environment and social atmosphere, and his open-ended and creative approach to theory, experimentation, and practical problem resolution affected and inspired the entire group.[21] More than anything, he inspired his colleagues and students to artfully reproduce theoretical variables abstracted from the dynamics of real-life social processes in carefully managed and controlled experiments. However, although most of this work was informed by the "assumptive framework" of field theory, field theory was sufficiently amorphous and open-ended to inspire a wide variety of individual research programs. Many of Lewin's students (and Lewin himself at times) treated field theory as a metatheoretical or methodological position and not a substantive theoretical program (Patnoe, 1988, p. 18).

It is thus not surprising that students such as Kelley admitted to being "vastly more influenced by Festinger than Lewin" (Kelley, in Patnoe, 1988, p. 64), or that, like Festinger, their later interests returned to the individual psychological. It was Festinger who directed the experimental work at MIT, and it was Festinger who extended Lewin's program of creative experimentation after Lewin's death, albeit in an increasingly individualistic direction. It was the Lewinian commitment to creative experimentation as the mark of a genuinely scientific social psychology that had attracted many of the MIT students to the program, and it continued to motivate

[21] Lewin effectively re-created at Iowa and MIT the research culture that he had originally created in Berlin (Ash, 1992). Many of his students tried to re-create this research culture at their own institutions in later years but had only limited success (Patnoe, 1988).

them in the years ahead as they developed their independent careers. In describing the legacy of the MIT program, John Darley aptly described Lewin as the "Christ figure" but maintained that "Festinger played the role of St Paul who made the thing happen" (Darley, in Patnoe, 1988, p. 209). As Darley put it,

> Festinger... demonstrated that we in social psychology could do as good, well-reasoned, methodologically sophisticated experiments as people working with infrahuman subjects... to show that we could be as rigorous as the others. (p. 209)

The early work of Festinger and Schachter on communication and cohesion in social groups (Festinger, 1950; Festinger et al., 1950, 1956; Schachter, 1951) was based on studies of genuine social groups and theories about the "anchoring" of attitudes and behavior to social reference groups. However, the later work of these two major figures become increasingly focused on individually engaged cognitive processes and interpersonal behavior, based upon studies of experimentally created and designated groups. Thus Festinger's (1957) theory of cognitive dissonance abandoned any link to the social,[22] and stimulated a research program that was exclusively concerned with individually engaged cognitive processes and was based upon studies of experimentally created groups. Schachter and Singer's (1962) classic experimental study of the "cognitive, social and physiological determinants" of emotional states did not involve any exploration of the social dimensions of emotion. It did not, for example, explore the question of how the content and conditions of emotions (or their expression) might be oriented to represented social groups. The Schachter-Singer experiment only explored interpersonal influences and cognitive inferences among experimental subjects who were strangers to each other.

Lewin's own position was somewhat ambiguous. He did creatively explore and practically exploit the social dimensions of human psychology and behavior. However, his theoretical orientation often remained individually psychological and interpersonal. As he acknowledged in *Principles of Topological Psychology* (1936), his primary interest was with the individual psychological "life space." Although Lewin's general theoretical perspective was originally gestaltist in orientation, he progressively distanced himself from gestaltism as his career developed.[23] Although he

[22] Except for the field studies of millennium cults reported in Festinger et al. (1956).

[23] Lewin remained agnostic on the question of whether psychological fields are isomorphic with neural fields (Ash, 1992), a central tenet of the Berlin School of Gestalt psychology.

sometimes treated the social field as the aspect of an individual's life space that is orientated to represented social groups (1946/1997d, p. 360), at other times he treated it in terms of the overlapping life spaces of individuals, in which the psychological fields of individuals include other individuals (i.e., in the individualistic fashion of Krech & Crutchfield, 1948).

The individuals of Lewinian theory often remain autonomous individuals, in tension or harmony with other autonomous individuals, unlike the social individuals of Asch (1952) or Sherif (1948), who were treated as intrinsically related to social groups (through the representation of their socially grounded individuality or identity). Although Lewin did recognize the essential psychological constitution of social groups (1940/1997b, p. 69) and abjured operational definitions of the meaning of theoretical constructs (1942/1997c, p. 13), he generally stuck to his own operational definition of social groups in terms of the dynamical "interdependence" of their parts (1939/1997a, p. 60). For Lewin, as for Krech and Crutchfield, the essential feature of a group is the dynamic interaction between the individual members. In consequence, Lewinian groups included Allportian co-acting and interacting groups as well as genuine social groups.

While some of Lewin's experimental studies employed genuine social groups and explored socially engaged attitudes and behavior, such as the Lewin et al. (1939) study of different "social atmospheres" in Boys Clubs, many involved experimentally assembled aggregate populations. This was true of the famous studies of "group decision and group change," for example, reported in Lewin (1947a). The pregnant housewives who made up the experimental groups in the study of nutritional habits were artificially assembled populations, originally strangers to each other. The "neighborhood groups" that formed the populations assembled for the milk study (selected via door-to-door solicitation and attendance at the local community center) are ambiguous: some of the participants knew each other, others did not. The experimental groups in the "intestinals" study were composed of Red Cross volunteers: while they were no doubt members of a genuine social group, this was not the variable explored in the experiment, since *all* the experimental groups were made up of Red Cross volunteers.

While he expressed his debt to the work of Wertheimer and Köhler in his first book, *A Dynamic Theory of Personality* (1935), he acknowledged that he was moving beyond their theoretical position in the open letter to Köhler that formed the preface of his second book, *Principles of Topological Psychology* (1936). Perhaps for this reason, Köhler would not support Lewin for an appointment at the New School for Social Research when Lewin was suggested as a candidate (Marrow, 1969, p. 159).

Because of this, it was very easy for those working within and later continuing the Lewinian theoretical and experimental program to simply focus on the individual psychological and the interpersonal. This was certainly the case in much of the work done within the Lewinian program during the 1950s and 1960s. As McGrath and Altman (1966) and Sherif (1977) complained, much of the work on small groups during the 1950s and 1960s focused on immediate interpersonal attraction and preferences (independently of any relation to represented social groups). Thibaut and Kelley, in *The Social Psychology of Groups* (1959), one of the most influential texts of this period, restricted their account of social groups to a cost-benefit analysis of individuals' motivation for behavior within dyadic interpersonal relationships, on the assumption that such an experimental analysis of the behavior of individuals in dyads could be extended to all forms of social cognition, emotion, and behavior:[24]

> In the analysis that follows we begin with the two-person relationship, the dyad. We so begin in order first to attempt to understand the simplest of social phenomena by endeavoring to be as clear and as explicit as we can about the conditions necessary for the formulation of a dyadic relationship and about the interpersonal relations manifested there. Our bias on this point is apparent: we assume that if we can achieve a clear understanding of the dyad we can subsequently extend our understanding to encompass the problems of larger and more complex social relationships. (p. 6)

In response to critics who complained that their work was too behaviorist in orientation, Thibaut and Kelley later claimed (Kelley & Thibaut, 1978)[25] that their work was a development of the Lewinian tradition of which Lewin himself would have approved, since it promised to resolve

[24] This is not to deny the significance of the work of Thibaut and Kelley or its relevance for a social psychology. Thibaut and Kelley described the interpersonal processes and individually engaged "cost-benefit" motives that promote the creation of elemental social groups (dyads), and they acknowledged that the "group norms" that emerge from such patterns of interpersonal action and individual motives come to *displace them* as the basis for social forms of cognition, emotion, and behavior. Insofar as their theory may be conceived as an attempt to develop a theory of the *formation* of certain kinds of social groups, it may be treated as a genuine attempt to "bring back groups," as Thibaut later insisted that it was (in Patnoe, 1988, p. 59). Thibaut and Kelley conceived of their experimental studies in contrast to Allport-type experiments of the "individual in relation to some fixed social stimuli that are under the experimenter's control," which Thibaut characterized as "individual psychology in a social setting" (in Patnoe, 1988, p. 59). Nonetheless, most of the experimental groups created were Allportian face-to-face groups.

[25] Significantly their second book was entitled *Interpersonal Relations* rather than *The Social Psychology of Groups*.

the problem of "overlapping life spaces" that Lewin was working on just before he died. Yet this is itself indicative of the frequently individualistic orientation of the Lewinian notion of psychological life space and his individualistic analysis of social groups in terms of the interdependence of individuals in "dynamic interaction" – an analysis of which Floyd Allport would have heartily approved.

Thibaut and Kelley's assumption that a "clear understanding of the dyad" could be extended to "encompass the problems of larger and more complex social relationships" essentially reprised Floyd Allport's (1924a) claim that experimental studies of small face-to-face groups could ground explanations of complex social psychological phenomena, and some such assumption runs through most of the experimental literature of this period. Thus Bales (1953), for example, followed Allport and Krech and Crutchfield in maintaining that interpersonal behavior is the "ultimate stuff" of social psychology, and he defined small groups in terms of interacting persons:

> A small group is defined as any number of persons engaged in interaction with one another in a single face to face meeting or a series of such meetings, in which each member receives some impression or perception from each other member distinct enough so that he can, either at the time or in later questioning, give some reaction to each of the others as an individual person, even though it be only to recall that the other was present. (p. 30)

Analogously, Stogdill (1950) offered M. Smith's (1945) definition of a social group as the "most satisfactory definition available at the present time." According to this definition, a social group is "a unit consisting of a plural number of organisms (agents) who have collective perception of their unity and who have the ability to act/or are acting in a unitary manner toward the environment" (p. 2). Stogdill also endorsed Krech and Crutchfield's definition of a social group as a "set of individuals" in "dynamic interaction with each other," claiming that it was similar to Smith's (which is doubtfully true, although both do follow Floyd Allport's definition).[26]

In any case, it was fairly easy for avowed practitioners of the Lewinian tradition to simply develop the experimental paradigm for the study of interpersonal interactions in small experimental groups that was established in the prewar years by Allport, Dashiell, and the Murphys. For

[26] The difference between the Smith definition and the Krech and Crutchfield definition is roughly the difference between Allport's (1924a) co-acting and face-to-face (interacting) groups.

example, Deutsch's (1949) studies of cooperation and conflict offered a game-theoretical analysis of individually engaged motives for behavior and their effect on small face-to-face group functioning, based upon studies of experimentally created groups of about six persons. Lewin's frequent focus on individually engaged "psychological fields" directed toward other persons made it easy for later practitioners to represent the Lewinian "situation" and "construal of the situation" in individual psychological and interpersonal terms. An example of this development of the Lewinian tradition, claimed to be an exemplar of that tradition by two of its avowed inheritors (L. Ross & Nisbett, 1991), is the series of laboratory and field experimental studies of "bystander apathy" developed by Latané and Darley (Darley & Latané, 1968; Latané & Darley, 1970). The "situations" investigated by Latané and Darley were interpersonal, not social: the "victims" and potential "helpers" studied were not presumed to be (and were not selected as) members of distinctive social groups. The situational "pressure" was operationally defined in terms of the number of other persons present, irrespective of their membership of any represented social group. On the basis of such studies, the phenomenon of bystander apathy was explained in terms of individually engaged cognitive states, such as "diffused responsibility" and "failure to represent the situation as an emergency."

## V

Franz Samelson (2000) has recently claimed that Floyd Allport's influence was waning in the 1950s and 1960s. While Allport may have been infrequently referenced as a theoretical precursor, his spirit clearly informed the experimental methodology of small-group research in the 1950s and 1960s, so much so that McGrath, in his 1978 review of the small-group experimental literature, could exhaustively classify the experimental research surveyed under the Allportian headings of "co-acting" and "interacting" (face-to-face) groups. Only a very few studies during this period employed preexisting social groups to study the social dimensions of cognition, emotion, and behavior. Subjects were rarely selected on the basis of their membership of established social groups (some notable exceptions were Charters & Newcomb, 1952; Festinger et al., 1950, 1956; French, 1941; Gorden, 1942; Kelley & Volkart, 1952; Siegel & Siegel, 1957). In most cases, the "social" groups studied were randomly selected sets of individuals externally designated (by experimenters) as groups for the purposes of experimental analysis. These experimental groups satisfied

Allport's definition of a group (and Lewin's operational definition of a group) but were only genuine social groups in rare cases, and then by accident: for example, when volunteer students just happened to be members of some independently constituted social group[27] (which would have been treated as a source of confounding in most experiments). While a few studies arguably touched on the formation of genuine social groups from experimentally assembled collections of strangers, very few studies were concerned with the dynamics of the formation, maintenance, and transformation of social attitudes, emotions, and behaviors among members of *preformed* social groups, which would have required the prior determination that the study subjects were indeed members of such groups (D. Katz & Schanck, 1938, p. 45).

The fertile tradition of research on persuasion and attitude change that grew out of wartime studies of propaganda and advertising (Hovland, Lumsdaine, & Sheffield, 1949) and the work of the interdisciplinary Communication and Attitude Change Program at Yale University (Hovland et al., 1953) mainly focused on controlled experimental studies of interpersonal communication, exploring the properties of the communicator, the communicatee, and the communication. This despite the fact that a number of studies suggested the powerful role played by the social grounding of attitudes – by their "anchoring" in social reference groups (Kelley & Volkart, 1952; Kelley & Woodruff, 1956).

This last point perhaps deserves further comment. While some theorists in the 1950s and 1960s were prepared to grant that attitudes and behaviors could be anchored in social groups, other theorists were almost as reluctant as Floyd and Gordon Allport to accept that attitudes and behaviors are ever engaged socially. In his development of "social comparison" theory, Festinger (1954) assumed that individuals only evaluate their opinions and abilities socially when "objective, non-social means" are unavailable and that they evaluate their opinions and abilities individually, and thus rationally, when such means are available.[28] In various subtle ways, socially engaged attitudes and behaviors were denied or deprecated as a modern (or postmodern) malaise: they were held to represent the psychology or psychopathology of "other directed people" (Reisman, 1950) or of

[27] As in the case of the subjects in Asch's (1951) experimental groups in his classic conformity studies. See Chapter 7.

[28] Although in practice the availability of "objective, non-social means" of evaluating opinions and abilities seems to be singularly ineffective in undermining or overriding socially engaged stereotypes and prejudices, a point recognized by many early American social psychologists (Faris, 1925; Horowitz, 1936; D. Katz & Braly, 1933; Lasker, 1929).

"group think" (Janis, 1968). Alternatively, they were explained away (or excused) by reference to interpersonal pressures to conform, conceived in terms of individual rewards and punishments for conforming or failing to conform. Aside from Sherif (1948) and Asch (1952), few theorists seemed to acknowledge that individuals might just accept and engage attitudes and behavior socially for no reason other than that such attitudes and behaviors are prescribed by the social groups to which their identities *as individuals* are oriented.[29]

While small-group research expanded dramatically in the 1950s and early 1960s,[30] it declined in the late 1960s and early 1970s (McGrath, 1978; Steiner, 1974). Whether this was due to the absence or impoverishment of theory and application (Helmreich, Bakeman, & Scherwitz, 1973; McGrath, 1978; McGrath & Altman 1966) or the decline in military funding (Cina, 1981) is not clear, but the contraction of small-group research signaled the end of the social in social psychology.

Despite the theoretical promise and confident anticipations of the 1950s, detailed research on reference groups was simply never developed. At the end of the 1960s, Hyman and Singer (1968) lamented the fact that the hope and promise of theory and research on reference groups had not been fulfilled. While theory and research on comparison groups (grounded in Festinger's 1954 study) continued apace, research on "normative" reference groups, that is, those groups that serve as the source of the individual's norms, attitudes, or values (Kelley, 1952), declined precipitously in psychological social psychology.

Things were not much better in sociological social psychology. Elenor Singer, in her chapter on "Reference Groups and Social Evaluations" in the 1988 sociological handbook of social psychology (*Social Psychology: Sociological Perspectives*, prepared by the Section on Social Psychology of the American Sociological Association), noted that the use of the concept had declined since 1968 and repeated the lament that the explanatory promise of the concept of reference groups had not been fulfilled. She recommended that the normative concept of a reference group be abandoned,

[29] Compare Pepitone (1999, p. 183), who notes that most people behaving uniformly by wearing blue jeans, driving sports utility vehicles while talking on cellular phones, and the like, do not generally look to others for the validation of their views, since they have "already decided that their way . . . is right." By wearing blue jeans or driving their sport utility vehicles they affirm their identities: they "show others, or some others, where they stand."

[30] Hare (1972, quoted in Sherif, 1977) reported a fourfold increase in studies during the period 1959–1969 in relation to all the decades before 1959.

and that reference group theory be restricted to "comparison processes" (p. 91).[31]

In any case, none of the post-1950 studies cited in Hyman and E. Singer (1968) or E. Singer (1988) relating to normative reference groups were published in mainstream social psychology journals. While there is some discussion of reference groups in the 1968 *Handbook of Social Psychology* (Lindzey & Aronson, 1968, in the chapters by Berkowitz, Collins, & Raven; De Vos & Hippler; Freeman & Giovannoni; Getzels, McGuire, Moore, Sarbin, & Allen; and Sears), there is none in the 1985 and 1998 editions. A search of *Psychological Info* from 1960 to 2000 yields around 750 entries on reference groups, only a mere handful of which were published in mainstream social psychology journals. As Jerome Singer (one of Schachter's students) noted, "The study of how a group shapes an individual's opinion... gradually ceased to be a major research question" (1980, p. 161).

[31] In contrast to Shibutani (1955), who proposed to restrict reference group theory to normative groups, that is, to those groups "whose perspective constitutes the frame of reference of the actor" (p. 563).

# 9

# Crisis

Farr (1996, p. 156) notes that historical contributors to the various editions of the *Handbook of Social Psychology* (G. W. Allport, 1954, 1968a, 1985; Jones, 1985, 1998) have come to represent American social psychology as having a "long past" but "short history" (echoing Gustav Fechner's famous claim about scientific psychology). Many represent the "short history" as beginning with the 1935 edition of the handbook, especially Dashiell's (1935) chapter on experimental social psychology,[1] and coming to full fruition with the explosive postwar development of the discipline. Indeed, some claim social psychology only really became a scientific discipline in the postwar period, when practitioners embraced a common experimental paradigm for social psychological investigation (Levine & Rodrigues, 1999). Many authors also note that the postwar development built upon the prewar foundations of the 1930s. As Cartwright (1979) put it, "social psychologists were well-prepared to respond" to the war (p. 84), and after the war they developed "a vast storehouse of well-established empirical findings" (p. 87).

This is important to stress. Although the immediate postwar years witnessed perhaps the high-water mark of social forms of social psychology, they also witnessed the explosive development of the asocial scientific and experimental tradition that originated in the 1920s and 1930s. Many of the founders of scientific and experimental social psychology in the 1920s and 1930s repudiated the competing theoretical paradigms prevalent at the time (appeals to instinct, the group mind, crowd theories,

[1] See, e.g., Fiske and Goodwyn (1996, p. xv), Lindzey and Aronson (1985a, p. iii), and E. E. Jones (1985, p. 77).

and so forth) and established a "new theoretical and methodological approach" (Cartwright, 1979, p. 83). Unfortunately, this theoretical and methodological approach was fundamentally asocial, and despite changes in fashion with respect to particular theories and research topics (social facilitation, group dynamics, conformity, cognitive dissonance, attribution, social cognition, and the like), this asocial paradigm remained entrenched throughout the rest of the twentieth century. In accord with this asocial paradigm, postwar American social psychology has remained focused on individually engaged and interpersonally directed psychological states and behavior investigated in studies of experimentally constituted "groups."

Many researchers in the postwar years simply developed the asocial theoretical research programs that originated in the prewar years. Thus the "social learning" theories of aggression promoted by Bandura (1962, 1973) and Berkowitz (1962; Berkowitz & Le Page, 1967) were developments of the earlier tradition of explaining aggression by reference to individually engaged motives and behaviors (Dollard et al., 1939; Miller & Dollard, 1941). "Social learning" in this tradition just meant "interpersonal learning": there was no theoretical requirement that the persons whose behavior is imitated be represented as members of a social group (serving as a reference group for socially engaged aggressive behavior, for example).

Analogously, Zajonc (1965) developed Allport's program of research on "social facilitation," perhaps better described as "interpersonal facilitation," since the audience or members of the co-acting experimental groups studied were rarely also members of distinctive social groups. Zajonc provided an individual psychological explanation of the facilitation of well-learned responses and impairment of the acquisition of new responses in terms of increases in "the individual's general arousal or drive level" (p. 273). As noted in Chapter 8, much of the so-called small-group research in the 1950s and 1960s continued the Allportian methodological tradition by focusing on co-acting or interacting (face-to-face) experimental groups, constituted by "sets of individuals" who were usually strangers to each other and were only rarely members of distinctive social groups.

While the 1950s perhaps represented the high-water mark for the advocacy of the original conception of the social dimensions of human psychology and behavior, many social psychologists were intent to develop the asocial metatheoretical and methodological program of Allport, Dashiell, and Murphy and Murphy. By the late 1960s, little of the social remained in the pages of the *Journal of Experimental Social Psychology* or the emancipated *Journal of Personality and Social Psychology*,[2] which now focused almost exclusively on individual psychological explanations of interpersonal behavior and individually engaged psychological processes. The dominant theoretical research programs were those developed from cognitive consistency theory (Abelson & Rosenberg, 1958; Festinger, 1957; Heider, 1958), exchange theory (Homans, 1961; Thibaut & Kelley, 1959), attribution theory (Bem, 1967; Heider, 1958; Kelley, 1967, 1971; Jones & Davis, 1965; Jones & Nisbett, 1972; L. Ross, 1977), interpersonal attraction (Kelley, 1950), and person perception (Heider, 1958; Tagiuri, 1968). The tradition of conformity studies remained interpersonal, with conformity explained in terms of individually engaged motives and interpersonal pressures (Milgram, 1963, 1974). The experimental literature on persuasion and communication continued to focus on interpersonal cognitive processes and individually engaged attitude change (Cohen, 1964).

By the late 1960s, social psychology was firmly committed to, and has since remained firmly committed to, the study of the interpersonal interactions of aggregations of individuals, of co-acting or interacting (face-to-face) groups, independently of their represented social group membership. During this period, social attitude research, now largely "public opinion" research, was devoted to surveying the common attitudes or opinions of dispersed aggregations of individuals (publics) independently of their represented social group membership. With the development of more sophisticated experimental methodologies and survey research techniques, things actually got worse. The absence of any prior social relation between subjects in experiments in social psychology came to be seen as a methodological desideratum because it was conceived as a way to avoid problems of experimental "contamination." Attitude research came to focus on the contents of individually engaged common attitudes to ensure that statistical assumptions about independence were not violated. This was particularly true of research on stereotypes, for example (Haslam, Turner, Oakes, McGarty, & Reynolds, 1998).

## I

A number of historians have laid part of the blame for the neglect of the social in American social psychology on the commitment by postwar social

[2] This journal, formerly titled the *Journal of Abnormal and Social Psychology*, abandoned its link with abnormal psychology in 1965.

psychologists to experimentation as the preferred method of research (Danziger, 2000; Gergen, 1978; MacMartin & Winston, 2000; Pepitone, 1999; Secord, 1990; Winston & Blais, 1996) and to associated statistical methods such as analysis of variance. Danziger, for example, maintains that "methodology is not ontologically neutral" (2000, p. 332) and suggests that the social was finally eliminated from social psychology when experiments came to be "conceptualized in terms of the demonstration of functional relationships between specific stimulus elements, now known as independent variables, and specific response elements, known as dependent variables" (Danziger, 2000, p. 342).[3]

On this account, the commonly accepted definition of psychology experiments in terms of the manipulation of independent variables functioned as a constraint on theoretical conceptions of the social psychological. As MacMartin and Winston (2000) put it, "Methodology served to constrain ontology" (p. 350). In consequence,

> The enshrinement of the manipulated independent variable helped to justify the laboratory study of attitude change, aggression, competition, moral development, in the ahistorical, acultural and decontextualized approach that was so common in the 1950s and 1960s. (Winston & Blais, 1996, p. 613)[4]

It is certainly true that there was an explosion in experimental studies in the postwar period, along with an increased employment of associated statistical techniques such as analysis of variance (Rucci & Tweney, 1980).

[3] Compare Pepitone (1999), who claims that the neglect of the social in the postwar years was at least in part "a by-product of what has become a principle feature of the lab experiment – the testing of hypothesis by creating the independent variable. That is, the reduction of the subject-matter reflects the fact that not all subjects of interest to social psychology can be represented by independent variables" (p. 181).

[4] Gardner Murphy made a similar point in the 1960s:

> The tools, of course, determined in considerable measure what the product would be.... What you conceptualize and what you measure are two aspects essentially of the same thing, the measuring tool itself. (1965, p. 23)

Murphy and Murphy (1931) had earlier warned of the danger of assuming "that the *one* inevitable method of social psychology was the laboratory control of discrete variables, the isolation and measurement of each variable in terms of its effects" (p. 12). Murphy et al. (1937) later noted,

> It has become very evident in recent years that the social psychologist has thrust many of his problems into the laboratory without adequate consideration of the matrix in which his most certain and valuable data lie. He has simplified his phenomena in such a way as to exclude essential facts necessary to the understanding of social life, and has succeeded in experimental and quantitative control by leaving out most of the variables about which we really need to know. (p. 10)

Christie (1965) contrasted the 30 percent of *Journal of Abnormal and Social Psychology* articles reporting experimental studies in 1948 with the 83 percent in 1958. He also noted the increasing use of college students as subject populations and analysis of variance as the preferred method of statistical analysis. He predicted that within a few years "the number of published articles in the *Journal of Abnormal and Social Psychology* should reach an asymptote with all articles reporting experiments on college students using analysis of variance designs" (p. 151).

While tongue in cheek, it was a remarkably prescient prediction.[5] Higbee and Wells (1972) compared articles published in the 1969 *Journal of Personality and Social Psychology* with Christie's study of the earlier *Journal of Abnormal and Social Psychology*, and they noted that 87 percent of the studies employed experimental manipulation and 76 percent used college students. The later study by Higbee, Millard, and Folkman (1982) confirmed that this trend continued throughout the 1970s. Helmreich (1975) noted an increase in the number of articles reporting experiments in the *Journal of Personality and Social Psychology* from 56 percent in 1961 to 84 percent in 1974. Sherman, Buddie, Dragan, End, and Finney's (1999) recent study of research methods represented in the *Journal of Personality and Social Psychology* and the *Personality and Social Psychology Bulletin* between 1976 and 1996 records the continuing dominance of experimentation over all other methods of research.

There is little doubt that social psychologists became increasingly committed to experimentation as the preferred method of research in an effort to attain scientific respectability among their psychological colleagues, just as psychologists in earlier years (from Wundt and Titchener to Watson and Hull) had committed themselves to experimentation as the preferred method of research to gain scientific respectability among their natural scientific colleagues (Whitely, 1984).[6] However, it is doubtful if the commitment to experimentation per se or the development of more sophisticated statistical techniques was responsible for the abandonment of the original conception of the social. As the rare experimental studies

[5] It was certainly more prescient than the prediction made by S. Stansfield Sargent (1965) in his discussion of Gardner Murphy's paper in the same volume. Sargent optimistically anticipated "a non-experimental trend in social psychology (or perhaps several nonexperimental trends) in addition to the established and accepted experimental emphasis" (p. 35).

[6] During World War II and immediately after, there was significant resistance to accepting social psychology as a genuine scientific discipline (Capshew, 1999). See also Sam Parkovnick's (1998) account of Gordon Allport's failure to set up a research advisory office in social psychology and to secure funding for it.

produced in the postwar period demonstrated (e.g., Charters & Newcomb, 1952; Kelley & Volkart, 1952; Siegel and Siegel, 1957), the social dimensions of human psychology and behavior could be studied experimentally so long as subjects in experimental groups were preselected as members of genuine social groups (as in the above studies) or they were exposed to information relating to potential reference groups (as in the studies by Asch et al., 1938, and Asch, 1951, for example).[7] While the employment of analysis of variance increased exponentially after the war, this statistical technique did not itself preclude the experimental analysis of the social dimensions of human psychology and behavior. In fact, the first use of this statistical technique in social psychology (according to Christie, 1965) was in Edwards's (1941) study of socially engaged attitudes oriented to the represented attitudes of members of political groups.

In the postwar years, social psychologists did regularly characterize experiments in terms of the manipulation of independent variables and did treat the experimental manipulation and control of variables as the royal road to causal judgment in social psychological research:

> A laboratory experiment may be defined as one in which the investigator creates a situation with the exact conditions he wants to have and in which he controls some, and manipulates other, variables. He is able then to observe and measure the effect of the manipulation of the independent variables on the dependent variables in a situation in which the operation of other relevant factors is held to a minimum. (Festinger, 1953, p. 137)

> In the laboratory experiment, sufficient control can be achieved to obtain definitive answers, and systematic variation of different factors is possible. As a result of this greater control, precision and manipulability, conclusive answers can be obtained and relatively precise and subtle theoretical points can be tested. (Festinger, 1953, p. 140)

The manipulation and control of variables was treated as the feature that distinguishes the experimental method from supposedly inferior

[7] It is doubtful if the original selection of those social groups that serve as reference groups can be explored experimentally, however, since it is difficult to reproduce the basic dimensions of "anticipatory socialization" in a laboratory experiment:

> In the laboratory, selection is ordinarily reduced to a binary choice: to affiliate with this group or that, that group or none, for a limited time and for a purpose that, from the subject's point of view, may be trivial; and acceptance is similarly measured with respect to groups that are ephemeral rather than enduring, peripheral rather than central to the subject's life. (E. Singer, 1988, pp. 72–73)

Of course, such phenomena can be studied nonexperimentally, and were in fact explored in a pioneering series of developmental and correlational studies by Ruth Hartley (1957, 1960a, 1960b, 1960c).

methods of social psychological research (such as field studies, survey research, and the like):

> An experiment differs from other research methods in that the experimenter has some degree of control over the variables involved and the conditions under which the variables are observed. (Edwards, 1954, p. 260)

These characterizations of experimentation in psychology did not originate in the postwar period. According to Winston (1990), the common definition of experimentation in psychology in terms of the manipulation of independent variables (while holding other variables constant and observing the effects upon dependent variables) was first introduced by E. G. Boring (1933, pp. 9–10) and popularized in psychology through Robert Woodworth's introductory text *Psychology* (1934) and his graduate text *Experimental Psychology* (1938) – the so-called Columbia Bible. It was also Woodworth who promoted the notion that experimentation is the best if not the only means of identifying causes, by distinguishing experimentation from other comparative and correlational methods. The use of this definition of experimentation in psychology texts increased dramatically from the 1930s to the 1970s (Winston & Blais, 1996), from around 5 percent in the 1930s to 95 percent in the 1970s (by contrast, around 5 percent of biology texts and 0 percent of physics texts included this definition in the 1970s!).[8] Moreover, the use of this definition increased earlier and more dramatically in social psychology than in general experimental psychology (Danziger & Dzinas, 1997).

By the postwar years, the concept of an "independent variable" had come to be restricted within social psychology to those variables manipulated by experimenters:

> In an experiment, however, the experimenter has control over certain variables. These variables are called the *independent* variables. Independent variables are variables which the experimenter himself manipulates or changes. (Edwards, 1954, p. 261)

Nonetheless, at least in the early days, independent variables were sometimes conceived as including "subject" or "organismic" variables such as the social group orientation of experimental subjects. Thus Edwards (1954, p. 260), who cited Woodworth (1938) as the source of his definition of an independent variable, allowed that the preselection of subjects

[8] This gives the lie to the common presumption that experimental social psychology modeled itself on the practice of hard sciences such as physics. See Danziger (1997, pp. 178–179) for a useful discussion of this point.

according to political party affiliation counted as the experimental manipulation of variables (as in Edwards, 1941). Analogously, Festinger (1953) treated the preselection of subjects according to religious affiliation in his own experimental study of "group-belongingness" and voting behavior (Festinger, 1947) as a form of experimental manipulation:

> Decisions about the kinds of persons to be used as subjects, how they are to be recruited, and what they are to be led to expect before they come to the experiment *provide important opportunities for the manipulation of variables*. (p. 146, emphasis added)

However, by the 1960s American social psychologists had largely abandoned the empirical study – including the experimental study – of the social dimensions of human psychology and behavior because by then they had almost completely abandoned the original conception of the social dimensions of human psychology and behavior, including the original conception of a social group. Once they had abandoned the original conception of the social, they naturally had no interest in conducting experiments that explored the social dimensions of human psychology and behavior, nor in treating the social dimensions of psychological states and behavior as independent variables, even as broadly conceived by Edwards (1954) and Festinger (1953).

The asocial prewar conception of social psychological phenomena, as individually engaged and interpersonally directed psychological states and behavior, originally developed by Floyd Allport, was certainly "tailor made for an experimental social science" (Danziger, 2000, p. 333). However, Allport's individualistic conception was primarily shaped by his empiricism and moral and political individualism, not by his commitment to experimentation per se. The postwar commitment to experimentation, conceived as the active manipulation of independent variables, certainly supported the individualistic theoretical conception of social psychological phenomena inherited from the prewar period but did not *presuppose* it.

One way of illustrating this point is by noting that one of the exemplars of the postwar experimental paradigm (cited by Danziger, 2000, as an exemplar of that paradigm) was Carl Hovland's research program on communication and attitude change at Yale (Hovland et al., 1953).[9] As Danziger (2000, p. 342) notes, this research program was "based upon

[9] This program grew out of research on propaganda and morale conducted for the Information and Education Section of the War Department (Hovland et al., 1949).

a model that had already been in vogue before World War II" and was part of the general behaviorist program associated with the Yale Institute of Human Relations. Although Hovland's asocial approach to attitudes and attitude change was also tailor-made for the postwar experimental paradigm, at least part of the Yale program was devoted to the experimental study of the social engagement of attitudes and the difficulties of changing such socially "anchored" attitudes (largely through the contributions of Harold Kelley, a student of Lewin's who joined the Yale program in 1950).[10] The experimental paradigm was itself essentially neutral and capable of supporting both social and individualistic approaches to human psychology and behavior.

This is important to stress, since part of the rhetorical point of explaining the postwar neglect of the social by American social psychology in terms of its commitment to experimentation is to champion the (supposedly) Wundtian view that experimentation is not (or frequently is not) an appropriate or possible means of exploring the social dimensions of human psychology and behavior. Perhaps the clearest and most articulate statement of this view is put forward by Danziger (1997), who claims that the standard conception of psychology experiments in terms of the manipulation of independent variables embodies the gratuitous assumption that social psychological phenomena are logically *atomistic*.

According to Danziger, this conception presupposes that "components" can be added or subtracted to experimental situations without altering the theoretical identity of the structure or process studied. Consequently, the "representation of situations and actions in terms of discrete, logically independent elements became the model for the conceptualization of all psychological reality" (Danziger, 1997, p. 168). According to this model,

> There could only be elements that retained their identity irrespective of the relationships into which they entered, all elements being logically independent. Such elements would vary only quantitatively, not qualitatively, between individuals and between situations, and the relationship between them would be essentially additive. (p. 177; cf. Danziger, 2000, p. 344)

This is a telling criticism of the assumptions that ground a great many actual experiments in social psychology. The principle of atomism, like

[10] As Kelley (1999) noted, however, the studies on these topics were not treated as an integral part of the Yale program. His "group focus" existed side by side with the "individualistic focus" of Hovland and Janis, and the two orientations "were never brought into confrontation" (p. 37).

the ideal of manipulative experimentation, is often associated with the idea of scientific social psychology (Greenwood, 1994). Yet although the relational nature of many social psychological phenomena creates very definite and distinctive problems for isolative experimentation, it does not preclude the experimental investigation of social psychological phenomena (Greenwood, 1989). After all, an electromagnetic field cannot be decomposed into its elements any more than a trial by jury, but the intrinsic dimensions and common concomitants of electromagnetic fields and jury trials can be experimentally varied and controlled.

It is possible to explore experimentally both the quantitative and qualitative dimensions of social psychological phenomena. Assuming that Milgram (1963, 1974) was successful in reproducing an authoritarian social structure in his studies on "destructive obedience" (which is reasonable, since psychology experiments are themselves instances of such structures), then Milgram's experimental setup can be employed (as in fact it was) to investigate how levels of obedience vary with the physical proximity of the experimenter and "learner" to the "teacher" *and* to explore how the experimental situation is transformed from an authoritarian social structure to something quite different when essential "components" are changed: for example, when subjects no longer represent the experimenter as a competent authority.[11] Analogously, one may investigate how the degree of conformity in Asch-type experiments varies as a function of the number of other persons present *and* explore the changes to the "psychological field" that are produced when the other persons are represented as strangers as opposed to members of a social group.[12]

## II

However, the increased sophistication of techniques of isolation and control and the employment of statistical techniques such as analysis of variance certainly exacerbated the neglect of the social that began with the

[11] This happens when two experimenters are employed and one expresses doubt about the safety of the procedures. See Milgram's (1974) Experiment 15.

[12] See Chapter 4, n. 25 for a discussion of this point. Such investigations might of course require the analysis of subject accounts of their representation of the experimental situation, a procedure that Asch himself routinely employed. However, this presents no special problem, since it only (reasonably) supposes that subjects are generally reliable authorities on their representation of experimental situations, not that they are authorities on the reasons or causes of their behavior in such situations (see Greenwood, 1989, chap. 11).

metatheoretical and methodological program initiated by Floyd Allport in the 1920s. When experimenters abandoned any attempt to explore the social dimensions of human psychology and behavior, their commitment to increased methodological and statistical rigor virtually ensured that experiments would exclude any residual social dimension. For when the social dimensions of human psychology and behavior are not themselves the object of experimental study (when they are not themselves treated as independent variables), then any form of "psychological connection" between subjects grounded in their orientation to social groups is a source of confounding, and violates assumptions about statistical independence (in field and survey studies of attitudes and stereotypes as much as in laboratory experiments). With the increased commitment to experimental and statistical rigor went the last vestige of the social.

What put the final nail in the coffin was not the commitment to experimentation per se but the commitment to an increasingly narrow conception of experimentation that developed in the postwar years. Winston and Blais (1996) note that the contemporary concept of experimentation in psychology "generally implies some notion of randomization, although this concept is rarely mentioned in textbook definitions" (p. 614). Although study designs involving the random assignment of subjects to experimental treatment groups had been employed since the 1930s and were increasingly utilized as the decades progressed, it was only in the postwar period that randomization came to be seen as an essential feature of experiments in social psychology.[13]

[13] Danziger (2000) describes this development in this way:

> Actual social groups were gradually replaced by hypothetical groups that had a purely statistical reality. The random assignment of individuals to different groups defined only by their experimental treatment constituted a fundamental and inescapable part of this methodological regime. (p. 344)

However, Danziger's claim appears overstated. No doubt the random assignment of subjects to experimental groups exacerbated the neglect of the social, but its neglect was widespread before randomization techniques were regularly employed in social psychology experiments. Moreover, neither the employment of experimental techniques per se nor the treatment of experimental groups as statistical groups mandated the random assignment of subjects to experimental conditions. As noted earlier, experiments could be and were performed employing social group affiliation as the manipulated independent variable (e.g., Charters & Newcomb, 1952; Kelley & Volkart, 1952; Siegel & Siegel, 1957), and statistical generalizations could be and were made about social groups based on the membership samples employed in experimental groups constituted according to social group affiliation. Thus the random assignment of individuals to experimental conditions was not an "inescapable part" of the experimental regime. Nonetheless, very few experiments managed to actually escape it as the postwar decades progressed.

Although Woodworth (1934, 1938) originally promoted the definition of psychological experiments in terms of the manipulation of independent variables, he never treated randomization as an essential feature of psychology experiments (and there is no discussion of randomization in the second and third editions of Woodworth's *Experimental Psychology* [Woodworth & Schlosberg, 1954; Kling & Riggs, 1972]). As late as 1975, Brenton Underwood, the author of a highly successful series of texts and manuals on experimental psychology (Underwood, 1949, 1966a, 1966b; Spatz & Underwood, 1970), maintained that "natural-group" designs exploring "subject variables" (such as social group orientation) are legitimate forms of psychological experimentation (Underwood, 1975).[14]

However, postwar social psychologists quickly elevated the random assignment of subjects to experimental groups to the status of a necessary condition for experimentation in social psychology. The first and second editions of *Research Methods in Social Relations* allowed that from a "logical point of view" the manipulation of variables is not necessary for experimental inquiry (Jahoda, Deutsch, & Cook, 1951, p. 59) and that the "equality" of experimental treatment groups can be attained via methods other than randomization, such as subject matching or frequency distribution control (Selltiz, Jahoda, Deutsch, & Cook, 1959, pp. 77–80). However, by the third and fourth editions, "true experiments" were defined as those in which potential confounding effects are excluded via randomization (Selltiz, Wrightsman, & Cook, 1976), and "subject" or "organismic" variables (such as religious or political group affiliation) were carefully distinguished from genuine "experimental" variables (Kidder, 1981, p. 19).[15] Analogously, Crano and Brewer's *Principles of Research in Social Psychology* (1973, p. 33) defined "true experiments" as manipulations involving the random assignment of subjects to treatment groups (compare Rosenblatt & Miller, 1972).

This restrictive conception of the experiment was in due course enshrined in the later editions of the *Handbook of Social Psychology*. The

[14] Commenting on such designs, Underwood (1975) noted the following:

> It might be argued that such procedures do not really constitute an experiment since the experimenter does not administer different treatments. Nonetheless, the thinking that is brought to such procedures and to their outcomes is the thinking that surrounds experiments. (p. 11)

[15] Quasi-experimental studies and nonexperimental studies were defined (following Campbell & Stanley, 1966, and Cook & Campbell, 1979) as studies "that do not have randomly assigned treatment and comparison groups" (Kidder, 1981, p. 43).

random assignment of subjects to experimental conditions, originally seen as a "major advantage" of experimentation (Aronson & Carlsmith, 1968, p. 7), came to be viewed as the "single essential attribute" of experiments in social psychology, for it ensures the "minimum guaranteed level of confidence that the treatments caused the observed differences in behavior" (Carlsmith, Ellsworth, & Aronson, 1976, p. vii). The random assignment of subjects to different experimental conditions came to be treated as "the criterial attribute for defining a study as an *experiment*" (Carlsmith et al., 1976, p. 15; the claim is repeated in Aronson, Brewer, & Carlsmith, 1985, p. 447, and Aronson, Wilson, & Brewer, 1998, p. 112: "the essence of an experiment is the random assignment of participants to experimental conditions").

In this fashion the elimination of the social from experimental social psychology was effectively institutionalized, since the randomization requirement left virtually no methodological space for the experimental study of the social dimensions of human psychology and behavior.[16] This is because the random assignment of subjects to experimental treatment groups effectively precludes the employment of genuine social groups in experimental social psychology.[17] Recent studies of research trends in social psychology from the 1960s to the 1990s document the increasing dominance of "true experiments" involving the random assignment of subjects to experimental treatment conditions (Reis & Stiller, 1992; Sherman et al., 1999; West, Newsom, & Fenaughty, 1992).[18]

[16] I say virtually no methodological space, because it is still in principle possible to manipulate social group orientation while employing randomization. For example, Catholics may be randomly assigned to different treatment groups in which Catholic orientation is made salient for one group only via different experimental instructions.

[17] Cf. Danziger (2000, p. 345):

> An experimental practice based on the randomized assignment of individuals to treatment groups has an implicit societal ontology, one that operates with populations rather than societal formations. (p. 345)

[18] It might be objected that this methodological commitment to exclude variables such as social group affiliation is a scientific virtue, since without the random assignment of subjects to treatment groups one cannot rule out the possibility that differences in experimental outcomes (differences in the measured dependent variable) may be due not to differences in social group affiliation but to subject variables such as personality traits associated with membership of different social groups. Thus differences in the experimental responses of Catholics as opposed to Protestants, or Democrats as opposed to Republicans, for example, may be due to personality factors associated with affiliation with these social groups (or to motives for affiliation with these groups).

This is of course true, but it is hardly a sufficient reason for abandoning the experimental investigation of socially engaged forms of cognition, emotion, and behavior,

Thus experimental social psychology in the postwar period effectively came to embrace the supposedly Wundtian position, that social – or "cultural" – variables cannot be explored experimentally but must be explored by alternative comparative or correlational methods. Since it was held that such variables cannot be experimentally investigated, "a social psychologist who has a question about such variables as these will be obliged to conduct a nonexperimental study" (Carlsmith et al., 1976, p. 35).[19]

## III

There were no doubt other reasons for the postwar abandonment of the original conception of the social dimensions of human psychology and behavior. Franz Samelson (2000) has recently suggested that many practitioners may have settled on the experimental study of individually engaged and interpersonal psychological states and behaviors of individuals in small artificially created experimental groups because of post-1960s pressures to "publish or perish."[20] They may have seen this as a more efficient means of generating published papers than the experimental study of social forms of human psychology and behavior, which generally requires the preselection of members of established social groups or the complicated methods of field experimentation and longitudinal analyses characteristic of Newcomb's (1943) Bennington study.[21] Experimental studies of individually engaged psychological states and behaviors of individuals in small artificially created experimental groups were generally

since such alternative explanations can be evaluated by additional experiments or by techniques such as subject matching or frequency distribution control. It is even less of a reason for abandoning the employment of social groups in conjunction with aggregate groups to experimentally isolate forms of socially engaged cognition, emotion, and behavior.

[19] Experimental social psychologists quickly embraced Woodworth's (1938) early distinction between experimental and correlational methods, but it was only in the postwar period that experimental methods came to be treated as superior (Stam, Radtke, & Lubek, 2000). However, the study of the social dimensions of human psychology and behavior was not assigned to correlational methods as a consequence of the general acceptance of Woodworth's distinction. As noted in Chapter 7, very few of the groups employed in either the experimental or correlational studies documented by Murphy et al. (1937) constituted genuine social groups.

[20] The complaint was anticipated by Gordon Allport in 1968: "May it be also that the goad to 'publish or perish' encourages swift, piecemeal, unread and unreadable publications, crammed with method but scant on meaning?" (1968b, p. 16).

[21] See Patnoe (1988, p. 88) for this sort of explanation (attributed to Pepitone): "Research on groups is not only expensive but time consuming."

represented as requiring fewer subjects and as less sensitive to confounding factors (although both assumptions are of doubtful validity).

In any case, by the 1960s social psychologists were no longer much concerned with the demarcation of social groups or the integration of social and psychological orientations. They now felt that social psychology had established itself as an experimental science and that they were free to develop it more or less as they saw fit: "In the garb of the lab coat, social psychology is becoming more and more 'scientifically respectable' and less and less viewed as the domain for soft-headed 'do-gooders'" (Deutsch & Krauss, 1965, p. 215). The hopeful visions of Newcomb (1951) and Krech and Crutchfield (1948) came to be replaced by fairly aggressive restatements of Floyd Allport's position.[22] Zajonc (1966) maintained that social psychology, like any other branch of psychology, is to be distinguished simply by reference to the causal variables studied. In the "social" psychology of humans, the relevant causal variable is the influence of other persons, as opposed to the "social" psychology of rats or cockroaches, where the relevant causal variable is the influence of other rats or cockroaches:[23]

For instance, the rat's response of "turning left in a T-maze" may be analyzed in terms of the number of reinforced trials that have been given to the animal (the psychology of learning); or in terms of the level of the animal's hunger (the psychology of motivation); or in terms of the physical properties of the right arm of the maze as opposed to those of the left arm (the psychology of perception). If all of the above variations – reinforcement, deprivation, and physical stimulation – are held constant, and if we observe the rat's responses of "turning left in the T-maze" when there happens to be one other rat in the right arm of the maze, we become social psychologists.

Social psychology deals with *behavioral dependence and interdependence among individuals*. By "behavioral dependence" we mean a relation between the

[22] It is instructive to compare the 1965 volume *Basic Studies in Social Psychology*, edited by Proshansky and Seidenberg, with its companion volume, *Current Studies in Social Psychology*, edited by Steiner and Fishbein (1965). These joint volumes replaced the 1947 *Readings in Social Psychology*, edited by Newcomb and Hartley (and the 1952 and 1958 revised editions). The *Basic Studies* volume was "designed to illustrate the type of research – mostly published before 1958 – in which current work in the field is rooted," and the *Current Studies* volume was "designed to exemplify very recent developments" (Kelman, 1965, p.v.). However, despite the optimistic intentions, there is little continuity between the two volumes. The *Basic Studies* volume includes a healthy sampling of theory and research on the social dimensions of human psychology and behavior, the *Current Studies* volume very little.

[23] For the social psychology of the cockroach, see Zajonc, Heingartner, and Herman (1969).

behavior of a number of individuals, such that a given behavior of one or more individuals is a cause or an occasion for change in the behavior of one or more other individuals. "Interdependence" simply means that the dependence is mutual and reciprocal. (Zajonc, 1966, p. 1)[24]

For Zajonc, as for Allport, social psychology was just a branch of general psychology concerned with individually engaged psychological states and behavior. Zajonc (1966) maintained that it is positively misleading to characterize social psychology as concerned with *social* cognition, emotion, and behavior, and he operationally defined the social in terms of causal dependencies between interpersonal behaviors without reference to represented social groups:

It is unfortunate that the field being introduced to the reader bears a "social" label – which, because it means so many different things, actually means very little. But even if the term "social" *explains* nothing specifically about man, it is still necessary for us to agree on what it *denotes*, for we shall have to use the word repeatedly. Since we define social psychology as the study of behavioral dependence and interdependence among individuals, "social" will mean a property of one organism's behavior which makes the organism vulnerable to the behavior of another organism. (p. 8)

As the sources of so-called social uniformity, Zajonc identified a variety of individually engaged reasons and causes, such as co-action, imitation, vicarious learning, communication, cooperation, and conflict (p. 87).

Indeed, as social psychology became more rigorously experimental (and self-confident), it was common for authors like Zajonc to abjure any attempt to define the social and to dismiss any attempt to offer an account of the distinctively social nature of social psychological phenomena as a sterile semantic quibble. Refusing to provide any contentful or connative characterization of social psychological phenomena, they proffered ostensive or denotive definitions by simply indicating the topics actually studied by social psychologists. Thus Aronson (1972), for example, in his enormously influential and popular *The Social Animal*, studiously avoided any explicit definition:

What is social psychology? There are almost as many definitions of social psychology as there are social psychologists. Instead of listing some of the definitions, it

[24] Compare this with the accounts of "mutual and reciprocal" dependence offered by Floyd Allport and Mead. See Chapter 4.

might be more informative to let the subject matter define the field. The examples presented in the preceding pages are all illustrations of sociopsychological situations. As diverse as these situations may be, they do contain one common factor: social influence. (p. 5)

"Social influence," however, was defined in purely interpersonal terms: "...this becomes our working definition of social psychology: the influences that people have upon the beliefs or behavior of others" (p. 6).

By the 1970s, the subject matter of social psychology had come to be prescriptively defined in terms of what people who called themselves "social psychologists" studied (at least in departments of psychology). Insko and Schopler (1972) noted that social psychologists originally borrowed their material from general psychology, sociology, and anthropology, but they suggested that

gradually, however, social psychologists began to evolve a subject matter more exclusively their own.... With tongue in cheek we can define social psychology as that discipline which people who call themselves social psychologists are interested in studying. (pp. xiii–xiv)

A few stalwarts still maintained a social and integrative vision throughout the 1960s. Works like Secord and Backman (1964), Secord, Backman, and Slavitt (1976), and Newcomb, Turner, and Converse (1965) made a brave attempt to integrate psychological social psychology with sociology, anthropology, and psychiatry. However, these works were more illustrations of the interdisciplinary scholarship of their authors than reflections of general interest among social psychologists. Moreover, even Newcomb et al. (1965), for example, although it included a survey of studies documenting the social dimensions of psychological states and behavior, still presented this survey in conformity with the now dominant interpersonal conception of social psychology as "the study of how people think, feel, and behave toward one another" (p. 1).[25]

[25] According to Newcomb, Turner, and Converse (1965), social psychologists "want to *understand* the conditions under which this particular sequence of behaviors is likely to occur, and when it is not" (p. 2). Social interaction, the basic subject matter of social psychology, was defined as

any set of observable behaviors on the part of two or more individuals when there is reason to assume that in some part these persons are responding to each other. What all these observable forms of interaction have in common is a *sequence of behaviors* on the part of two or more persons. (p. 3)

## IV

To be sure, not everyone was happy with this situation. For example, Volkart (1971, quoted in J. M. Jackson, 1988, p. 79), commenting on the 1968 five-volume *Handbook of Social Psychology*, complained that social psychology in the 1960s represented

> a rather vast umbrella under which a number of different theorists and researchers huddle together for protective reasons; but few of them know how to talk to each other. They may all be "social psychologists"; but they speak different languages, have different assumptions, and use quite different methods. (p. 899)[26]

D. Katz (1967) lamented the abandonment of the integrative theoretical and disciplinary visions of the 1950s and complained about "the continuing and growing fragmentation of the discipline" (p. 341). Similar complaints were voiced by Argyris (1967), Kruglanski (1975), Ring (1967), and Tajfel (1972).

A number of theorists identified the source of the problem as the increasingly asocial nature of the discipline of social psychology (Harré & Secord, 1972; Moscovici, 1972; Pepitone, 1976; Sampson, 1977; Sherif, 1977; Tajfel, 1972). However, the specifically social focus of these critical voices was almost obliterated by the cacophony of metatheoretical, methodological, and moral critiques of the discipline that constituted the 1970s "crisis" in social psychology. Despite the explosive growth of social psychology in the postwar decades, not everyone was convinced that social psychology in the 1970s could be characterized as having "better methods for collecting and analyzing data, a vast storehouse of well-established empirical findings, more rigorous conceptual models, and much more sophisticated theory" (Cartwright, 1979, p. 87).

Throughout the late 1960s and 1970s, social psychology was assailed by a wide variety of critiques that were advanced in the main by practitioners themselves.[27] Over and above complaints about the impoverished and fragmented nature of social psychological theory, there were serious doubts raised about the putative empirical progress of social psychology despite the plethora of experimental studies. Gordon Allport (1968b, p. 3) talked of the "slender achievements" of social psychology, and Sherif

[26] This strikingly echoes Durkheim's (1901/1982c) complaint at the turn of the century that social psychology "is hardly more than a term which covers all kinds of general questions, various and imprecise, without any definite object" (p. 41).

[27] Although some of the more extreme critiques were advanced by nonpractitioners (see Greenwood, 1989).

(1977, pp. 368–389) alleged that only a few "golden kernels" could be extracted from the experimental "chaff" of the preceding decades. More radically, critics such as Gergen (1973, 1976, 1978), Israel and Tajfel (1972), and Triandis (1976a) complained about the culturally and historically bounded nature of the experimental findings of the past decades, maintaining that they were restricted to a temporary phase of American culture and not generalizable beyond it. For example, theories of competition and conflict were doubtfully generalizable beyond the individualistic and capitalist culture of America (Plon, 1974), and conformity studies in the Asch paradigm were doubtfully generalizable beyond the McCarthy era in the United States (Gergen, 1973).[28]

Doubts were raised about the social relevance of much research (D. Katz, 1978; Ring, 1967), and moral and methodological concerns expressed about the use of deception with respect to experimental subjects (Baumrind, 1964; Kelman, 1967). However, perhaps the most common complaint focused on the artificiality of the laboratory experiment in social psychology (Argyris, 1975; Babbie, 1975; Borgetta & Bohrnstedt, 1974; Campbell & Stanley, 1966; Chapanis, 1967; Harré & Secord, 1972) and questioned the generalizability and real-world relevance of isolative laboratory studies: "The greatest weakness of laboratory experiments lies in their artificiality. Social processes observed to occur within a laboratory setting might not necessarily occur within more natural social settings" (Babbie, 1975, p. 254).

Concerns were expressed about the various "interaction effects" that appear to plague experimentation in social psychology (A. G. Miller, 1972a). As Beloff (1973) put it,

> A necessary precondition of the experimental method is that the phenomenon being investigated should not be materially affected by the procedure used to investigate it. . . . No such assumption is possible, unfortunately, with human subjects. (p. 11)

These interaction effects included contaminating variables such as "experimenter effects" (Rosenthal, 1966), "demand characteristics" (Orne, 1962; Orne & Holland, 1968), and "evaluation apprehension" (M. J. Rosenberg, 1969). Further doubts were raised about the generalizability of findings based on experimental populations of volunteers (Rosenthal & Rosnow, 1969, 1975), especially as they were largely (psychology) student volunteers (Higbee et al., 1982; Higbee & Wells, 1972). These complaints

[28] A complaint apparently supported by Bond and Smith (1996), Larson (1990), and Perrin and Spencer (1980).

led many to call for alternatives to deceptive laboratory experiments, such as field experiments (McGuire, 1967; Silverman, 1977) or role-playing experiments (Forward, Canter, & Kirsch, 1976; Jourard, 1968; Kelman, 1967; Ring, 1967).

While many of these critiques were legitimate, others were misdirected. The complaint about the historically and culturally bounded nature of social psychological phenomena generated a fierce debate (Greenwald, 1976; Jahoda, 1976; Manis, 1975; Secord, 1976; Triandis, 1976b; Wolff, 1977) but was somewhat of a red herring. Fundamental social psychological phenomena may vary cross-culturally and transhistorically (nothing in the nature of reality or science precludes it) but cannot be presumed to do so. As Simmel noted in 1908, the basic forms of sociation may be universal, even if their contents are historically and culturally diverse.[29] For example, it is an empirical question whether Lewin et al.'s (1939) findings about the relation of aggression to "social atmosphere" among American boys in the 1930s apply to Indian and Chinese boys in the 1980s.[30] It is not a question that can be determined by metatheoretical or methodological prescription.

## V

Most of the discussion about the artificiality of experiments focused on the technical "contamination" problems and on the predictability of psychological states and behavior beyond the confines of the laboratory experiment. Some social psychologists explored technical means of addressing interaction effects (Suls & Rosnow, 1988), and others questioned the generalizability of such effects themselves (X. T. Barber, 1978; X. T. Barber et al., 1969). Defenders of laboratory experimentation noted that the theoretical adequacy of explanations of human psychology and behavior tested by reference to isolated and controlled "closed-system" laboratory studies does not depend on their ability to license predictions about human psychology and behavior in "open systems" beyond the laboratory (Berkowitz & Donnerstein, 1982; Henschel, 1980; Kruglanski, 1976; Mook, 1983). Just as theories about superconductivity can be tested via predictions about how superconductors behave in isolation from local magnetic fields, which can interfere with superconductivity, so too theories about aggression can be tested via predictions about how volunteers

[29] See the discussion at the end of Chapter 2.

[30] Meade (1986), cited in Collier, Minton, and Reynolds (1991), suggests that they do not.

(including psychology student volunteers) behave in isolation from potentially interfering parents and policemen.

However, these legitimate defenses generally ignored the critical problem of the *identity* of the phenomena created in laboratory experiments in social psychology (Secord, 1982): the problem of whether, for example, instances of behavior in which subjects inflict electric shocks on others under laboratory conditions count as genuine instances of aggression (especially when subjects represent their behavior as beneficial to others, having been provided with a learning-theory rationale; see Baron & Eggleston, 1972; Kane, Joseph, & Tedeschi, 1976). A special instance of this problem, almost universally ignored by both critics and defenders of the experimental method, is the problem of employing randomly selected experimental groups of co-acting or interacting strangers to explore the social dimensions of human psychology and behavior. The problem was almost universally ignored because by this time the investigation of the social dimensions of human psychology and behavior had been effectively abandoned.

Social psychologists had been aware of the problem in the early days. Murphy and Murphy had warned about the difficulties of reproducing the "meaning" of everyday situations in social psychology experiments, especially when attempts were made to model them on experiments in physics (Murphy et al., 1937, p. 12).[31] Lewin (1935), in a famous paper contrasting Aristotelean and Galilean modes of thought, had noted that experimental situations represent specially created idealizations of everyday situations that enable the causal dynamics of such situations to be identified. Thus the Galilean law of free-falling bodies ($s = 1/2gt^2$) is experimentally determined via the "frictionless rolling of an ideal sphere down an absolutely straight and hard plane," a state of affairs that rarely occurs and only approximates the "unimpeded fall" of a body described by the law. However, although the causal dynamic described by the law is not manifest in the everyday motions of bodies, which are generally impeded, the (experimentally confirmed) law can be employed in the explanation and prediction of the everyday motions of bodies (taking into account impeding forces) because the motions of bodies *retain their identity as the motions of bodies* under artificial conditions of experimental isolation.

Analogously, although the types of groups with homogeneous memberships but different and distinctive styles of leadership employed in

[31] See Murphy and Murphy (1931):

> Much of the social behavior which is the actual marrow of the social sciences would not or could not occur in the artificial situation, in which the conditions were determined by the experimenter. (p. 22)

the Lewin et al. (1939) study of social atmospheres and aggression, for example, might rarely occur in the everyday social world (where individual reasons for joining such groups might generate heterogeneity in memberships), the experimentally identified relationships between forms of leadership and patterns of aggression can be employed to explain everyday patterns of aggression in terms of differences in leadership styles, given that *aggression and leadership styles retain their identity as aggression and leadership styles* under the artificial conditions of experimental isolation. That is, while Lewin recognized the fundamental differences between the types of situations "artificially constructed" in experiments and the adulterated and multifarious situations of everyday life, he insisted on the fundamental *theoretical identity* of the basic dimensions of the phenomena isolated in experiments and the everyday phenomena to which explanations and predictions based on such experiments are applied.[32]

Although his discussion of the problem was less sophisticated than Lewin's (and neglected Lewin's emphasis on the theoretically constructive nature of experiments), Festinger (1953) maintained essentially the same position. He noted that the main point and purpose of laboratory (or field) experimentation is to construct specially created and controlled situations in which causal relationships between variables (broadly conceived) can be discriminated via the elimination or attenuation or control of potentially interfering and confounding variables:

> The laboratory experiment should be an attempt to create a situation in which the operation of variables will be clearly seen under special identified and defined conditions. It matters not whether such a situation would ever be encountered in real life. In most laboratory experiments such a situation would certainly *never* be encountered in real life. In the laboratory, however, we can find out exactly how a certain variable affects behavior or attitudes under special, or "pure," conditions. (p. 139)

Festinger did not mean that experimental social psychologists should not concern themselves with what happens in the real world. The whole point of the experiment is to identify causal relations and deploy references

[32] While Lewin (1935) contrasted the Galilean and Aristotelian conception of causal explanatory laws, he might just as well have contrasted the modern scientific with the Humean conception of causal laws, since he identified the Aristotelian position with the view that causal laws in science express "regularity in the sense of frequency" (p. 13). This is significant, because most contributors to the artificiality of experiments debate in social psychology misconstrued the external validity (Campbell & Stanley, 1966) of experiments as the ability to license predictions about regularities in everyday situations based on regularities demonstrated in laboratory experiments – as if the explanatory relevance of Galileo's law of falling bodies depended on its ability to license predictions about everyday instances of unimpeded fall.

to them in explanations and predictions of behavior in real-life situations, which requires a basic theoretical identity between the "pure" or "ideal" situations explored in the laboratory and everyday situations:

> The possibility of application to a real-life situation arises when one knows enough about these relationships to be able to make predications concerning a real-life situation. . . . there should be an active interrelation between laboratory experimentation and the study of real-life situations. (pp. 139–140)

The fact that laboratory experiments are artificial or ideal or pure,[33] in the sense that (ideally) everyday forms of potential interference with an investigated causal sequence are absent or attenuated, is thus an epistemic advantage and not a disadvantage of the laboratory experiment, so long as the basic theoretical identity of experimental and everyday situations is maintained. Without this form of isolation and control, there would be no point in engaging in experimental studies of the causal dimensions of everyday situations; one might as well study everyday situations directly. As Festinger (1953) put it, the common complaint about the artificiality of experiments

> probably stems from an inaccurate understanding of the purposes of a laboratory experiment. A laboratory experiment need not, and should not, be an attempt to duplicate a real-life situation. If one wanted to study something in a real-life situation, it would be rather foolish to go to the trouble of setting up a laboratory duplicating the real-life situation. Why not simply go to the real-life situation and study it? (p. 139)

It is a critical assumption of the logic of experimentation that potential interfering and confounding factors can be eliminated or attenuated without altering the theoretical identity of the structural factors or causal processes isolated in experiments. This assumption is relatively unproblematic in the natural sciences, but it represents a real problem for experimental social psychology (see Greenwood, 1989). As Chapanis (1967) put it, with respect to social psychological phenomena, "The very act of bringing a variable into the laboratory usually changes its nature" (p. 566). While Festinger and Lewin were aware of the necessity of the basic theoretical identity of experimental and everyday situations for explanatory and predictive inference, they were rather too quick to assume

33 It is important to stress that "ideal" or "pure" in this context just means "isolated from normal forms of interference with the investigated causal processes." The processes themselves are equally real instances of causal processes whether they occur in experimental or nonexperimental contexts.

that it was satisfied in their own experiments and mistakenly equated (in Festinger's case, at least) the *intensity* of subject engagement with theoretical identity. For example, serious doubts may be raised about the presumed identity of the causal dynamics of everyday prison incarceration and Hanay, Banks, and Zimbardo's (1973) experimental exploration of these dynamics employing student volunteers (see Yardley, 1982, for a detailed critique), although there is little doubt about the intensity of the experience for the experimental subjects (in fact, the experiment had to be prematurely abandoned because it got too intense for the subjects). Analogously, one may reasonably doubt the theoretical identity of the social psychological phenomena reproduced in Milgram's (1963, 1974) studies of obedience and the social psychological phenomena grounding the behavior of concentration camp guards and the soldiers in the My Lai massacre (Orne & Holland, 1968), even though there is little doubt about the intensity of the experiences undergone by Milgram's subjects.

Yet by the 1970s concern about the theoretical identity of experimental and everyday situations had almost completely vanished. The problems associated with the artificiality of experiments were dismissed by distinguishing between *mundane* and *experimental* realism and maintaining that experimenters should only concern themselves with the latter form of realism. An experiment was defined as mundanely realistic insofar as there was a "similarity of events occurring in a laboratory setting to those likely to occur in the 'real world'" (Carlsmith et al., 1976, p. 81). For the reasons just noted, there is no special need for experiments to be mundanely realistic (and most good experiments are not), so long as the experimental situations created retain their theoretical identity with the everyday situations that are the putative objects of explanatory and predictive inference. Yet experimental realism as it came to be conceived in the 1970s was no guarantor of theoretical identity, because it was simply equated with the intensity of the experimental manipulation as experienced by subjects.[34] An experimental situation was held to be experimentally realistic "if the situation is realistic to the subjects – if they believe it, if they are forced to attend to it and take it seriously – in short, if it has impact on them" (Carlsmith et al., 1976, p. 81; cf. Aronson & Carlsmith, 1968, p. 22; Aronson et al., 1985, p. 482; Aronson et al., 1998, p. 131).

[34] Aronson, Wilson, and Brewer (1998, p. 132) appear to acknowledge this by noting that experimental studies may be low on both mundane and experimental realism but high on *psychological realism*, defined in terms of identity of psychological processes occurring in the experiment and everyday life.

Such "technical" responses enabled many to dismiss the more radical complaints as exaggerated and even as tantamount to the betrayal of the precepts of a science of social psychology. Gergen's radical position was regularly dismissed as unscientific (Jackson, 1988; Schlenker, 1974). The crisis itself was quickly followed by a spate of denials of a crisis. Elms (1975) claimed that the crisis was a crisis of confidence only, and many others claimed that the various critiques were overstated (Deutsch, 1976; McGuire, 1967). Some maintained that metatheoretical and methodological disputes, like regular theoretical disputes, are signs of scientific health and vitality and grounds for optimism so long as a pluralistic tolerance is maintained (Deutsch, 1976; Kiesler & Lucke, 1976).

In the meantime, the metatheoretical and methodological juggernaut that was scientific and experimental social psychology lumbered on largely unmodified. Mainstream social psychologists were dismissive of role-playing as an experimental strategy (Aronson & Carlsmith, 1968; Cooper, 1976; A. G. Miller, 1972b), generally endorsing Freedman's (1969) characterization of role-playing experiments as "unrealistic":[35]

> The data from role playing are people's guesses as to how they would behave if they were in a particular situation.... Role playing tells us what men think they would do. It does not tell us what men would actually do in the real situation. (p. 114)

Many concurred with A. G. Miller's (1972b) judgment that deception experiments remained "unfortunate but necessary" (p. 636).[36] In any case, the various methodological critiques had little effect on practice. In the ensuing years, there was an *increase* in laboratory experimentation, including deception experimentation, in relation to alternative forms of empirical enquiry, such as field experiments or role-playing studies (Fried, Gumper, & Allan, 1973; Higbee et al., 1982; Suls & Gastorf, 1980).

[35] Despite the fact that many studies attested to the high degree of realism that can be attained in role-playing experiments (Hanay, Banks, & Zimbardo, 1973; Janis & Mann, 1965; Mixon, 1972; Olson & Christiansen, 1966), at least in terms of the intensity of the experimental manipulations experienced by subjects.

[36] Experimental studies of role-playing as an alternative strategy to deception produced ambiguous results. Greenberg (1967), I. A. Horowitz and Rothschild (1970), and Willis and Willis (1970) reported qualified success, whereas Darroch and Steiner (1970), Holmes and Bennett (1974), and Simons and Pilliavin (1972) reported qualified failure. However, the methodological strategy of experimentally evaluating role-playing experiments via their ability to reproduce the results of deception experiments was itself problematic, given the original methodological (as well as moral) concerns about deception experiments (Greenwood, 1983).

## VI

The crisis in social psychology was effectively resolved for many by the adoption of the "social cognition" paradigm in the late 1970s and 1980s, propelled by the dramatic success of the "cognition revolution" in general psychology (Baars, 1986; Gardner, 1985; Lachman, Lachman, & Butterfield, 1979). It was generally agreed that the crisis could be resolved and the scientific promise of social psychology fulfilled by "getting inside the head" (Taylor & Fiske, 1981) to experimentally study social forms of cognition: that is, cognition *directed toward other persons and social groups*. The social cognition paradigm appeared to satisfy both those committed to an experimental psychology and those who wished to have a more cognitive and "social" social psychology.

Social cognition was the dominant topic of conferences in the late 1970s, and of edited collections in the early 1980s. Fiske and Taylor's definitive text *Social Cognition* came out in 1982. In the early 1980s the journal *Social Cognition* was instituted, along with the "Attitudes and Social Cognition" section of the *Journal for Personality and Social Psychology*.[37] The social cognition paradigm was so popular in the 1980s that Ostrom (1984) came to talk about the "sovereignty" of social cognition. Markus and Zajonc (1985) claimed that social psychology and cognitive psychology are "nearly synonymous" and reaffirmed that social psychology is just a branch of individual psychology. E. E. Jones (1985) critically dismissed talk of a crisis as not merely overstated but damaging and downright disloyal to the scientific institution of social psychology. McGuire (1985) reaffirmed the commitment of social psychology to the asocial theoretical and methodological paradigm that began with Floyd Allport's (1924a) individualistic and experimental reorientation of social psychology.

It should be stressed that the 1980s conception of social cognition was no more social than Floyd Allport's (1924a) conception of "social consciousness" or Gordon Allport's (1935) conception of "social attitudes." Focusing on cognition rather than behavior and calling cognition "social" did not return the social to social psychology. Social cognition was (and continues to be) conceived and defined in terms of cognitive states and processes directed toward other persons and social groups, not in terms of cognitive states and processes oriented to the represented

[37] According to S. E. Taylor (1998), at one point in the 1980s about 80 percent of the submissions to the *Journal for Personality and Social Psychology* were on social cognition.

psychology and behavior of members of social groups.[38] The only difference recognized between social cognition and nonsocial cognition is in terms of the types of objects toward which cognitive states and processes are directed, not in terms of differences between the states and processes themselves. It is presumed that the same individually engaged modes of information processing that underlie our perception and cognition of nonsocial objects (tables, trees, and tarantulas) also underlie our perception and cognition of social objects (other persons or social groups):

As one reviews research on social cognition, the analogy between the perception of things and the perception of people becomes increasingly clear. The argument is made repeatedly: the principles that describe how people think in general also describe how people think about people. (Fiske & Taylor, 1991, p. 18)

This is simply a cognitive restatement of the presumption that underlay Dashiell's review of the studies of the interpersonal behavior of individuals in small groups in his chapter on experimental social psychology in the 1935 *Handbook of Social Psychology* (which continued the asocial program of experimental research laid out by Floyd Allport in the 1920s):

Particularly is it to be borne in mind that in this objective stimulus-response relationship of an individual to his fellows we have to deal with no radically new concepts, no principles essentially additional to those applying to non-social situations. (p. 1097)

It might be objected that social cognition theorists do recognize possible differences between cognitive states and processes relating to social objects and nonsocial objects. Ostrom (1984), for example, reiterated the question raised earlier by Krech and Crutchfield (1948): "One frequently asked question is whether social knowledge differs in any significant way from knowledge about nonsocial objects" (p. 6). While Krech and Crutchfield maintained that differences between social and nonsocial cognition were only differences in degree but not in kind (in particular, quantitative

[38] The editorial introduction to the first edition of the journal *Social Cognition* endorsed this definition: "Any article that focuses on the perception of, memory of, or processing of information *involving people or social events* falls within our purview." (1982, p. i, emphasis added). For this issue, Miller and Cantor (1982) were commissioned to review a work held to partially set the research agenda for social psychology and to sketch the "current state of the art in cognitive social psychology" (p. 88): Nisbett and Ross's (1980) *Human Inference: Strategies and Shortcomings of Social Judgment*, which dealt exclusively with individually engaged forms of reasoning and inference.

differences in the properties of their objects), Ostrom acknowledged possible differences in the cognitive states and processes themselves:[39]

> The two forms of knowledge differ in four ways: properties of the object, properties of the perceiver, properties of contingencies in human action, and properties of the perceiver as a participant in interaction. (p. 9)

Yet this does not recognize any distinctively social dimension of social cognition. It is not that social cognition is conceived as being socially as opposed to individually engaged, it is simply that individually engaged cognitive states and processes are conceived to differ as a function of the types of objects to which they are directed. In any case, socially and individually engaged forms of cognition do not differ in terms of the types of objects to which they are directed. As Thomas and Znaniecki (1918) noted long ago, socially engaged forms of cognition may be directed toward both social and nonsocial objects: toward the weather and rivers as well as toward teachers and the unemployed.

## VII

One might have thought (or hoped) that the crisis literature would have stimulated a critical reappraisal of the direction of social psychology, which might in turn have encouraged a retrieval of the earlier social tradition. However, this was not to be. Sherif's (1977) suggested way out of the crises – that social psychology rejuvenate itself by returning to its original concern with social norms – was largely ignored. At the end of the day, the critical focus of the crisis period only exacerbated the historical neglect of the social, in two fundamental ways.

First, many of the original critics were dissatisfied with the merely technical responses to the experimental crisis and with the increasingly cognitive orientation of the discipline. Kenneth Gergen, one of the most vocal and radical critics of mainstream social psychology, developed what became known as the "social constructionist" movement in social psychology (Gergen, 1985, 1989; Gergen & Davis, 1985). Those who remained critical of the theory and practice of social psychology tended to align themselves with this position, or were seen by mainstream social psychologists as aligned with this position (which, for present purposes, amounted

[39] Ostrom (1984) suggested that social and nonsocial perception may involve different but nonetheless individually engaged encoding and retrieval mechanisms: "May it be that persons are best represented by schemas and nonsocial objects by prototypes?" (pp. 23–24).

to the same thing). Advocates of the social constructionist movement in social psychology (Gergen, 1985, 1989; Gergen & Davis, 1985; Kitzinger, 1987; Parker, 1989; Parker & Shotter, 1990; Potter & Wetherell, 1987; Shotter, 1987, 1992) advanced a number of radical epistemological theses encompassing all forms of knowledge (natural scientific as well as psychological and social psychological) that were unlikely (and remain unlikely) to recommend themselves to mainstream social psychologists committed to a scientific and experimental social psychology.

Ironically, the conception of the social deployed within social constructionism is almost as impoverished as that of mainstream social psychology. The social dimensions of the phenomena studied by social psychology are not held to be intrinsic properties of the phenomena themselves but rather a function of the social dimensions of our linguistic or cognitive constructs of them. This is because, according to social constructionism, thought or discourse putatively "about the world" does not function "as a reflection or map of the world, but as an artifact of communal interchange" (Gergen, 1985, p. 266). The "objects" of social psychological theory and discourse, like the "objects" of natural scientific theory and discourse, are socially constructed: whatever social properties they have are a constitutive product of the social dimensions of our cognitive and linguistic constructions of them.[40]

Most of the intellectual work of this metatheoretical position is done by the notion that knowledge, including social psychological knowledge, is cognitively constructed, or theoretically informed (Feyerabend, 1975; Kuhn, 1970). Social psychological theories are not subject to empirical evaluation, since what counts as empirical confirmation is itself cognitively or theoretically constituted:

> To count something as a fact or a datum one requires a forestructure of theoretical understanding.... These orienting terms are embedded within more elaborated theories, whether implicit or explicit. Thus, what counts as observation is determined by preexisting theoretical commitments, and these commitments do not themselves spring from the soil of observation.
>
> ...Once formulated, the theory will determine what counts as evidence, confirmation and disconfirmation. Competing theories, implying alternative ontologies, are thus incommensurable.(Gergen, 1988, p. 2)

This naturally led Gergen (1982) to the relativist conclusion, anathema to most social psychologists, that social psychological theories are radically

40 These claims can of course be seriously questioned. I have critically discussed the limitations of social constructionism elsewhere (Greenwood, 1992, 1994).

underdetermined by empirical data and that as a consequence "*Virtually any experimental result used as support for a given theory may be used as support for virtually any alternative theory*" (p. 72).

This association of the advocacy of the social in social psychology with a movement that denies the possibility of a scientific and experimental psychology (at least as ordinarily conceived) has not recommended the notion of a distinctive social psychology to mainstream practitioners. The notion of a distinctive social psychology has come to be dismissed along with the notion that all knowledge is just a social construction, just as it was earlier dismissed along with the notion of the social mind. The purely cognitive aspects of the constructionist position (shorn of its radical epistemology) have been assimilated within the mainstream program of cognitive social psychology. Stroebe and Kruglanski (1989, p. 488), for example, maintain that social constructionism and cognitive social psychology are "complementary" rather than "antagonistic."

Second, the methodological focus on interaction effects in laboratory experiments simply led many social psychologists to develop increasingly rigorous strategies for eliminating or alleviating the effects of all variables other than those produced by the manipulated independent variable(s) under study (Suls & Rosnow, 1988). In consequence, influences on the psychology and behavior of subjects grounded in social group orientation were methodologically excluded as potentially contaminating or confounding variables. For example, in eliminating or attenuating the "demand characteristics" of laboratory experiments, experimenters also eliminated or attenuated the "demand characteristics" of real-life situations that are oriented to the represented psychology and behavior of members of social groups (such as the appropriate attitudes and behavior of fellow Catholics, fellow Democrats, or family members in such situations).[41] As noted earlier, by this time "true experiments" had come to be conceived as designed to exclude the influence of all subject variables (including the orientation of subjects to different social groups) via the random assignment of subjects to manipulated treatment groups. The

[41] As I have argued elsewhere (Greenwood, 1989), the demand characteristics of laboratory experiments in social psychology are only problematic when they are the *wrong* demand characteristics: when they are artifactual products of the laboratory experiment rather than the social demands of the social situation putatively studied via laboratory experiments. One of the virtues of experimental role-playing – or as I prefer to call it, experimental simulation – is its potential ability to reproduce the social demands of everyday life rather than the peculiar demands of ambiguous laboratory experiments employing deception (Greenwood, 1983).

methodological reaction to the crisis only exacerbated the elimination of the social by reinforcing the increasingly narrow and restrictive conception of experimentation adopted in the postwar period.

## VIII

Here is where we find American psychological social psychology at the beginning of the 21st century. Virtually no trace remains of the original conception of the social dimensions of cognition, emotion, and behavior. Things appear to be little different in so-called sociological social psychology. The 1995 sociological handbook of social psychology, *Sociological Perspectives in Social Psychology* (Cook, Fine, & House, 1995), which was sponsored by the Social Psychology Section of the American Sociological Association and presented as the successor to the 1986 handbook, *Social Psychology: Sociological Perspectives* (M. Rosenberg & Turner, 1986), has little to distinguish it from the fourth edition of the psychological *Handbook of Social Psychology* (Gilbert, Fiske, & Lindzey, 1998).

The social interactionist tradition remains focused on the interpersonal rather than the social. There are no entries for "reference groups" listed in the index; Hyman is cited only once, and only for his methodological critique of *The Authoritarian Personality* (Adorno, Frenkel-Brunswik, Levinson, & Nevitt Sanford, 1950). Schuman's chapter on attitudes, belief, and behavior is full of references to psychological social psychologists such as Abelson, G. W. Allport, Bem, Hovland, and McGuire. Howard's chapter on social cognition is similarly crowded with references to Bem; Fiske and Taylor; Heider, Jones, and Davis; Kelley; Markus and Zajonc; and Nisbett and Ross. Meeker and Leik's chapter on experimentation in sociological social psychology follows Aronson et al. (1985, 1998) (in the third and fourth editions of the psychological *Handbook of Social Psychology*) in maintaining that socially engaged beliefs and attitudes, such as religious beliefs and attitudes, "cannot be studied experimentally," since experimentation requires the random assignment of subjects to experimental conditions (Meeker & Leik, 1995, p. 641).

# 10

# The Rediscovery of the Social?

In this work I have argued that the distinctive conception of the social dimensions of human psychology and behavior and of the discipline of social psychology recognized by early theorists such as Wundt, James, Durkheim, Weber, and Simmel and developed by early American social psychologists such as Ross, McDougall, Dunlap, Judd, Kantor, Schanck, Wallis, Bernard, Bogardus, Ellwood, Faris, Thomas, and Young was progressively neglected from the 1930s onward and virtually abandoned by the 1960s, to such a degree that scarcely a trace of the social remains in contemporary American social psychology. I have charted this decline and suggested a variety of explanations for it.

However, it might be objected that while this account may apply with some justice to much of the period beginning in the 1930s, things have considerably improved in the past fifteen years or so. There appears to have been a recent revival of interest in the social within social psychology, as evidenced by a spate of books with titles like *Perspectives on Socially Shared Cognition* (Resnick, Levine, & Teasley, 1991), *What's Social About Social Cognition?* (Nye & Brower, 1996), and *Group Beliefs* (Bar-Tel, 1990). American social psychology now appears to be recognizing and embracing the contribution of the more distinctively social tradition of European social psychology, particularly the important work of Moscovici[1] and his colleagues on "social representations" (Farr & Moscovici, 1984; Moscovici, 1961, 1981) and of Tajfel and

[1] Moscovici (1981) explicitly acknowledges Durkheim's theory of "collective representations" as the intellectual ancestor of his theory of "social representations."

Turner on "social identity" theory (Tajfel, 1978, 1981; Tajfel & Turner, 1986; Turner, 1987). Proponents of the new discipline of "cultural psychology" (Cole, 1996; D'Andrade, 1981; Shweder, 1990; Shweder & Le Vine, 1984) seem to promise a renewed concentration on the "sociocultural" dimensions of many psychological and behavioral processes and explicitly represent themselves as reconstituting the original discipline developed by Durkheim, Wundt, and McDougall (Fiske, Kitayama, Markus, & Nisbett, 1998).

These sorts of developments are encouraging and provide some grounds for optimism. However, as answers to the historical critique developed in this work, they are not very convincing and remind one of Christine Keeler's remark during the "Profumo affair" when she heard that British Cabinet ministers had denied her allegations of sexual liaisons: "Of course that's what they would say!"

## I

In the face of complaints about the individualistic cognitive psychological emphasis of social cognition research (Graumann & Somner, 1984; Sampson, 1977, 1981), researchers in the field of social cognition have come to recognize the limitations of the standard approach to social cognition, which simply applies the information-processing paradigm of cognitive psychology to the study of the perception and cognition of other persons (independently of the social orientation of the cognitive agent). Thus Fiske and Taylor (1991), for example, have acknowledged that "the social perceiver . . . has been viewed as something of a hermit, isolated from the social environment. Missing from much research on social cognition have been other people in a status other than that of stimulus" (p. 556).

However, one might be forgiven for remaining skeptical when one looks at the proposed solutions for returning the social to social psychology. In a volume entitled *What's Social About Social Cognition?* Fiske and Goodwyn (1996) stress the need to forge a link between social cognition and small-group research: "Small group research and social cognition research need each other" (p. xiii). Yet small-group research in the 1990s represents a tradition of research that is no more social than the tradition of research on social cognition: it represents a tradition of research devoted to the study of co-acting and interacting (face-to-face) experimental groups as defined by Floyd Allport. The tradition appealed to by Fiske and Goodwyn (1996) in their proffered "solution" is the interpersonal tradition as defined by Gordon Allport:

> The earliest and most current definitions of the field unanimously endorse the need to study interaction as well as perceptions and interpretations. Allport (1954) defined social psychology as "an attempt to understand and explain how the thought, feeling, and behavior of individuals are influenced by the actual, imagined, or implied presence of other human beings" (p. 5). Prominent in this definition are both actual interactions and cognitively mediated interactions (imagined, implied). (p. xiv)

One way of making this point is by recalling that the primary problem with research on social cognition is that it is based on the assumption that there is no fundamental difference between social cognition (cognitive states and processes relating to social objects, such as other persons and groups) and nonsocial cognition (cognitive states and processes relating to nonsocial objects, such as tables, trees, and tarantulas). Both forms of cognitive states and processes are conceived as *individually* engaged (without orientation to represented social groups):

> As one reviews research on social cognition, the analogy between the perception of things and the perception of people becomes increasingly clear. The argument is made repeatedly: the principles that describe how people think in general also describe how people think about people. (Fiske & Taylor, 1991, p. 18)

As noted earlier, this was precisely the assumption made by Dashiell (1935) in his chapter on experimental social psychology in the 1935 *Handbook of Social Psychology*, which continued the asocial tradition of experimental research introduced by Floyd Allport:

> Particularly is it to be borne in mind that in this objective stimulus-response relationship of an individual to his fellows we have to deal with no radically new concepts, no principles essentially additional to those applying to non-social situations. (p. 1097)

Thus, one can hardly be optimistic about the prospects for a more social psychology when the tradition of small-group research with which Fiske and Goodwyn think that social cognition research ought to be integrated turns out to be precisely the asocial research tradition effectively inaugurated by Dashiell's chapter in the 1935 *Handbook*:

> Whether social psychology is viewed as social influences, mental interaction, or conversation (interstimulation), each of these implies people together, and each implies that people are interpreting each other. The foundational definitions suggest a central place for face-to-face interaction, as studied, for example, within

small groups, research on which has been present since the beginning of American social psychology. For example, the index to Karpf's (1932) *American Social Psychology* contains numerous references to the "group approach"; Dashiell's (1935) survey of experiments examining the influence of social situations on individuals explicitly defines itself in terms of person-to-person relationships and the effects of the group on the individual. (Fiske & Goodwyn, 1996, p. xv)

This is where we came in. It represents the problem not the solution.[2]

## II

Ostensibly more promising but ultimately as disappointing is the "social identity" theoretical approach originally developed by Tajfel and Turner (Tajfel, 1978, 1981; Tajfel & Turner, 1986; Turner, 1987) and represented by the chapter authored by Haslam, McGarty, and Turner (1996) in *What's Social About Social Cognition?*[3] Haslam et al. justly complain about the "failure to take into account the *distinct* nature of social cognition" and correctly diagnose this failure as due to the "relatively straightforward borrowing (or extension) of paradigms and theories from the non-social (i.e. experimental cognitive) domain over the past two or three decades" (p. 29). They claim that, in contrast, some European theorists, notably social identity theorists such as Tajfel and Turner, have bucked this individualistic trend by developing

an analysis of the psychological substrates of inter-group behavior that incorporates an appreciation of the distinct role of the group in determining individual cognition and behavior. These ideas, and the central concept of social identity, were subsequently developed in self-categorization theory... in an attempt to provide *inter alia* a non-individualistic analysis of social influence. (p. 30)

[2] In a slightly more promising vein, Higgins (2000) comes close to acknowledging the social dimensions of cognition by distinguishing between the "cognition of social psychology" and the "social psychology of cognition": "Social psychology is social because *what* is learned concerns the social world, and *where* the learning takes place is in the social world" (p. 3). As examples of the social psychology of cognition, Higgins appropriately cites Festinger (1950), Sherif (1935, 1936), Asch (1948, 1952), Charters and Newcomb (1952), and Hyman (1942), Merton and Kitt (1952), Newcomb (1952), and Siegel and Siegel (1957) on reference groups.

However, there are no cited studies of the social dimensions of cognition beyond the 1950s. All the post-1950s studies cited are interpersonal, in line with Higgin's endorsement of Gordon Allport's (1954) interpersonal definition of the social, interpreted by Higgins as equivalent to Weber's interpersonal definition of the social (Higgins, 2000, p. 4).

[3] The chapter is entitled "Salient Group Memberships and Persuasion: The Role of Social Identity in the Validation of Beliefs."

Unfortunately, although the social identity approach appears to be very promising and close to the original conception of social forms of cognition, emotion, and behavior (as forms of psychology and behavior oriented to the represented psychology and behavior of members of social groups), this appearance turns out to be illusory. What Haslam et al. offer and investigate is yet another individual cognitive explanation of persuasive force and general interpersonal influence in terms of perceived "psychological equivalence":

> Influence and persuasion represent social forms of action that are mediated by perceptions of shared group membership (or social categorical identity). In particular, it is argued that people are more likely to be persuaded and positively influenced by others with whom they recognize a shared identity. The theory proposes that these others are persuasive because their psychological equivalence to self is seen to qualify them to validate self-relevant aspects of reality. In other words, we come to believe what others tell us when we categorize them as similar to us in relevant ways, and we cease to believe them when we categorize them as different. (p. 30)

The problem with this approach is that it appeals to the same impoverished conception of a social group that informs the individualistic experimental tradition originated by Floyd Allport. Haslam et al. make no distinction between social groups and aggregate groups of individuals who merely share, or are represented as sharing, some common property or properties. For Haslam et al., "shared group membership" is equivalent to "social category" identity. In the experimental research grounding "social-identity" theory, only the categorical identity of the source of a communication or potential influence is manipulated, not information about the beliefs, attitudes, emotions, and behavior of members of social reference groups (as in the original studies by Sherif and Asch approvingly cited by the authors). The sorts of group memberships explored in these studies are memberships of aggregate groups, whose members exemplify, or are represented as exemplifying, some common property or properties, such as being male, Australian, of teenage years, female, or African-American. These are just the sort of aggregate groups or "category groups" that Newcomb (1951), for example, quite properly distinguished from genuine social groups:

> I should like to outline what seem to me the necessary distinctions between groups, as amenable to social psychological study, and other forms of human collectivities. For social psychological purposes at least, the distinctive thing about a group is that its members share norms about something.... They serve to distinguish, for social psychological purposes at least, a group from a number of persons at a

street intersection at a given moment, and also from a mere category, such as all males in the State of Oklahoma between the ages of 21 and 25. (p. 38)[4]

As noted earlier, the fact that a set of individuals hold a certain belief is not sufficient to constitute that set of individuals as a social group or generally even incline them to represent themselves as members of a social group. The population of persons who (correctly or incorrectly) believe that today is their birthday, who think that motorcyclists should wear safety helmets, and who believe in déjà vu do not constitute social groups and do not generally represent themselves as members of social groups. Yet the represented groups that are the putative source of influence in the experimental studies documented by Haslam et al. (1996) are of precisely this aggregate nature: the sets of individuals "who want to outlaw the sale and consumption of alcohol" and the sets of individuals "who want to improve road safety."

That is, Haslam et al. do not really provide "a non-individualistic analysis of social influence": what they provide is an account of *interpersonal influence* independent of the orientation of beliefs and attitudes to the represented beliefs and attitudes of members of social groups.[5] No doubt we do tend to find the communications of those persons whose views we agree with more persuasive than those whose views we disagree with, but there need be nothing social about this (and there is no reason to conclude this based on the experimental studies described by Haslam et al.). The social-identity theory of Tajfel and Turner, like other "self-labeling" theories of identity (Breakwell, 1983; Deaux, 1993; McCall & Simmons, 1978; S. Rosenberg & R. A. Jones, 1972; Weinreich, 1983; Zavalloni, 1971), does not provide a theoretical account of the social dimensions of identity – that is, of the psychological orientation of identity formation, achievement, and development to represented social groups. At best social-identity theory provides a theoretical account of individually engaged psychological processes grounded in interpersonal *identification*:

[4] Compare Asch (1952), who distinguished social groups from mere category groups, such as "persons who are five years old or the class of divorced persons" (p. 260).

[5] Furthermore, one may seriously wonder if Haslam et al. (1996) really do grasp the nature of socially engaged forms of cognition, emotion, and behavior when they maintain that their research demonstrates that "the contribution of groups to the persuasion process is far from peripheral or unthinking but rather is absolutely central to these and other related forms of *rational cognitive activity*" (p. 30, emphasis added). Many of the early advocates of a distinctive social psychology (and many of their critics, such as Floyd Allport) stressed the irrational nature of social psychological states and behavior. See Chapters 6 and 7.

For example, if in a particular situation we define ourselves in terms of group membership such as "male" or "Australian," this means that at that time we perceive ourselves as sharing identity with other members of these social categories and less as unique individuals. (Haslam et al., 1996, p. 36)

Perhaps the clearest way of illustrating this point is by noting that, for Haslam et al., sociality is treated as antagonistic to individuality rather than as the social medium within which individuality is developmentally constituted:

The theory postulates that at different times we perceive ourselves as unique individuals and at other times as members of groups and that the two are equally valid expressions of self. That is, it is proposed that our *social identities* (deriving from the groups we perceive ourselves to be members of in particular contexts) are as true and basic to self as personal identity (derived from views of oneself as a unique individual) and that the extent to which we define ourselves at either the personal or social level is both flexible (context dependent) and *functionally antagonistic.* (pp. 35–36 emphasis added)

Yet while socially and individually engaged beliefs (and emotions and behavior) may sometimes be antagonistic (one belief may be held socially and a contrary belief individually), they need not be (one and the same belief can be held both socially and individually), and there are no compelling grounds for treating sociality and individuality as antagonistic – as if a congenital Robinson Crusoe was better placed to develop an individual identity than a Catholic psychologist mother.[6]

Haslam et al. approvingly cite the work of Sherif and Asch, but their conception of social identity is a long way from Sherif's (1936, 1948) psychology of ego-involvements and Asch's (1952) intrinsically social beliefs and attitudes. Whatever the avowed intellectual ancestry, the form and content of their version of social-identity theory locates them unambiguously in the individualistic psychological tradition of Floyd and Gordon Allport, in which sociality or even "commonality" is seen as antithetical to the development of individuality – as opposed to the tradition represented by McDougall, Sherif, and Asch (and by Baldwin, Cooley, Mead, Blumer, and Goffman, 1959, 1961), in which sociality is seen as the developmental medium of individuality.

[6] Although Gordon Allport (1935) explicitly maintained something close to this view: "By no means do all attitudes have social reference. Personality cannot then be regarded as completely dependent upon culture" (p. 838). Isolates such as Robinson Crusoe "are deficient or else completely lacking in cultural contexts, and yet have diverse and intricate personalities" (p. 838).

## III

The contributions to *Perspectives on Socially Shared Cognition* (Resnick, Levine, & Teasley, 1991), sponsored and published by the American Psychological Association, appear more fundamentally social but turn out to focus on interpersonally distributed aspects of knowledge systems ("shared" knowledge in the sense of sharing parts of a cake or a lottery win) and the ("often idiosyncratic") cognitive processes involved in the construction of knowledge:

> The metaphor of cognitive systems as social systems in connectionist and blackboard models of thinking (cf. Minsky, 1986) makes the entire cognitive science community more open than it was a decade ago to the idea of knowledge as distributed across several individuals whose interactions determine decisions, judgments and problem solutions. (Resnick, 1991, p. 3)

While probably unjustified, such remarks give the collection a distinctly supraindividual impression, and the generally holistic rhetoric of the collection indicates that it is not likely to recommend itself to mainstream social psychologists:

> This volume is about a phenomenon that seems almost a contradiction in terms: cognition that is not bounded by the individual brain or mind.... For cognition is, by past consensus and implicit definition, an individual act bounded by the physical facts of brain and body. (Resnick, 1991, p. 1)

In fact, this is just how the "socially shared cognition" perspective tends to be represented in the mainstream social psychological literature. Old associations die hard, and talk of the distinctively social dimensions of cognition recall objections to the group mind:

> The recent interest among many social psychologists in socially shared cognition seems likely to make this issue more salient by raising the specter of a "group mind." How can cognition, which presumably occurs only within individual minds, possibly be viewed as a *group* activity? (Levine & Moreland, 1998, p. 417)

The specter is banished, as usual, by maintaining that socially shared cognition can be reductively analyzed in terms of individually engaged forms of cognition (in much the way that Gordon Allport banished the specter of the social mind via reductive accommodation in his 1929 review of McDougall's 1928 reissue of *The Group Mind*):

> At first glance, analyses such as these may seem encouraging to people who believe in the reality of groups. But a closer examination often reveals subtle forms of reductionism. In many of these analyses, for example, group phenomena are

explained entirely in terms of individual thoughts and feelings (see Tajfel, 1979; Taylor & Brown, 1979). Even when explanations are offered at the group level, the individual almost always serves as an explicit or implicit model for the group, which is then analyzed through metaphor or analogy. Most analyses of group decision making, for example, seem to assume that groups make decisions in much the same way as individuals do. (p. 418)

This response reiterates Giddings (1896, p. 151) negative answer to the question raised by Durkheim in 1901/1982c (p. 42) as to whether the processes underlying socially and individually engaged forms of cognition, emotion, and behavior are distinct. Once again, the social is analytically reduced out of existence.[7]

To be fair to the contributors to *Perspectives on Socially Shared Cognition*, they also focus on "the various and often hidden ways in which the social permeates thinking, especially by shaping the forms of reasoning and language... available to members of a community" (Resnick, 1991, p. 3). However, these schemas for reasoning and language are held to be derived from the general *cultural* and not the local social context: "Human cognition is so varied and so sensitive to cultural context that we must also seek mechanisms by which people actively shape each other's knowledge and reasoning processes"(p. 2). That is, the theoretical perspective advocated by Resnick and the other contributors purports to occupy the same conceptual space as what has come to be known as "cultural psychology," and consequently it inherits all the problems of this "once and future discipline" (Cole, 1996).[8]

## IV

The emerging discipline of cultural psychology appears to mark a more serious attempt to reestablish the earlier social tradition of social psychology. Michael Cole's *Cultural Psychology: The Once and Future Discipline* (1996), for example, explicitly locates itself within the tradition of a

[7] Interestingly enough, although many of the studies documented in Resnick et al. (1991) are doubtfully studies of socially engaged cognition, they do appear to delineate genuinely supraindividual forms of "group cognition": that is, knowledge or expertise that is distributed among sets of individuals and of which no individual person has full mastery. Such groups seem to function as genuine "entitative" groups (T. Jones, 1999) or "thing-like" groups (Durkheim, 1901/1982c, p. 35): they have properties or powers – in this case, knowledge or expertise – that cannot be attributed to any of the individuals that compose them.

[8] Michael Cole, one of the foremost proponents of cultural psychology, contributes the closing commentary to the Resnick et al. (1991) volume.

distinctively social form of psychology. The discipline that once was, and which Cole aims to reinstate, is Wundt's *Völkerpsychologie* (although Cole also cites Cooley, Mead, and Vygotsky as intellectual precursors).[9] Cole complains about the neglect of culture in American psychology in much the same fashion as others have complained about the neglect of the social (Farr, 1996; Graumann, 1986), and he tries to develop an objective psychology that is hermeneutically sensitive to the cultural and historical context in which cognitive (and emotive and behavioral) processes are embedded. In this fashion, he envisions cultural psychology as transcending the traditional bifurcation between the causal-experimental and historical-cultural methods of the *Naturwissenschaften* and *Geisteswissenschaften*, in much the same fashion as Wundt and Vygotsky (1934/1986, 1978) conceived of their psychologies as transcending traditional disciplinary divides.

Cultural psychology is to be carefully distinguished from "cross-cultural" psychology. In Cole's view, traditional cross-cultural psychology fundamentally misrepresents the relation between psychological processes and culture: it treats culture as an independent variable, merely modifying to some degree psychological processes presumed to be universal. In contrast, Cole insists that culture is the medium in which psychological development takes place and through which psychological processes are enacted via practical everyday activities situated in historically conditioned contexts. According to Cole (1996), psychological processes are partially constituted by cultural processes, in conjunction with ontogeny and phylogeny. Culturally and historically conditioned ontogenesis is "intertwined" with evolution, transforming biologically evolved cognitive rudiments into *culturally and historically local forms of psychological functioning* (p. 331).[10]

Now this sort of commitment to distinctive forms of human psychology restricted to different cultures and historical periods does clearly locate cultural psychology within the tradition from which Wundt's *Völkerpsychologie* derived. It is also true that many early American social psychologists tended to associate and sometimes identify cultural and social forms of psychology. Titchener (1910), for example, clearly equated

[9] Analogously, Fiske et al. (1998) cite Durkheim, Wundt, and McDougall (along with Dewey, Baldwin, Mead, Le Bon, and Tarde) as "the researchers who founded what are now social psychology and cultural psychology" (p. 917).

[10] Compare Fiske et al. (1998), who claim that "culture, psyche, and evolutionary biology constitute one another" (p. 916).

social psychology of the form advocated by Wundt with the comparative study of cultural differences in human psychology:

> Just as the scope of psychology extends beyond man to the animals, so too does it extend from the individual man to groups of men, to societies.... The study of the collective mind gives us a psychology of language, a psychology of myth, a psychology of custom, etc.; it also gives us a differential psychology of the Latin mind, of the Anglo-Saxon mind, of the Oriental mind, etc. (p. 28)

Many of the papers in Murchison's 1935 *Handbook of Social Psychology* documented the cultural history and social psychology of the negro (Herskovits, 1935), the red man (Wissler, 1935), the white man (Wallis, 1935b) and the yellow man (Harvey, 1935), and the early twentieth century editions of the *Journal of Abnormal and Social Psychology* and the *Journal of Social Psychology* contained many comparative psychological papers (the *Journal of Social Psychology* was originally subtitled *Racial, Political and Differential Psychology*).[11]

However, the contemporary conception of a cultural psychology remains problematic. In the first place, there is serious ambiguity about what it amounts to. "Cultural psychology" means quite different things to different advocates of the discipline. In contrast to Cole's (1996) integrative version of Wundt's "second psychology" (Cahan & White, 1992), mainstream social psychologists such as Ross and Nisbett (1991) treat cultural psychology as just an extended form of Wundt's "first psychology" (individual or experimental psychology): "cultural factors" are conceived in terms of Lewinian "situational" variables and individual subject "construals" of situations. For Richard Shweder (1990), in contrast, cultural psychology is, or is better conceived as, a constructivist and relativist branch of interpretive social science and the humanities.

More deeply problematic is the supposed relation between the social and the cultural. Cole (1996) often talks of "socio-cultural processes" without any attempt to explicate the implied difference between the social and the cultural. Cole seems to recognize (implicitly at least) that they are not the same, since he acknowledges similar social formations within different cultures (p. 61). So also did Wundt, who noted that beliefs or practices or languages that were once socially engaged by members of restricted social groups can become distributed and sedimented over wider populations, to become part of the common conceptual, emotive, and behavioral repertoire of whole classes, societies, or nations. These

[11] The subtitle was dropped in 1949.

may be no longer engaged socially but simply accepted (or taken for granted) as a consequence of cultural-historical transmission. Thus, although the "mental products" that form the objective observation base of *Völkerpsychologie*, such as language, myth, and custom, may have their origins in "social community," they may also become sedimented and distributed culturally as enduring "thing-like" products.[12]

I do not wish to engage the critical question of how the cultural is to be distinguished from the social, other than to note these speculative Wundtian suggestions. It is perhaps sufficient to point out at this historical juncture that the question is scarcely addressed in the cultural psychology literature and not at all in the social psychological literature. Nonetheless, the distinction between social and cultural psychology is important to stress, for the following reason. The notion of a distinctive social psychology, unlike the notion of a distinctive cultural psychology, is not intrinsically tied to the notion that the psychologies of different social groups in different times and places are themselves different.

The difference between social and individual cognition, emotion, and behavior is a difference in the manner in which psychological states and behaviors are engaged: socially (because and on condition that other members of a social group are represented as engaging these psychological states and behavior) as opposed to individually (for reasons or causes independent of whether any member of a social group is represented as engaging these psychological states and behavior). While there is little doubt that wide differences can exist in the psychological states and behaviors engaged socially by members of different social groups in different places and times or in the same place and time (Catholics, psychologists, and Hell's Angels, for example), there is no special reason for supposing that the basic processes underlying the social engagement of cognition, emotion, and behavior vary cross-culturally or transhistorically or between different social groups in the same culture and historical period. The same is of course true for the basic processes underlying the *individual* engagement of cognition, emotion, and behavior, even though persons do believe different things and behave differently in different places and times, and in the same place and time.[13] As Simmel (1894) stressed, the

[12] This was of course precisely Wundt's reason for treating them as objective indicators of underlying psychological processes.

[13] Behaviorists and cognitivists working within individual psychology recognize this variance but accommodate it in terms of differences in stimulus and reinforcement conditions or differences in stimuli, memory, learning, and so forth.

study of social forms of human psychology and behavior is normally focused on the "forms of sociation itself, as distinct from the individual interests and contents in and through which sociation is realized" (p. 272), although of course any particularly empirical study has to be focused on some particular realization.

This is not to deny that the basic psychological processes underlying socially engaged psychological states and behavior *may* vary cross-culturally and transhistorically: that is an empirical matter to be determined empirically. It is just to recognize that the justification of a social psychology distinct from individual psychology (based on the distinction between socially and individually engaged psychological states and behavior) does not depend on the assumption that they do. The case is quite different with cultural psychology, whose very rationale is based on the assumption that fundamental psychological processes vary cross-culturally and transhistorically. That is simply a logical consequence of the theoretical commitment by cultural psychologists to the notion that psychological processes are partially constituted by culture: if they really are partially constituted by culture, significant differences in cultural (or historical) context entail local differences in fundamental psychological processes.

It would also be fair to say that cultural psychology, at least as advocated by Cole, fails to live up to its revolutionary and integrative promise, like Wundt's *Völkerpsychologie* and the cultural-historical psychology of Vygotsky and Luria (1931, 1976, 1979). Cole has done important work demonstrating that the empirical estimation of human psychological capacities in different cultures must employ measures adapted to the local context ("culturally contextualized" measures) to avoid distortion and devaluation of the psychological capacities of persons from different cultures. Cole's own entry into the study of cultural factors evolved out of his attempt to understand the mathematical failures of Kpelle schoolchildren in Liberia while he was serving as a psychological advisor to overseas development projects in the 1960s. Unwilling to dismiss their poor school performance as entirely resulting from perceptual-cognitive deficits, especially in the face of their demonstrated street smarts in the marketplace, he developed what he considered to be a more "culturally appropriate" measure of their mathematical concepts and skills. This involved the estimation of volumes of rice by reference to the Kpelle "standard unit" of one kopi (a tin cup that holds one dry pint), which was better suited to the significant practices of their particular form of life. Yet his research in this area did not demonstrate a distinctively local form of cognitive processing bound to that particular culture or its equivalents: it demonstrated that "*cultural differences in cognition reside more in the situations to which particular cognitive processes are applied than in the existence of a process in one cultural group and its absence in another*" (Cole, Gay, Glick, & Sharp, 1971, p. 233).

While admirable in its development of effective psychological measures and pedagogical innovations, none of Cole's work demonstrates culturally or historically local forms of psychological functioning.[14] It may be that Cole's reluctance to embrace this implication of cultural psychology is a function of his commitment to the principles of "first psychology," with its core commitment to objectivity, and his mistaken equation of objectivity with universality (along with generations of post-Newtonian psychologists who aimed to create a properly scientific and objective discipline of psychology). Other cultural psychologists, such as D'Andrade (1981) and Shweder (1990), are less enamored of the "first psychology" paradigm and treat cultural psychology as essentially a branch of interpretative social science or the hermeneutical humanities. Following colleagues in these disciplines down familiar relativist and social constructionist roads

[14] The early Russian work in this area is equally disappointing. Alexander Luria's (1931, 1976) comparative studies in the Soviet Republics of Central Asia related intellectual development to educational, economic, and industrial changes but did not demonstrate any culturally or historically local forms of psychological functioning. Luria learned in the 1930s the lesson that Cole learned in the 1960s: that one is liable to seriously underestimate and distort the cognitive abilities of members of a cultural group if one employs culturally inappropriate instruments. Luria abandoned the use of the Stanford-Binet test of intelligence when it generated retarded levels of problem solving for children in remote Siberian villages, who, like Cole's Kpelle rice estimators, seemed to have no special difficulty solving everyday problems in their form of cultural life. However, Luria never unambiguously demonstrated (and, as far as I know, never maintained that he had demonstrated) distinctively Siberian forms of psychological functioning – as opposed to universal forms of psychological functioning adapted to the cultural and historical vagaries of the local context.

It perhaps ought to be stressed that the demonstration of genuinely local forms of psychological functioning restricted to particular cultures or historical periods is no easy task and faces formidable methodological problems. It is far from clear, for example, how one is supposed to distinguish between a universal form of cognitive processing differentially embedded in a different cultural context (the types of situations usually postulated and studied by Cole) and a genuinely local form of cognitive processing. The problem is compounded in Cole's own case by his central claim that the notion of "normal" or "proper" cognitive functioning is normatively grounded in local historical and cultural contexts, since Cole also maintains that some anomalous performances of persons in some cultures (e.g., on Piagian tasks) cannot be explained away in terms of culturally inappropriate measures. This makes it very difficult to understand how one could discriminate between a historically and culturally local form of psychological processing and a simple deficit (since both involve deviation from some standard).

(Geertz, 1973; Lutz, 1982, 1988), they appear to treat every difference in the cultural or historical *content* of cognition (or emotion or behavior) as representing distinctively local forms of psychological processes. Shweder (1990), for example, treats cognitive processing as "content-driven, domain specific, and constructively stimulus-bound" (p. 13). This is scarcely likely to recommend cultural psychology to the dedicated empiricists and experimentalists of American social psychology at the beginning of the 21st century. Cole's own "first psychology" brand of cultural psychology can be easily accommodated (and arguably already has been accommodated) by mainstream social psychology (see, e.g., Fiske et al., 1998).

## V

The position of Moscovici and the European tradition of research on "social representations" is more ambiguous, although it manifests some of the same tensions as cultural psychology and recent more avowedly "social" approaches to social cognition. The influence this tradition has had (or is likely to have) on American social psychology is difficult to estimate.[15] However, it is fair to say that, as originally advanced by Moscovici (1961), it does represent a development of the conception of the social dimensions of human psychology and behavior found in Durkheim and early American social psychology.

In the first place, in this tradition, social representations are held to be intrinsically social: that is, they are characterized as social by reference to the avowed social dimensions of representations themselves, not derivatively (as in the social cognition literature) by reference to social objects of representation (other persons or groups). In this European tradition of research, the forms of social representation studied are as frequently directed to nonsocial objects (such as health, Paris, and the physical environment) as social objects (such as women, minorities, and the British Royal Family). The fact that nonsocial objects can be socially represented was also explicitly recognized by Durkheim and by early American social psychologists such as Thomas and Znaniecki (1918).

In the second place, at least in his earlier work, Moscovici followed Durkheim in laying great stress on the social orientation of social representations and their association with distinct and restricted social groups.

[15] At a recent conference in New York exploring the relations between European "social representations" research and American "social cognition" research (Deaux & Philogene, 2000), there appeared to be much mutual respect and collegiality but little theoretical intersection or communication.

Thus Moscovici's (1961) *La psychoanalyse: Son image et son public* focused on the different conceptions of psychoanalysis held by members of different social groups, such as working-class men as opposed to professionals, including professional psychoanalysts. Moscovici originally maintained that he was developing Durkheim's concept of collective representation by focusing on prescriptive and conventionalizing forms of social representation, which he held to be more fluid and plastic than the static and concrete "collective representations" that were the primary object of Durkheim's concern. Indeed, this was Moscovici's avowed reason for employing the term "social" rather than "collective" in describing the types of representations that interested him.

Unfortunately, Moscovici's vigorous efforts to dissociate the study of social representations from the individualist program of experimental social psychology has led him to indulge in the same sort of "holistic" rhetorical excesses to which Durkheim was prone, which has tended to alienate even the more intellectually accommodating social cognition theorists. For example, according to Moscovici (1998a), social representations "have a life of their own, communicating between themselves, opposing each other and changing in harmony with the course of life; vanishing, only to re-emerge under new guises" (p. 410). This sort of rhetoric has also sometimes led Moscovici, like Durkheim, to focus more upon the "externality" and "constraint" of social representations, or their thing-like nature, than their social nature per se:

> Through their autonomy, and the constraints they exert (even though we are perfectly aware that they are "nothing but ideas") it is, in fact as unquestionable realities that we are led to envisage them. The weight of their history, custom and cumulative content confronts us with all the resistance of a material object. (1998a, p. 412)

Nonetheless, like Durkheim, Moscovici maintains that social representations are dynamically grounded in the psychological representations of individuals. These representations, when socially engaged by members of a social group, exercise a form of constraint on members of that social group:

> With Serge Galam, a physicist, I have demonstrated that individual representations at the first stage may create a new social representation at the second stage, which constrains the individual ones that do not disappear or become inactive. And this finding excludes an often made criticism, i.e. that our conception entails something like a group mind or collective consciousness. (1998b, p. 9)

Moscovici, like Durkheim and early American social psychologists, is one of the few contemporary theorists to recognize that the fundamental difference between social and individual representations lies in the manner in which they are engaged. Representations are engaged socially when they are oriented to the represented psychology and behavior of members of social groups, and they are engaged individually when they are engaged independently of any social group orientation:

> Therefore the only factor that makes possible the disciplined character of representations, their stability, impersonality and normativity, is their being produced by regular interactions between individuals, resting upon a stable, impersonal institution that is relatively independent from members' preferences or choices. A social representation is compulsive for everyone because it is based upon the norms and practices of the community. Hence social representations cannot be reduced to cognitive processes alone, nor explained in the same way as individual representations lacking those qualities. (1998b, p. 10)

Unfortunately, as this quotation indicates, the general trend of social representation research in the past few decades has been to focus on the stable, sedimented, and widely distributed forms of representation that constitute communal, common-sense, or cultural forms of representation. For this Moscovici has been roundly criticized (Parker, 1987), but his response to such criticisms appears to move him closer to contemporary cultural psychology and further away from the social form of social psychology characteristic of the early American social psychological tradition:

> I do believe that social psychology is a sort of anthropology of our culture.... At any rate the concept of culture would be better suited to define the phenomena that delimit our sphere of responsibility and the problems for which we should endeavor to find an answer. (1987, p. 527)

## VI

On a more positive note, the very fact that such positions and alternatives are being canvassed and discussed in the literature is grounds for some optimism. And some recent discussions come very close to recognizing the social dimensions of human psychology and behavior. Perhaps the closest approximation to the original social tradition in early American social psychology is Daniel Bar-Tel's (1990) treatment of group beliefs in *Group Beliefs*. Bar-Tel defines group beliefs as follows: "Group beliefs are defined as convictions that group members (a) are aware that they share and (b) consider as defining their 'groupness'" (p. 36). A group itself is defined as

> *a collective of individuals with a defined sense of membership and shared beliefs, including group beliefs; which regulate their behavior at least in matters relating to the collective.* This definition encompasses groups of various sizes and kinds including small groups, associations, organizations, political parties, interest groups, religious denominations, ethnic groups, and even nations. *The three necessary and sufficient conditions for a collective to be a group are: (a) Individuals in the collective should define themselves as group members; (b) they should share beliefs, including group beliefs; and (c) there should be some level of coordinated activity.* (p. 41)

This definition of a group has close affinities with the definitions of social groups to be found in early American social psychology. Unfortunately, like the "social identity" approach, this analysis makes the fundamental error of treating social and individual beliefs (and emotions and behavior) as falling into exclusive categories based on the manner in which they are conceived by the individual. Thus Bar-Tel (1990) distinguishes between personal, common, and group beliefs independently of how they are engaged:

> The present conception differentiates among three types of beliefs: *personal beliefs*, *common beliefs*, and *group beliefs*. *Personal beliefs* are those beliefs that individuals perceive as being uniquely their own. These beliefs are not perceived as being shared. Rather, they are believed to be formed by the individuals themselves, and as long as they are not shared, they are considered to be private repertoire. Personal beliefs distinguish individuals from one other by characterizing them as unique persons.
>
> Beliefs that are shared are called *common beliefs*. In these cases, individuals believe that their beliefs are also held by other individuals. Common beliefs can be shared by a small group of family members, friends, members of an organization, members of a society, members of a religion, and even by the majority of human beings. From a specific individual's perspective, common beliefs can be acquired from external sources or formed by himself/herself and later disseminated among other people. (p. 35)

Group beliefs are defined, as above, in terms of the awareness that they are shared and their constitutive role in defining the social group.

Yet as noted in Chapter 1, certain beliefs and attitudes (and emotions and behaviors) can be held both socially and individually, and both sorts of beliefs and attitudes can be common or uncommon to members of a social group. Furthermore, the fact that certain beliefs are represented by persons as "uniquely their own" or "formed by the individuals themselves" is of course no guarantee that they are in fact unique or individually engaged. As Faris noted (1952),

> We are often entirely unaware of the influence of social attitudes upon our own judgments and activity; our subjective feeling is that we make up our own minds and freely choose our activity, even though detached observation shows that we are clearly dominated by collective definitions. (pp. 205–206)

More fundamentally, as was noted earlier, it is quite wrong to suppose that only personally or individually engaged beliefs and attitudes ("formed by the individuals themselves") support the development of individuality and that socially engaged beliefs and attitudes somehow represent impediments to the development of individual identity. Different members of the same social groups (Catholic psychologist mothers) can determine their unique identities by their different social trajectories: by the characteristic ways in which (and the different degrees to which) they satisfy or fail to satisfy the norms of reputation and honor prescribed by these social groups.

It would not be hard to modify Bar-Tel's analysis to bring it closer to the original conception of the social in American social psychology, and the fact that it comes so close is grounds for optimism. Whether this sort of analysis marks a new beginning remains to be seen. For any optimism must be tempered with a healthy dose of realism (or skepticism), since there are also reasons to doubt whether many contemporary American social psychologists really do take the social dimensions of cognition, emotion, and behavior seriously.

McGuire (1986), for example, one of the senior figures of American social psychology, objects to the way in which the term "social" is applied to "cognition, representation, etc," reprising Zajonc's 1966 complaint about the use of the term "social" to describe the field of social psychology. McGuire distinguishes six different senses in which representations are characterized as "social" and complains about the resulting confusion, since "where a given representation falls on one of these dimensions says little about its location on the other five" (p. 102). The six meanings of "social," as applied to representations, are as follows:

> The most defensible but least interesting reason for adding the "social" modifier is to limit the discussion to the subset of representations that deal with social objects, for example, perceptions of other people, of interpersonal relations, of social institutions etc. More often "social" is added rather to specify representations

that are shared by members of the given society as in Durkheim's (1898) notion of "collective representations" in contrast to heterogeneous attitudinal positions that differentiate members of a group. A third usage of "social" refers to the extent to which the representation originates, not in the knower's genes or private experiences, but through interacting with other people, as in Blumer's (1969) symbolic interactionist adaptation of Mead's (1934) Chicago school of social behaviorism. In a fourth usage "social" refers to the extent to which representations are interpersonally communicable (by being phenomenologically accessible and verbally expressible) rather than being implicit deep structures that guide the person's experiences and actions even though he/she is unaware of them or can at most visualize them only as unverbalizable images. A fifth use of "social" is to distinguish those representations and inferences that serve to maintain the current social system and cultural forms from those not having such a function. A sixth usage is to impute a transcendental quality to representations such that they have an existence outside of individual heads, as in language structures that transcend individual speakers and hearers. (pp. 102–103)

The first of these meanings, which McGuire considers the "least interesting," exactly captures the sense in which human psychology and behavior have been generally defined as social from the 1930s onward. Psychological states and behaviors have been defined as social by reference to their "social objects," namely, other persons and social groups, independently of their orientation to the represented psychology and behavior of members of social groups. None of the other meanings cited by McGuire capture the sense in which social forms of cognition, emotion, and behavior were held to be social by early American social psychologists. There are no real grounds for optimism here.

## VII

In this work I have documented and tried to explain the historical neglect of the distinctive conception of the social dimensions of cognition, emotion, and behavior embraced by early American social psychologists. In conclusion, I want to stress that none of the reasons offered as partial explanation of this neglect – the fear of supraindividual social minds, the apparent threat posed to autonomy and rationality, the assimilation of crowd and social psychology, and the narrowly restrictive conception of experimentation embraced in the postwar period – were particularly good reasons for neglecting the social dimensions of human psychology and behavior. Consequently, there are no impediments in principle to the objective and experimental study of the social dimensions of cognition, emotion, and behavior. What once was neglected can again be revived, given the will and the institutional ways.

While the history of social psychology has generally been represented as a progressive triumph of the individualistic experimental approach, there have always been a vocal minority who have resisted this approach and defended their alternative visions (Pandora, 1977; Stam et al., 2000). Although some of the representatives of the individualistic tradition may lament the apparent weakening of the commitment to this paradigm by the new generation of social psychologists (Aronson, 1999; Zimbardo, 1999), others may see it as a liberating opportunity to explore new alternatives – including the rediscovery of the social in social psychology.[16]

[16] This of course presumes that there remains a social dimension to be discovered. It might be argued that the social dimensions of human psychology and behavior, or at least American psychology and behavior, were themselves a casualty of the increasingly individualistic orientation of American society. Such concern appears to be expressed in postmodern angst about the limits of autonomy individualism, for example, concerning the "emptiness of a life without sustaining social commitments" (Bellah et al., 1985, p. 151). Thus the social dimensions of American psychology may have been eliminated along with a distinctively social psychology. I personally think this is rather doubtful, although it is an empirical matter that can only be addressed by a social psychology that acknowledges the social engagement of psychological states and behavior.

# References

Abelson, R. P., & Rosenberg, M. J. (1958). Symbolic psychology: A model of attitudinal cognition. *Behavioral Science, 3*, 1–13.

Ach, N. (1905). *Über die Willenstätigkeit und das Denken*. Göttingen: Vandenhoeck and Ruprecht.

Adorno, T. W., Frenkel-Brunswik, E., Levinson, D. J., & Nevitt Sanford, R. (1950). *The authoritarian personality*. New York: Harper & Row.

Allport, F. H. (1919). Behavior and experiment in social psychology. *Journal of Abnormal Psychology, 14*, 297–306.

Allport, F. H. (1920). The influence of the group upon association and thought. *Journal of Experimental Psychology, 3*, 159–182.

Allport, F. H. (1924a). *Social psychology*. Boston: Houghton Mifflin.

Allport, F. H. (1924b). The group fallacy in relation to social science. *Journal of Abnormal and Social Psychology, 19*, 60–73.

Allport, F. H. (1932). Psychology in relation to social and political problems. In P. S. Achilles (Ed.), *Psychology at work*. New York: McGraw-Hill.

Allport, F. H. (1933). *Institutional behavior*. Chapel Hill, NC: University of North Carolina Press.

Allport, F. H. (1934). The J-curve hypothesis of conforming behavior. *Journal of Social Psychology, 5*, 141–181.

Allport, F. H. (1955). *Theories of perception and the concept of structure*. New York: Wiley.

Allport, F. H. (1961). The contemporary appraisal of an old problem. *Contemporary Psychology, 6*, 195–196.

Allport, F. H. (1962). A structuronomic conception of behavior: Individual and collective: I. Structural theory and the master problem of social psychology. *Journal of Abnormal and Social Psychology, 64*, 3–30.

Allport, F. H. (1974). Floyd H. Allport. In E. G. Boring & G. Lindzey (Eds.), *A history of psychology in autobiography* (Vol. 6). Englewood Cliffs, NJ: Prentice-Hall.

Allport, G. W. (1929). Review of *The Group Mind* (2nd ed.) by William McDougall. *Journal of Abnormal and Social Psychology, 24*, 126.
Allport, G. W. (1935). Attitudes. In C. Murchison (Ed.), *Handbook of social psychology*. Worcester, MA: Clark University Press.
Allport, G. W. (1939). *Personality: A psychological interpretation*. New York: Henry Holt.
Allport, G. W. (1954). The historical background of modern social psychology. In G. Lindzey (Ed.), *Handbook of social psychology*. Reading, MA: Addison-Wesley.
Allport, G. W. (1967). Gordon Allport. In G. Lindzey (Eds.), *A history of psychology in autobiography* (Vol. 5). New York: Appleton Century.
Allport, G. W. (1968a). The historical background of modern social psychology. In G. Lindzey & E. Aronson (Eds.), *The handbook of social psychology* (2nd ed.). Reading, MA: Addison-Wesley.
Allport, G. W. (1968b). Six decades of social psychology. In S. Lundstedt (Ed.), *Higher education in social psychology*. Cleveland: Case Western Reserve University Press.
Allport, G. W. (1985). The historical background of modern social psychology. In G. Lindzey & E. Aronson (Eds.), *The handbook of social psychology* (3rd ed.). Reading, MA: Addison-Wesley.
Altman, I. (1987). Centripedal and centrifugal trends in psychology. *American Psychologist, 42*, 1058–1069.
Amundson, R. (1985). Psychology and epistemology: The place versus response controversy. *Cognition, 20*, 127–153.
Angell, J. R. (1908). *Psychology*. New York: Henry Holt & Co.
Apfelbaum, E., & Lubek, I. (1976). Resolution versus revolution: The theory of conflicts in question. In L. H. Strickland, K. J. Gergen, & F. E. Abound (Eds.), *Social psychology in transition*. New York: Plenum.
Argyris, C. (1967). The incompleteness of social-psychological theory: Examples from small group, cognitive consistency, and attribution research. *American Psychologist, 24*, 893–908.
Argyris, C. (1975). Dangers in applying results from experimental social psychology. *American Psychologist, 30*, 469–485.
Arieli, Y. (1964). *Individualism and naturalism in American ideology*. Cambridge, MA: Harvard University Press.
Aronson, E. (1972). *The social animal*. San Francisco: Freeman.
Aronson, E. (1999). Adventures in experimental social psychology: Roots, branches and sticky new leaves. In A. Rodrigues & R. V. Levine (Eds.), *Reflections on 100 years of experimental social psychology*. New York: Basic Books.
Aronson, E., & Carlsmith, J. M. (1968). Experimentation in social psychology. In G. Lindzey & E. Aronson (Eds.), *The handbook of social psychology* (2nd ed.). Reading, MA: Addison-Wesley.
Aronson, E., Brewer, M. B., & Carlsmith, J. M. (1985). Experimentation in social psychology. In G. Lindzey & E. Aronson (Eds.), *The handbook of social psychology* (3rd ed.). New York: Random House.

Aronson, E., Wilson, T. D., & Brewer, M. B. (1998). Experimentation in social psychology. In D. T. Gilbert, S. T. Fiske, & G. Lindzey. (Eds.), *The handbook of social psychology* (4th ed.). New York: Oxford University Press.

Asch, S. E. (1948). The doctrine of suggestion, prestige, and imitation in social psychology. *Psychological Review, 55*, 250–276.

Asch, S. E. (1951). Effects of group pressure upon the modification and distortion of judgements. In H. Guetzkow (Ed.), *Groups, leadership, and men*. Pittsburgh: Carnegie Press.

Asch, S. E. (1952). *Social psychology*. Englewood Cliffs, NJ: Prentice-Hall.

Asch, S. E. (1986). *Social psychology* (2nd ed.). Oxford: Oxford University Press.

Asch, S. E., Block, H., & Hertzman, M. (1938). Studies in the principles of judgements and attitudes: I. Two principles of judgement. *Journal of Psychology, 5*, 219–251.

Ash, M. (1992). Cultural contexts and scientific change in psychology: Kurt Lewin in Iowa. *American Psychologist, 47*, 198–207.

Astington, J. W., Olson, D., & Harris, P. (Eds.). (1988). *Developing theories of mind*. Cambridge: Cambridge University Press.

Baars, B. J. (1986). *The cognitive revolution in psychology*. New York: Guilford Press.

Babbie, E. R. (1975). *The practice of social research*. Belmont, CA: Wadsworth.

Backman, C. W. (1983). Towards an interdisciplinary social psychology. In L. Berkowitz (Ed.), *Advances in experimental social psychology* (Vol. 16). New York: Academic Press.

Bacon, F. (1620). *Novum organum*. London.

Bagehot, W. (1884). *Physics and politics*. New York: D. Appleton & Co.

Baldwin, J. M. (1895). *Mental development in the child and the race*. New York: Macmillan.

Baldwin, J. M. (1897). *Social and ethical interrelations in mental development*. New York: Macmillan.

Baldwin, J. M. (1911). *The individual and society*. Boston: Gorham Press.

Baldwin, J. M. (1913). *History of psychology*. New York: G. P. Putnam's Sons.

Bales, R. F. (1953). A theoretical framework for interaction process analysis. In D. Cartwright & A. Zander (Eds.), *Group dynamics: Research and theory*. Evanston, IL: Row, Peterson & Co.

Bandura, A. (1962). Social learning through imitation. In M. R. Jones (Ed.), *Nebraska symposium on motivation: 1962*. Lincoln: University of Nebraska Press.

Bandura, A. (1973). *Aggression: A social learning analysis*. Englewood Cliffs, NJ: Prentice-Hall.

Barber, B. (1961). Resistance by scientists to scientific discovery. *Science, 134*, 596–602.

Barber, X. T. (1978). *Pitfalls of human research: Ten pivotal points*. New York: Pergamon.

Barber, X. T., Calverly, D. S., Forgione, A., McPeake, J. D., Chaves, J. F., & Brown, B. (1969). Five attempts to replicate the experimenter bias effect. *Journal of Consulting and Clinical Psychology, 33*, 1–10.

Barker, E. (1915). *Political thought in England from Herbert Spencer to the present day*. London: Home University Library.

Barnes, B. (1977). *Interests and the growth of knowledge*. London: Routledge & Kegan Paul. (Second edition, 1982, Routledge & Kegan Paul)

Baron, R. A., & Eggleston, R. J. (1972). Performance on the "aggression machine": Motivation to help or harm? *Psychonomic Science, 26*, 321–322.

Bar-Tel, D. (1990) *Group beliefs*. New York: Springer-Verlag.

Bartlett, F. C. (1932). *Remembering: A study in experimental and social psychology*. New York: Macmillan.

Bauer, R. A. (1952). *The new man in Soviet psychology*. Cambridge, MA: Harvard University Press.

Baumrind, D. (1964). Some thoughts on ethics of research: After reading Milgram's "Behavioral Study of Obedience." *Journal of Abnormal and Social Psychology, 51*, 616–623.

Bellah, R. N., Madsen, R., Sullivan, W. M., Swidler, A., & Tipton, S. M. (1985). *Habits of the heart: Individualism and commitment in American life*. Berkeley: University of California Press.

Beloff, J. (1973). *Psychological sciences*. London: Crosby Lockwood Staples.

Bem, D. J. (1967). Self-perception: An alternative interpretation of cognitive dissonance phenomena. *Psychological Review, 74*, 183–200.

Berelson, B. (1954). Content analysis. In G. Lindzey (Ed.), *Handbook of social psychology*. Cambridge, MA: Addison-Wesley.

Berkowitz, L. (1962). *Aggression: A social psychological analysis*. New York: McGraw-Hill.

Berkowitz, L. (1968). Social motivation. In G. Lindzey & E. Aronson (Eds.), *The handbook of social psychology* (2nd ed.). Reading, MA: Addison-Wesley.

Berkowitz, L., & Donnerstein, E. (1982). External validity is more than skin deep: Some answers to criticisms of laboratory experiments. *American Psychologist, 37*, 245–257.

Berkowitz, L., & Le Page, A. (1967). Weapons as aggression-eliciting stimuli. *Journal of Personality and Social Psychology, 7*, 202–207.

Bernard, L. L. (1921). The misuse of instinct in the social sciences. *Psychological Review, 28*, 96–119.

Bernard, L. L. (1924). *Instinct: A study in social psychology*. New York: Holt.

Bernard, L. L. (1926a). *An introduction to social psychology*. New York: Holt.

Bernard, L. L. (1926b). Review of Floyd Allport's *Social Psychology*. *Psychological Bulletin, 23*, 285–289.

Bernard, L. L. (1931). Attitudes and the redirection of behavior. In K. Young (Ed.), *Social attitudes*. New York: Henry Holt & Co.

Bloor, D. (1976). *Knowledge and social imagery*. London: Routledge. (Second edition, 1991, Chicago Universty Press)

Blumenthal, A. L. (1970). *Language and psychology: Historical aspects of psycholinguistics*. New York: Wiley.

Blumenthal, A. L. (1975). A re-appraisal of Wilhelm Wundt. *American Psychologist, 30*, 1081–1088.

Blumenthal, A. L. (1979). The founding father we never knew. *Contemporary Psychology, 24*, 547–550.

Blumenthal, A. L. (1985). Wilhelm Wundt: Psychology as the propaedeutic science. In C. E. Buxton (Ed.), *Points of view in the modern history of psychology*. New York: Academic Press.

Blumer, H. (1937). Social psychology. In E. P. Schmidt (Ed.), *Man and society*. New York: Prentice-Hall.

Blumer, H. (1969). *Symbolic interactionism: Perspective and method*. Englewood Cliffs, NJ: Prentice-Hall.

Blumer, H. (1984). *Symbolic interactionism*. Englewood Cliffs, NJ: Prentice-Hall.

Boas, F. (1934). Aryans and non-Aryans. In *Race and democratic society*. New York: Augustin.

Bogardus, E. S. (1918). *Essentials of social psychology*. Los Angeles: University of Southern California Press.

Bogardus, E. S. (1922). *A history of social thought*. Los Angeles: University of Southern California Press.

Bogardus, E. S. (1924a). The occupational attitude. *Journal of Applied Sociology, 8*, 171–177.

Bogardus, E. S. (1924b). *Fundamentals of social psychology*. New York: Century.

Bond, M. H., & Smith, P. B. (1996). Culture and conformity: A meta-analysis of studies using Asch's line judgement task. *Psychological Bulletin, 119*, 111–137.

Borgetta, E. E., & Bohrnstedt, G. (1974). Some limits on generalizability from social psychological experiments. *Social Methods and Research, 3*, 111–120.

Boring, E. G. (1933). *The physical dimensions of consciousness*. New York: Century.

Boring, E. G., & Lindzey, G. (Eds.). (1967). *A history of psychology in autobiography*. New York: Appleton Century.

Boring, E. G., & Lindzey, G. (Eds.). (1974). *A history of psychology in autobiography* (Vol. 6). Englewood Cliffs, NJ: Prentice-Hall.

Bosanquet, B. (1899). *Philosophical theory of the state*. New York: Macmillan.

Breakwell, G. (1983). *Threatened identities*. New York: Wiley.

Bringmann, W., & Tweney, R. (Eds.). (1980). *Wundt studies*. Toronto: Hogrefe.

Brock, A. (1992). Was Wundt a "Nazi"?: Völkerpsychologie, racism, and anti-Semitism. *Theory and Psychology, 2*, 205–223.

Brooks, D. H. M. (1986). Group minds. *Australasian Journal of Philosophy, 64*, 456–470.

Brown, J. R. (Ed.). (1984). *Scientific rationality: The sociological turn*. Dordrecht: Reidel.

Bühler, K. (1908). Nachtrag: Antwort auf die von W. Wundt erhobenen Einwande gegen die Methode der Selbstbeobachtung an experiementell erzeugten Erlebnissen. *Archiv für die gesamte Psychologie, 12*, 93–112.

Burgess, R. (1977). The withering away of social psychology. *American Sociologist, 12*, 12–13.

Burrow, T. (1924). Social images versus reality. *Journal of Abnormal and Social Psychology, 19*, 230–235.

Buss, A. R. (1975). The emerging field of the sociology of psychological knowledge. *American Psychologist, 30*, 988–1002.

Butterfield, H. (1951). *The Whig interpretation of history*. London: Bell.

Cahan, E. D., & White, S. H. (1992). Proposals for a second psychology. *American Psychologist, 47*, 224–235.
Calhoun, J. H. (1838). Remarks on the state rights' resolutions in regard to abolition, Jan. 12, 1838. In J. H. Calhoun, *Works*. New York.
Campbell, D. T., & Stanley, J. C. (1966). *Experimental and quasi-experimental designs for research*. Chicago: Rand McNally.
Cantril, H. (1941). *The psychology of social movements*. New York: Wiley.
Capshew, J. H. (1999). *Psychologists on the march: Science, practice, and professional identity in America*. New York: Cambridge University Press.
Carlsmith, J. M., Ellsworth, P. C., & Aronson, E. (1976). *Methods of research in social psychology*. Reading, MA: Addison-Wesley.
Carlson, H. B. (1934). Attitudes of undergraduate students. *Journal of Social Psychology, 5*, 202–212.
Cartwright, D. (1949). Some principles of mass persuasion: Selected findings of research on the sale of U.S. War Bonds. *Human Relations, 2*, 253–267.
Cartwright, D. (1973). Determinants of scientific progress: The case of research on the risky shift. *American Psychologist, 28*, 222–231.
Cartwright, D. (1979). Contemporary social psychology in historical perspective. *Social Psychology Quarterly, 42*, 82–93.
Cartwright, D., & Zander, A. (Eds.). (1953). *Group dynamics: Research and theory*. Elmsford, NY: Row, Peterson & Co.
Cartwright, D., & Zander, A. (Eds.). (1968). *Group dynamics: Research and theory* (3rd ed.). London: Tavistock.
Chang, W. C., Lee, L., & Koh, S. (1996). The concept of self in a modern Chinese context. Paper presented at the 50th Anniversary of the Korean Psychological Association, Seoul, Korea.
Chapanis, A. (1967). The relevance of laboratory studies to practical situations. *Ergonomics, 10*, 557–577.
Charters, W. W., Jr., & Newcomb, T. M. (1952). Some attitudinal effects of experimentally increased salience of a membership group. In T. M. Newcomb & E. L. Hartley (Eds.), *Readings in social psychology* (2nd ed.). New York: Henry Holt & Co.
Chowdhry, K., & Chowdhry, T. M. (1952). The relative abilities of leaders and non-leaders to estimate opinions of their own groups. *Journal of Abnormal and Social Psychology, 47*, 51–57.
Christie, R. (1965). Some implications of recent trends in social psychology. In O. Klineberg & R. Christie (Eds.), *Perspectives in social psychology*. New York: Holt, Rinehart & Winston.
Cialdini, R. B., & Trost, M. R. (1998). Social influence: Social norms, conformity and compliance. In D. T. Gilbert, S. T. Fiske, & G. Lindzey (Eds.), *The handbook of social psychology* (4th ed.). New York: Oxford University Press.
Cina, C. (1981). *Social science for whom? A structural history of social psychology*. Doctoral dissertation, State University of New York at Stony Brook.
Cohen, A. R. (1964). *Attitude change and social influence*. New York: Basic Books.
Cole, M. (1996). *Cultural psychology: The once and future discipline*. Cambridge, MA: Harvard University Press.

Cole, N., Gey, J., Glick, J. A., & Sharp, D. W. (1971). *The cultural context of learning and thinking.* New York: Basic Books.

Collier, G., Minton, H. L., & Reynolds, G. (1991). *Currents of thought in American social psychology.* Oxford: Oxford University Press.

Collins, B. E., & Raven, B. H. (1968). Group structure: Attraction, coalitions, communications and power. In G. Lindzey & E. Aronson (Eds.), *The handbook of social psychology* (2nd ed.). Reading, MA: Addison-Wesley.

Collins, H. (1985). *Changing order.* London: Sage.

Comte, A. (1830–1842). *Cours de philosophie positive.* Paris. (Published as *The positive philosophy of Auguste Comte,* translated and condensed by Harriet Martineau, 1853, London, J. Chapman)

Comte, A. (1851–1854). *Systeme de politique positive.* Paris. (Published as *System of positive polity,* translated by J. H. Bridges et al., 1875–1877, London, Longmans, Green & Co.)

Converse, P., & Campbell, A. (1953). Political standards in secondary groups. In D. Cartwright & A. Zander (Eds.), *Group dynamics.* New York: Harper & Row.

Cook, K. S., Fine., G. A., & House, J. S. (Eds.). (1953). *Sociological perspectives in social psychology.* Boston: Allyn & Bacon.

Cook, T. D., & Campbell, D. T. (1979). *Quasi-experimentation: Design analysis for field settings.* Boston: Houghton Mifflin.

Cooley, C. H. (1902). *Human nature and the social order.* New York: Charles Scribner's Sons.

Cooley, C. H. (1909). *Social organization.* New York: Scribner's.

Cooper, J. (1976). Deception and role-playing: On telling the good guys from the bad guys. *American Psychologist, 31,* 605–610.

Crano, W. D., & Brewer, M. B. (1973). *Principles of research in social psychology.* New York: McGraw-Hill.

D'Andrade, R. G. (1981). The cultural part of cognition. *Cognitive Science, 5,* 179–195.

Danziger, K. (1983). Origins and basic principles of Wundt's Völkerpsychologie. *British Journal of Social Psychology, 22,* 303–313.

Danziger, K. (1988). On theory and method in psychology. In W. J. Baker, L. P. Mos, H. V. Rappard, & H. J. Stam (Eds.), *Recent trends in theoretical psychology.* New York: Springer-Verlag.

Danziger, K. (1997). *Naming the mind: How psychology found its language.* London: Sage.

Danziger, K. (2000). Making social psychology experimental: A conceptual history, 1920–1970. *Journal of the History of the Behavioral Sciences, 36,* 329–347.

Danziger, K. (2001). Wundt and the temptations of psychology. In R. W. Rieber & D. K. Robinson (Eds.), *Wilhelm Wundt in history: The making of a scientific psychology.* New York: Kluwer Academic/Plenum.

Danziger, K., & Dzinas, K. (1997). How psychology got its variables. *Canadian Psychology, 38,* 43–48.

Darley, J. M., & Latané, B. (1968). Bystander intervention in emergencies: Diffusion of responsibility. *Journal of Personality and Social Psychology, 8,* 377–383.

Darroch, R. K., & Steiner, I. D. (1970). Role-playing: An alternative to laboratory research? *Journal of Personality, 38*, 302–311.

Darwin, C. (1859). *On the origin of species*. London: J. Murray.

Dashiell, J. F. (1930). An experimental analysis of some group effects. *Journal of Abnormal and Social Psychology, 25*, 190–199.

Dashiell, J. F. (1935). Experimental studies of the influence of social situations on the behavior of individual human adults. In C. A. Murchison (Ed.), *Handbook of social psychology*. Worcester, MA: Clark University Press.

Dawkins, R. (1976). *The selfish gene*. Oxford: Oxford University Press.

Deaux, K. (1993). Reconstructing social identity. *Personality and Social Psychology Bulletin, 19*, 4–12.

Deaux, K., & Philogene, G. (Eds.). (2000). *Representing the social: Bridging social traditions*. Oxford: Basil Blackwell.

Deutsch, M. (1949). A theory of cooperation and competition. *Human Relations, 2*, 129–152.

Deutsch, M. (1976). Theorizing in social psychology. *Personality and Social Psychology Bulletin, 2*, 134–141.

Deutsch, M., & Gerard, H. B. (1955). A study of normative and informational social influences upon individual judgement. *Journal of Personality and Social Psychology, 51*, 629–636.

Deutsch, M., & Krauss, R. M. (1965). *Theories in social psychology*. New York: Basic Books.

Devine, P. G., Hamilton, D. L., & Ostrom, T. M. (1994). *Social cognition: Impact on social psychology*. San Diego: Academic Press.

De Vos, G. A., & Hippler, A. A. (1968). Cultural psychology: Comparative study of human behavior. In G. Lindzey & E. Aronson (Eds.), *The handbook of social psychology* (2nd ed.). Reading, MA: Addison-Wesley.

Dewey, J. (1896). The reflex arc concept in psychology. *Psychological Review, 3*, 357–370.

Dewey, J. (1917). The need for social psychology. *Psychological Review, 24*, 266–277.

Dewey, J. (1927). *The public and its problems*. New York: Henry Holt & Co.

Dewey, J. (1967). Soul and body. In *Early works of John Dewey, 1882–1898* (Vol. 1). Carbondale: Southern Illinois University Press. (Original work published 1886)

Diamond, S. (1974). *The roots of psychology: A sourcebook in the history of ideas*. New York: Basic Books.

Dion, K. L., Barry, R. S., & Miller, N. (1970). Why do groups make riskier decisions than individuals? In L. Berkowitz (Ed.), *Advances in experimental social psychology* (Vol. 11). New York: Academic Press.

Dolbeare, K. M. (1984). *American political thought* (Rev. ed.). Chatham, NJ: Chatham House.

Dollard, J., Doob, L. W., Miller, N. E., Mowrer, O. H., & Sears, R. R. (1939). *Frustration and aggression*. New Haven, CT: Yale University Press.

Doob, L. W. (1938). Review of *Experimental Social Psychology* (revised edition). *Psychological Bulletin, 35*, 112–115.

Du Bois-Reymond, E. (1842). *Zwei grosse Naturforscher des 19 Jahrhunderts: Ein Briefwechsel zwischen Emil Du Bois-Reymond and Karl Ludwig*. Leipzig: Barth.

Dudycha, G. J. (1937). An examination of the J-curve hypothesis based upon punctuality distributions. *Sociometry, 1*, 144–154.

Dunlap, K. (1919). Are there any instincts? *Journal of Abnormal and Social Psychology, 14*, 307–311.

Dunlap, K. (1925). *Social psychology*. Baltimore: Williams & Wilkins.

Durkheim, E. (1951). *Suicide* (J. A. Spaulding & G. Simpson, Trans.). New York: The Free Press. (Original work published 1897)

Durkheim, E. (1897b). Review of A. Labriola, *Essays on the Materialist Conception of History*. *Revue Philosophique*.

Durkheim, E. (1947). *The division of labor in society* (G. Simpson, Trans.). New York: Free Press. (Original work published 1893)

Durkheim, E. (1982a). *The rules of sociological method*. In S. Lukes (Ed.), W. D. Halls (Trans.), *Durkheim: The rules of sociological method and selected texts on sociology*. New York: Macmillan. (Original work published 1895)

Durkheim, E. (1982b). The psychological character of social facts and their reality. In S. Lukes (Ed.), W. D. Halls (Trans.), *Durkheim: The rules of sociological method and selected texts on sociology*. New York: Macmillan. (Original work published 1895)

Durkheim, E. (1982c). Preface to the second edition of *The rules of sociological method*. In S. Lukes (Ed.), W. D. Halls (Trans.), *Durkheim: The rules of sociological method and selected texts on sociology*. New York: Macmillan. (Original work published 1901)

Editor's preface. (1998). *Asian Journal of Social Psychology, 1*, iii–iv.

Editorial. (1982). *Social Cognition, 1*, i–ii.

Edwards, A. L. (1941). Political frames of reference as a factor influencing recognition. *Journal of Abnormal and Social Psychology, 36*, 34–61.

Edwards, A. L. (1954). Experiments: Their planning and execution. In G. Lindzey (Ed.), *The handbook of social psychology*. Reading, MA: Addison-Wesley.

Ellwood, C. A. (1917). *An introduction to social psychology*. New York: Appleton & Co.

Ellwood, C. A. (1924). The relations of sociology and social psychology. *Journal of Abnormal and Social Psychology, 19*, 3–12.

Ellwood, C. A. (1925). *The psychology of human society*. New York: Appleton & Co.

Elms, A. C. (1975). The crisis of confidence in social psychology. *American Psychologist, 30*, 967–976.

Emerson, R. W. (1912). *Journals* (Vol. 3). Boston. (Original work published 1834)

Emerson, R. W. (1983). Self reliance. In *Essays and lectures*. New York: Library of America. (Original work published 1841)

Espinas, A. (1877). *Des sociétés animales*. Paris: G. Baillière.

Fabian, R. (1997). The Graz school of gestalt psychology. In W. G. Bringmann, H. E. Lück, R. Miller, & C. E. Early (Eds.), *A pictorial history of psychology*. Chicago: Quintessence Publishers.

Faris, E. (1921). Are instincts data or hypotheses? *American Journal of Sociology, 27*, 184–196.
Faris, E. (1925). The concept of social attitudes. *Journal of Applied Sociology, 9*, 404–409.
Faris, E. (1952). *Social psychology*. New York: Ronald Press Co.
Farr, R. M. (1996). *The roots of modern social psychology 1872–1954*. Oxford: Basil Blackwell.
Farr, R. M. (1998). Preface. In I. U. Flick (Ed.), *The psychology of the social*. Cambridge: Cambridge University Press.
Farr, R. M., & Moscovici, S. (Eds.). (1984). *Social representations*. Cambridge: Cambridge University Press.
Fearing, F., & Krise, E. M. (1941). Conforming behavior and the J-curve hypothesis. *Journal of Social Psychology, 14*, 109–118.
Festinger, L. (1947). The role of group-belongingness in a voting situation. *Human Relations*, 154–180.
Festinger, L. (1950). Informal social communication. *Psychological Review, 57*, 271–282.
Festinger, L. (1953). Laboratory experiments. In L. Festinger & D. Katz (Eds.), *Research methods in the behavioral sciences*. New York: Holt, Rinehart & Winston.
Festinger, L. (1954). A theory of social comparison processes. *Human Relations, 7*, 117–140.
Festinger, L. (1957). *A theory of cognitive dissonance*. Stanford, CA: Stanford University Press.
Festinger, L., Riecken, H. W., & Schachter, S. (1956). *When prophecy fails*. Minneapolis: University of Minnesota Press.
Festinger, L., Schachter, S., & Back, K. (1950). *Social pressures in informal groups: A study of human factors in housing*. New York: Harper & Brothers.
Feyerabend, P. K. (1975). *Against method*. London: New Left Books.
Finney, R. L. (1926). The unconscious social mind. *Journal of Applied Sociology, 10*, 357–367.
Fiske, A. P., Kitayama, S., Markus, H. R., & Nisbett, R. E. (1998). The cultural matrix of social psychology. In D. T. Gilbert, S. T. Fiske, & G. Lindzey (Eds.), *The handbook of social psychology* (4th ed.). New York: Oxford University Press.
Fiske, S. T., & Goodwyn, S. A. (1996). Introduction. Social cognition research and small group research, a West Side Story or...? In J. L. Nye & A. M. Brower (Eds.), *What's social about social cognition? Research on socially shared cognition in small groups*. Thousand Oaks, CA: Sage.
Fiske, S. T., & Taylor, S. E. (1982). *Social cognition*. New York: Random House.
Fiske, S. T., & Taylor, S. E. (1991). *Social cognition* (2nd ed.). New York: McGraw-Hill.
Fitzhugh, G. (1854). *Sociology for the South, or the failure of free society*. Richmond, VA.
Flew, A. (1985). *Thinking about social thinking: The philosophy of the social sciences*. Oxford: Basil Blackwell.

Forgas, J. (Ed.). (1981) What is social about social cognition? In J. Forgas (Ed.), *Social cognition: Perspectives on everyday understanding*. New York: Academic Press.

Forward, J., Canter, R., & Kirsch, N. (1976). Role enactment and deception methodologies: Alternative paradigms? *American Psychologist, 31*, 595–604.

Fouillée, A. (1885). *Les science sociale contemporaine*. Paris: L. Hachette.

Fox, R. (1977). The inherent rules of violence. In P. Collett (Ed.), *Social rules and social behavior*. Oxford: Basil Blackwell.

Frager, R. (1970). Conformity and anticonformity in Japan. *Journal of Personality and Social Psychology, 15*, 203–210.

Freedman, J. L. (1969). Role-playing: Psychology by consensus. *Journal of Personality and Social Psychology, 13*, 107–114.

Freeman, H. E., & Giovannoni, J. M. (1968). Social psychology of mental health. In G. Lindzey & E. Aronson (Eds.), *The handbook of social psychology* (2nd ed.). Reading, MA: Addison-Wesley.

French, J. R. P., Jr. (1941). The disruption and cohesion of groups. *Journal of Abnormal and Social Psychology, 36*, 361–377.

French, J. R. P., Jr. (1944). Organized and unorganized groups under fear and frustration. *University of Iowa Studies in Child Welfare, 20*, 299–308.

Freud, S. (1955). *Massenpsychologie und Ich-Analyse* [*Group psychology and the analysis of the ego*]. In J. Strachey (Ed. & Trans.), *The Standard Edition of the Complete Psychological Work of Sigmund Freud* (Vol. 18). London: Hogarth Press. (Original work published 1921)

Fried, S. B., Gumper, D. C., & Allan, J. C. (1973). Ten years of social psychology: Is there a growing commitment to field research? *American Psychologist, 28*, 155–156.

Fuller, S. (1988). *Social epistemology*. Bloomington: Indiana University Press.

Furumoto, L. (1989). The new history of psychology. In I. S. Cohen (Ed.), *The G. Stanley Hall Lecture Series* (Vol. 9). Washington, DC: American Psychological Association.

Gardner, H. (1985). *The mind's new science: A history of the cognitive revolution*. New York: Basic Books.

Gates, G. S. (1923). The effect of an audience upon performance. *Journal of Abnormal and Social Psychology, 18*, 334–344.

Gault, R. H. (1921). The standpoint of social psychology. *Journal of Abnormal and Social Psychology, 21*, 41–46.

Geertz, C. (1973). *The interpretation of cultures*. New York: Basic Books.

Gergen, K. J. (1973). Social psychology as history. *Journal of Personality and Social Psychology, 26*, 309–320.

Gergen, K. J. (1976) Social psychology, science and history. *Personality and Social Psychology Bulletin, 2*, 373–383.

Gergen, K. J. (1978). Experimentation in social psychology: A reappraisal. *European Journal of Social Psychology, 8*, 507–527.

Gergen, K. J. (1982). *Towards transformation in social knowledge*. New York: Springer-Verlag.

Gergen, K. J. (1985). The social constructionist movement in modern psychology. *American Psychologist, 40*, 266–275.

Gergen, K. J. (1988). The concept of progress in psychological theory. In W. J. Baker, L. P. Moss, H. V. Rappard, & H. J. Stam (Eds.), *Recent trends in theoretical psychology*. New York: Springer-Verlag.

Gergen, K. J. (1989). Social psychology and the wrong revolution. *European Journal of Social Psychology, 19*, 463–484.

Gergen, K. J. (1994). *Towards transformation in social knowledge* (2nd ed.). London: Sage.

Gergen, K. J., & Davis, K. E. (Eds.). (1985). *The social construction of the person*. New York: Springer-Verlag.

Getzels, J. W. (1968). A social psychology of education. In G. Lindzey & E. Aronson (Eds.), *The handbook of social psychology* (2nd ed.). Reading, MA: Addison-Wesley.

Giddings, F. H. (1896). *The principles of sociology*. New York: Macmillan.

Giddings, F. H. (1924). Stimulation ranges and reaction areas. *Psychological Review, 31*, 449–455.

Gilbert, D. T., Fiske, S. T., & Lindzey, G. (Eds.). (1998). *The handbook of social psychology* (4th ed.). New York: Oxford University Press.

Gilbert, M. (1991). *On social facts*. Princeton, NJ: Princeton University Press.

Goffman, E. (1959). *The presentation of self in everyday life*. New York: Doubleday.

Goffman, E. (1961). *Asylums*. New York: Doubleday.

Gopnick, A. (1993). How we know our minds: The illusion of first-person knowledge of intentionality. *Behavioral and Brain Sciences, 16*, 1–14.

Gorden, R. M. (1942). Interaction between attitude and the definition of the situation in the expression of opinion. *American Sociological Review, 17*, 50–58.

Gordon, R. M. (1986). Folk psychology as simulation. *Mind and Language, 1*, 156–171.

Gorman, M. (1981). Pre-war conformity research in social psychology: The approaches of Floyd H. Allport and Muzafer Sherif. *Journal of the History of the Behavioral Sciences, 17*, 3–14.

Graumann, C. F. (1986). The individualization of the social and the desocialization of the individual: Floyd H. Allport's contribution to social psychology. In C. F. Graumann & S. Moscovici (Eds.), *Changing conceptions of crowd mind and behavior*. New York: Springer-Verlag.

Graumann, C. F., & Somner, M. (1984). Schema and inference: Models in cognitive social psychology. In J. R. Royce & L. P. Mos (Eds.), *Annals of theoretical psychology* (Vol. 1). New York: Plenum.

Green, T. H. (1900). Lectures on the principles of political obligation. In *Collected works* (Vol. 2). London: Longmans, Green.

Greenberg, M. S. (1967). Role-playing: An alternative to deception? *Journal of Personality and Social Psychology, 7*, 235–254.

Greenwald, A. G. (1976). Transhistorical lawfulness of behavior: A comment on two papers. *Personality and Social Psychology Bulletin, 2*, 391.

Greenwood, J. D. (1983). Role playing as an experimental strategy in social psychology. *European Journal of Social Psychology, 13*, 235–254.

Greenwood, J. D. (1989). *Explanation and experiment in social psychological science*. New York: Springer-Verlag.

Greenwood, J. D. (1991). Introduction: Folk psychology and scientific psychology. In J. D. Greenwood (Ed.), *The future of folk psychology*. Cambridge: Cambridge University Press.

Greenwood, J. D. (1992). Realism, empiricism, and social constructionism: Psychological theory and the social dimensions of mind and action. *Theory and Psychology*, 2, 131–151.

Greenwood, J. D. (1994). *Realism, identity and emotion*. London: Sage.

Greenwood, J. D. (2003). Social facts, social groups, and social explanation. *Nous*, 37, 93–112.

Gumplowitz, L. (1885). *Grundriss der Sociologie*. Vienna. (Published as *The outlines of sociology*, translated by F. W. Moore, 1899, Philadelphia, American Academy of Political and Social Science)

Gurnee, H. (1936). *Elements of social psychology*. New York: Farrar & Rinehart.

Haines, H., & Vaughan, G. M. (1979). Was 1898 a "great date" in the history of experimental social psychology? *Journal of the History of the Behavioral Sciences*, 15, 323–332.

Hanay, C., Banks, W. C., & Zimbardo, P. G. (1973). Interpersonal dynamics in a simulated prison. *International Journal of Criminology and Penology*, 1, 69–97.

Hare, A. P. (1972). Bibliography of small group research. *Sociometry*, 35, 111–150.

Harré, R. (1983a). *Personal being*. Oxford: Basil Blackwell.

Harré, R. (1983b). Identity projects. In G. Breakwell (Ed.), *Threatened identities*. New York: Wiley.

Harré, R., & Secord, P. F. (1972). *The explanation of social behavior*. Totowa, NJ: Rowman & Littlefield.

Harris, A. J., Remmers, H. H., & Ellison, C. E. (1932). The relationship between liberal and conservative attitudes in college students, and other factors. *Journal of Social Psychology*, 3, 320–335.

Harris, P. L. (1991). The work of the imagination. In A. Whiten (Ed.), *Natural theories of mind*. Oxford: Basil Blackwell.

Hartley, E. L. (1951). Psychological problems of multiple group membership. In J. H. Rohrer & M. Sherif (Eds.), *Social psychology at the crossroads*. New York: Harper.

Hartley, R. (1957). Personal characteristics and acceptance of secondary groups as reference groups. *Journal of Individual Psychology*, 13, 45–55.

Hartley, R. (1960a). Relationships between perceived values and acceptance of a new reference group. *Journal of Social Psychology*, 51, 181–190.

Hartley, R. (1960b). Personal needs and the acceptance of a new group as a reference group. *Journal of Social Psychology*, 51, 349–358.

Hartley, R. (1960c). Norm compatibility, norm preference, and the acceptance of new reference groups. *Journal of Social Psychology*, 52, 87–95.

Harvey, E. D. (1935). Social history of the yellow man. In C. Murchison (Ed.), *Handbook of social psychology*. Worcester, MA: Clark University Press.

Haskell, T. L. (1977). *The emergence of professional social science: The American Social Science Association and the nineteenth century crisis of authority*. Chicago: University of Illinois Press.

Haslam, A. S., McGarty, C., & Turner, J. C. (1996). Salient group memberships and persuasion: The role of social identity in the validation of beliefs. In J. L.

Nye & A. M. Brower (Eds.), *What's social about social cognition? Research on socially shared cognition in small groups*. Thousand Oaks, CA: Sage.

Haslam, A. S., Turner, J. C., Oakes, P. J., McGarty, C., & Reynolds, K. J. (1998). The group as a basis for emergent stereotype consensus. *European Review of Social Psychology, 9*, 203–239.

Hayes, J. R., & Petras, J. W. (1974). Images of persons in early American sociology: 3. The social group. *Journal of the History of the Behavioral Sciences, 10*, 391–396.

Heelas, P., & Lock, A. (Eds.). (1981). *Indigenous psychologies*. London: Academic Press.

Hegel, G. W. F. (1910). *The Phenomenology of Mind* (Trans J. B. Baille). London: G. Allen and Unwin. (Original worked published 1807)

Hegel, G. W. F. (1953). *The Philosophy of Right* (T. M. Knox, Trans.). Oxford: Clarendon Press. (Original work published 1821)

Heider, F. (1944). Social perception and phenomenal causality. *Psychological Review, 51*, 358–374.

Heider, F. (1946). Attitudes and cognitive organization. *Journal of Psychology, 21*, 107–112.

Heider, F. (1958). *The psychology of interpersonal relations*. New York: Wiley.

Helmreich, R. (1975). Applied social psychology: The unfulfilled promise. *Personality and Social Psychology Bulletin, 1*, 548–560.

Helmreich, R., Bakeman, R., & Scherwitz, L. (1973). The study of small groups. *Annual Review of Psychology*.

Henschel, R. L. (1980). The purposes of laboratory experimentation and the virtues of deliberate artificiality. *Journal of Experimental Social Psychology, 16*, 466–478.

Herbart, J. F. (1816). *Lehrbuch zur Psychologie*. Königsberg: Unzer.

Herder, J. G. (1969). Ideas for a philosophy of history of mankind. In F. M. Bernard (Trans. & Ed.) *J. G. Herder on social and political culture*. Cambridge: Cambridge University Press. (Original work published 1784)

Herskovits, M. J. (1935). Social history of the negro. In C. Murchison (Ed.), *Handbook of social psychology*. Worcester, MA: Clark University Press.

Higbee, K. L., Millard, R. J., & Folkman, J. R. (1982). Social psychology research during the 1970's: Predominance of experimentation and college students. *Personality and Social Psychology Bulletin, 8*, 180–183.

Higbee, K. L., & Wells, M. G. (1972). Some research trends in social psychology during the 1960's. *American Psychologist, 27*, 963–966.

Higgins, E. T. (2000). Social cognition: Learning about what matters in the social world. *European Journal of Social Psychology, 30*, 3–39.

Higgins, E. T., Ruble, D. N., & Hartup, W. W. (Eds.). (1983). *Social cognition and social development*. New York: Cambridge University Press.

Himmelfarb, D. (1988). Freedom, virtue and the founding fathers: A review essay. *Public Interest, 90*, 115–120.

Hirschberg, G., & Gilliland, A. R. (1942). Parent-child relationships in attitudes. *Journal of Abnormal and Social Psychology, 37*, 125–130.

Hobhouse, L. T. (1904). *Mind in evolution*. New York: Macmillan.

Hobhouse, L. T. (1913). *Development and purpose*. New York: Macmillan.

Holmes, D. S., & Bennett, D. H. (1974). Experiments to answer questions raised by the use of deception in psychological research: 1. Role-playing as an alternative to deception; 2. Effectiveness of debriefing after deception; 3. Effect of informed consent upon deception. *Journal of Personality and Social Psychology, 29*, 358–367.

Homans, G. C. (1961). *Social behavior: Its elementary forms*. New York: Harcourt Brace Jovanovich.

Hoppe, F. (1930). Erfolg und Misserfolg. *Psychologische Forschung, 14*, 1–62.

Horowitz, E. L. (1947). Development of attitude towards negros. In T. M. Newcomb & E. L. Hartley (Eds.), *Readings in social psychology*. New York: Holt. Reprinted from *Archives of Psychology, 194*.

Horowitz, I. A., & Rothschild, B. H. (1970). Conformity as a function of deception and role-playing. *Journal of Personality and Social Psychology, 14*, 224–226.

House, J. S. (1977). The three faces of social psychology. *Sociometry, 40*, 161–177.

Hovland, C. I., Janis, I., & Kelley, H. (1953). *Communication and persuasion*. New Haven, CT: Yale University Press.

Hovland, C. I., Lumsdaine, A. A., & Sheffield, F. D. (Eds.). (1949). *Experiments in mass communication* (Studies in Social Psychology in World War II No. 3). Princeton, NJ: Princeton University Press.

Howard, J. A. (1995). Social cognition. In K. S. Cook, G. A. Fine, & J. S. House (Eds.), *Sociological perspectives in social psychology*. Boston: Allyn & Bacon.

Hronszky, I., Feher, M., & Dajka, B. (Eds.). (1984). *Scientific knowledge socialized*. Dordrecht: Kluwer.

Hull, C. L. (1943). *Principles of behavior*. New York: Appleton-Century-Crofts.

Hull, C. L. (1952). C. L. Hull. In E. G. Boring, H. S. Langfield, & R. M. Yerkes (Eds.), *A history of psychology in autobiography* (Vol. 4). Worcester, MA: Clark University Press.

Hume, D. (1739). *Treatise of human nature*. Edinburgh.

Humphrey, G. (1951). *Thinking: An introduction to its experimental psychology*. New York: Methuen.

Hyman, H. (1942). The psychology of status. *Archives of Psychology*, No. 269.

Hyman, H., & Singer, E. (Eds.). (1968). *Readings in reference group theory and research*. New York: The Free Press.

Insko, C., & Schopler, J. (1972). *Experimental social psychology*. New York: Academic Press.

Israel, J., & Tajfel, H. (Eds.). (1972). *The context of social psychology: A critical assessment*. London: Academic Press.

Jackson, F., & Pettit, P. (1992). Structural explanation in social theory. In D. Charles & K. Lennon (Eds.), *Reductionism and anti-reductionism*. Oxford: Oxford University Press.

Jackson, J. M. (1988). *Social psychology, past and present: An integrative orientation*. Hillsdale, NJ: Lawrence Erlbaum Associates.

Jaensch, E. A. (1938). *Der Gegentypus*. Leipzig: Barth & Co.

Jahoda, G. (1976). Critique: On Triandis's "Social Psychology and Cultural Analysis." In L. H. Strickland, F. E. Aboud, & K. J. Gergen (Eds.), *Social psychology in transition*. New York: Plenum Press.

Jahoda, G. (1997). Wundt's "*Volkerpsychologie*". In W. G. Bringmann, H. E. Lück, R. Miller, & C. E. Early (Eds.), *A pictorial history of psychology*. Chicago: Quintessence Publishers.

Jahoda, M., Deutsch, M., & Cook, S. W. (1951). *Research methods in social relations*. New York: The Dryden Press.

James, W. (1890). *The principles of psychology*. New York: Holt.

Janis, I. L. (1951). *Air war and emotional stress: Psychological studies of bombing and civilian defense*. New York: McGraw-Hill.

Janis, I. L. (1968). *Victims of groupthink*. New York: Harcourt Brace & Jovanovich.

Janis, I. L., & Mann, L. (1965). Effectiveness of emotional role-playing in modifying smoking habits and attitudes. *Journal of Experimental Research on Personality, 1*, 84–90.

Jarvie, I. (1959). *Universities and left review*.

Jaspars, J. M. F., & Fraser, C. (1984). Attitudes and social representations. In R. M. Farr & S. Moscovici (Eds.), *Social representations*. Cambridge: Cambridge University Press.

Jehlen, M. (1986). *American incarnation: The individual, the nation, and the continent*. Cambridge, MA: Harvard University Press.

Jones, E. E. (1985). Major developments in social psychology during the past five decades. In G. Lindzey & E. Aronson (Eds.), *The handbook of social psychology* (3rd ed.). New York: Random House/Erlbaum.

Jones, E. E. (1998). Major developments in five decades of social psychology. In D. T. Gilbert, S. T. Fiske, & G. Lindzey (Eds.), *The handbook of social psychology* (4th ed.). New York: Oxford University Press.

Jones, E. E., & Davis, K. E. (1965). From acts to dispositions: The attribution process in person perception. In L. Berkowitz (Ed.), *Advances in experimental social psychology* (Vol. 2). New York: Academic Press.

Jones, E. E., & Nisbett, R. E. (1972). The actor and the observer: Divergent perceptions of the causes of behavior. In E. E. Jones, D. E. Kanouse, H. H. Kelley, R. E. Nisbett, S. Valins, & B. Weiner (Eds.), *Attribution: Perceiving the causes of behavior*. Morristown, NJ: General Learning Press.

Jones, T. (1999). FIC descriptions and interpretative social science: Should philosophers raise their eyes? *Journal for the Theory of Social Behaviour, 29*, 337–369.

Jourard, S. M. (1968). *Disclosing man to himself*. New York: Litton.

Judd, C. H. (1925). The psychology of social institutions. *Journal of Abnormal and Social Psychology, 20*, 151–156.

Judd, C. H. (1926). *The psychology of social institutions*. New York: Macmillan.

Kane, T. R., Joseph, J. P., & Tedeschi, J. T. (1976). Person perception and the Berkowitz paradigm for the study of aggression. *Journal of Personality and Social Psychology, 6*, 663–673.

Kant, I. (1974). *Anthropology from a pragmatic point of view* (M. J. Gregor, Trans.). The Hague: M. Nijhoff. (Original work published 1798)

Kantor, J. R. (1922). How is social psychology possible? *Journal of Abnormal and Social Psychology, 17*, 62–78.

Kantor, J. R. (1923). The problem of instincts and its relation to social psychology. *Journal of Abnormal and Social Psychology, 18*, 50–77.

Karpf, F. B. (1932). *American social psychology*. New York: McGraw-Hill.
Karsten, A. (1928). Psychische Sattigung. *Psychologische Forschung, 10*, 142–154.
Katz, D. (1967). Editorial. *Journal of Personality and Social Psychology, 7*, 341–344.
Katz, D. (1978). Social psychology in relation to the social sciences: The second social psychology. *American Behavioral Scientist, 21*, 779–792.
Katz, D. (1991). Floyd Henry Allport: Founder of social psychology as a behavioral science. In G. A. Kimble, M. Wertheimer, & C. White (Eds.), *Portraits of pioneers in psychology* (Vol. 1). Hillsdale, NJ: Lawrence Erlbaum Associates, 1991.
Katz, D., & Allport, F. (1931). *Students' attitudes*. Syracuse, NY: Craftsman Press.
Katz, D., & Braly, K. (1933). Racial stereotypes of one hundred college students. *Journal of Abnormal and Social Psychology, 28*, 280–290.
Katz, D., & Schanck, R. (1938). *Social psychology*. New York: Wiley.
Katz, E. (1957). The two-step flow of communication: An up-to-date report on a hypothesis. *Public Opinion Quarterly, 21*, 61–78.
Kelley, H. H. (1950). The warm-cold variable in first impressions of persons. *Journal of Personality, 18*, 431–439.
Kelley, H. H. (1952). Two functions of reference groups. In G. E. Swanson, T. M. Newcomb, & E. L. Hartley (Eds.), *Readings in social psychology* (2nd ed.). New York: Henry Holt & Co.
Kelley, H. H. (1955). Salience of membership and resistance to change of group-anchored attitudes. *Human Relations, 3*, 275–289.
Kelley, H. H. (1967). Attribution theory in social psychology. In D. Levine (Ed.), *Nebraska symposium on motivation*. Lincoln: University of Nebraska Press.
Kelley, H. H. (1971). Causal schema and the attribution process. In E. E. Jones, D. Kanouse, H. H. Kelley, R. E. Nisbett, S. Valins, & B. Weiner (Eds.), *Attribution: Perceiving the causes of behavior*. Morristown, NJ: General Learning.
Kelley, H. H. (1999). Fifty years in social psychology: Some reflections on the individual-group problem. In A. Rodrigues & R. V. Levine (Eds.), *Reflections on 100 years of experimental social psychology*. New York: Basic Books.
Kelley, H. H., & Thibaut, J. (1978) *Interpersonal relations: A theory of interdependence*. New York: Wiley.
Kelley, H. H., & Volkart, E. H. (1952). The resistance to change of group anchored attitudes. *American Sociological Review, 19*, 453–465.
Kelley, H. H., & Woodruff, C. L. (1956). Members' reactions to apparent group approval of a counternorm communication. *Journal of Abnormal and Social Psychology, 52*, 67–74.
Kelman, H. C. (1965). Preface to H. Proshansky & B. Seidenberg (Eds.), *Basic Studies in Social Psychology* (1965), and I. D. Steiner & M. Fishbein (Eds.), *Current Studies in Social Psychology* (1965). New York: Holt, Rinehart and Winston.
Kelman, H. C. (1967). Human use of human subjects: The problem of deception in social psychological experiments. *Psychological Bulletin, 67*, 1–11.
Kendler, H. H. (1952). What is learned? A theoretical blind alley. *Psychological Review, 59*, 269–277.

Kidder, L. H. (1981). *Selltiz, Wrightsman and Cook's research methods in social relations* (4th ed.). New York: Holt, Rinehart & Winston.
Kiesler, C. A., & Kiesler, S. B. (1969). *Conformity*. Reading, MA: Addison-Wesley.
Kiesler, C. A., & Lucke, J. (1976). Some metatheoretical issues in social psychology. In L. H. Strickland, F. E. Aboud, and K. J. Gergan (Eds.), *Social psychology in transition*. New York: Plenum Press.
Kimble, G. A. (1989). Psychology from the point of view of a generalist. *American Psychologist, 44*, 491–499.
Kimble, G. A. (1995). Discussant's remarks: From chaos to coherence in psychology. *International Newsletter of Uninomic Psychology, 15*, 34–38.
King, E. G. (1990). Reconciling democracy and the crowd in turn-of-the-century American social psychological thought. *Journal of the History of the Behavioral Sciences, 26*, 335–343.
Kinoshita, T. (1964). Shudan no gyoshusei to kadai no juyosei no docho kodo ni oyobuso koka. [The effects of cohesiveness and task importance on conformity behavior]. *Shinrigaku Kenkyu, 35*, 181–193.
Kitzinger, C. (1987). *The social construction of lesbianism*. London: Sage.
Kling, J. W., & Riggs, L. A. (1972). *Woodworth and Schlosberg's experimental psychology*. New York: Holt, Rinehart & Winston.
Klotz, I. M. (1980). The N-ray affair. *Scientific American, 242*, May, 122–131.
Kolstad, A. (1933). *A study of opinions on some international problems* (*Teacher's College Contributions to Education No. 555*).
Krantz, D. L. (1972). The mutual isolation of operant and non-operant psychology as a case study. *Journal of the History of the Behavioral Sciences, 8*, 86–102.
Krech, D., & Crutchfield, R. S. (1948). *Theory and problems of social psychology*. New York: McGraw-Hill.
Krech, D., Crutchfield, R. S., & Ballachey, E. L. (1962). *Individual in society: A textbook of social psychology*. New York: McGraw-Hill.
Kruglanski, A. W. (1975). Theory, experiment, and the shifting publication scene in social psychology. *Personality and Social Psychology Bulletin, 1*, 489–492.
Kruglanski, A. W. (1976). On the paradigmatic objections to social psychology. *American Psychologist, 31*, 655–663.
Kuhn, T. (1970). *The structure of scientific revolutions* (2nd. ed.). Chicago: Chicago University Press.
Kulp, D. H., II. (1934). Prestige as measured by single experience changes and their permanency. *Journal of Educational Research, 27*, 663–672.
Külpe, O. (1895). *Outlines of psychology* (E. B. Titchener, Trans.). New York: Macmillan.
Kuo, Z. Y. (1921). Giving up instincts in psychology. *Journal of Philosophy, 18*, 645–666.
Kusch, M. (1999). *Psychological knowledge: A social history and philosophy*. New York: Routledge.
Lachman, R., Lachman, J., & Butterfield, E. (1979). *Cognitive psychology and information processing*. Hillsdale, NJ: Erlbaum.
Lamarck, J. (1914). Zoological philosophy (H. Elliot, Trans.). London: Macmillan. (Original work published 1809)
La Piere, R. T. (1934). Attitudes versus action. *Social Forces, 13*, 230–237.

La Piere, R. T. (1938). *Collective behavior.* New York: McGraw-Hill.

Larson, K. S. (1990). The Asch conformity experiment: Replication and transhistorical comparisons. *Journal of Social Behavior and Personality, 5,* 163–168.

Lasker, B. (1929). *Race attitudes in children.* New York: Holt.

Latané, B., & Darley, J. M. (1970). *The unresponsive bystander: Why doesn't he help?* New York: Appleton-Century-Crofts.

Latour, B., & Woolgar, S. (1979). *Laboratory life.* London: Sage.

Lazarus, M. (1851). On the concept and possibility of a *Völkerpsychologie.*

Leahey, T. H. (1979). Something old, something new: Attention in Wundt and modern cognitive psychology. *Journal of the History of the Behavioral Sciences, 15,* 242–252.

Leahey, T. H. (1981). The mistaken mirror: On Wundt's and Titchener's psychologies. *Journal of the History of the Behavioral Sciences, 17,* 273–283.

Leary, D. E. (1979). Wundt and after: psychology's shifting relations with the natural sciences, social sciences and philosophy. *Journal of the History of the Behavioral Sciences, 15,* 231–241.

Le Bon, G. (1896). *The crowd: A study of the popular mind* [translation of *La psychologie des foules*]. (1895). London: T. Fisher Unwin.

Lévi-Bruhl, L. (1923). *Primitive mentality* (L. A. Clare, Trans.). New York: Macmillan.

Levine, J. M., & Moreland, R. L. (1998). Small groups. In D. T. Gilbert, S. T. Fiske, & G. Lindzey (Eds.), *The handbook of social psychology.* (4th ed.) Oxford: Oxford University Press.

Levine, R. V., & Rodrigues, A. (1999). Afterward: Reflecting on reflections. In A. Rodrigues & R. V. Levine (Eds.), *Reflections on 100 years of experimental social psychology.* New York: Basic Books.

Lévi-Strauss, C. (1960). The family. In H. Shapiro (Ed.), *Man, culture and society.* New York: Oxford University Press.

Lewin, K. (1935). The conflict between Aristotelian and Galilean modes of thought. In K. Lewin, *A dynamic theory of personality.* New York: McGraw-Hill.

Lewin, K. (1936). *Principles of topological psychology.* New York: McGraw-Hill.

Lewin, K. (1939). Field theory and experiment in social psychology: Concepts and methods. *American Journal of Sociology, 44,* 868–896.

Lewin, K. (1947a). Group decision and social change. In T. M. Newcomb & E. L. Hartley (Eds.), *Readings in social psychology.* New York: Holt.

Lewin, K. (1947b). Frontiers in group dynamics: Concept, method and reality in science: Social equilibria and social change. *Human Relations, 1,* 2–38.

Lewin, K. (1947c). Frontiers in group dynamics: II. Channels of group life: Social planning and action research. *Human Relations, 1,* 143–153.

Lewin, K. (1948). *Resolving social conflicts: Selected papers on group dynamics.* New York: Harper & Row.

Lewin, K. (1951). *Field theory in social science.* New York: Harper & Brothers.

Lewin, K. (1997a). Experiments in social space. In K. Lewin, *Resolving social conflicts and field theory in social science.* Washington, DC: American Psychological Association. Reprinted from *Harvard Educational Review,* 4, 21, 21–32, 1939.

Lewin, K. (1997b). The background of conflict in marriage. In K. Lewin, *Resolving social conflicts and field theory in social science.* Washington, DC: American Psychological Association. Reprinted from M. Jung (Ed.). (1940), *Modern marriage.* New York: F. S. Crofts.

Lewin, K. (1997c). Field theory and learning. In K. Lewin, *Resolving social conflicts and field theory in social science.* Washington, DC: American Psychological Association. Reprinted from *Yearbook of the National Society for the Study of Education,* 1942, 41, 215–242.

Lewin, K. (1997d). Behavior and development as a function of the total situation. In K. Lewin, *Resolving social conflicts and field theory in social science.* Washington, DC: American Psychological Association. Reprinted from L. Carmichael (Ed.), *Manual of child psychology.* New York: John Wiley.

Lewin K., Dembo, T., Festinger, L., & Sears, P. L. (1944). Level of aspiration. In J. M. Hunt (Ed.), *Personality and the behavior disorders.* New York: Ronald Press.

Lewin, K., Lippitt, R., & White, R. K. (1939). Patterns of aggressive behavior in experimentally created "social climates." *Journal of Social Psychology, 1,* 271–299.

Likert, R. (1932). A technique for the measurement of attitudes. *Archives of Psychology, 140,* 1–55.

Lindzey, G. (Ed.). (1954). *Handbook of social psychology* (2 vols.). Reading, MA: Addison-Wesley.

Lindzey, G., & Aronson, E. (Eds.). (1968). *The handbook of social psychology* (2nd ed., 5 vols.). Reading, MA: Addison-Wesley.

Lindzey, G., & Aronson, E. (1985). Preface. In G. Lindzey & E. Aronson (Eds.), *A handbook of social psychology* (3rd ed.). New York: Random House.

Liska, F. (1977). The dissipation of sociological social psychology. In L. H. Strickland, F. E. Aboud, & K. J. Gergen (Eds.), *Social psychology in transition.* New York: Plenum Press.

Lissner, K. (1935). Die Entspannung von Beduerfnissen durch Ersatzhandlungen. *Psychologische Forschung, 18,* 218–250.

Longino, H. (1990). *Science as social knowledge: Values and objectivity in scientific inquiry.* Princeton, NJ: Princeton University Press.

Lowry, R. T. (1987). *The evolution of psychological theory.* New York: Aldine.

Lubek, I. (1979). Aggression. In A. R. Buss (Ed.), *Psychology in social context.* New York: Irvington.

Lubek, I. (1986). Fifty years of frustration and aggression: Some historical notes on a longheld hypothesis. In K. S. Larsen (Ed.), *Dialectics and ideology in psychology.* Norwood, NJ: Ablex.

Lukes, S. (1968). Methodological individualism reconsidered. *British Journal of Sociology, 19,* 119–129.

Lukes, S. (1973a). *Emile Durkheim: His life and work.* London: Allen Lane.

Lukes, S. (1973b). *Individualism.* New York: Harper & Row.

Luria, A. R. (1931). Psychological expedition to Central Asia. *Science, 74,* 383–384.

Luria, A. R. (1976). *Cognitive development: Its cultural and social foundations* (Trans. M. Lopez-Morillas & L. Solotaroff). Cambridge, MA: Harvard University Press. (Original work published 1974)

Luria, A. R. (1979). *The making of mind: A personal account of Soviet psychology.* Cambridge, MA: Harvard University Press.
Lutz, K. (1982). The domain of emotion words in Ifaluk. *American Ethnologist, 9,* 113–128.
Lutz, K. (1988). *Unnatural emotions.* Chicago: University of Chicago Press.
Lynd, R. S., & Lynd, H. M. (1929). *Middletown.* New York: Harcourt Brace.
Lynd, R. S., & Lynd, H. M. (1937). *Middletown in transition.* New York: Harcourt Brace.
Lyons, J. O. (1978). *The invention of the self.* Carbondale: Southern Illinois University Press.
Maccoby, E. E., Newcomb, T. M., & Hartley, E. L. (Eds.). (1958). *Readings in social psychology* (3rd ed.). New York: Henry Holt & Co.
Maccoby, E. E., & Wilson, W. (1957). Identification and observational learning from films. *Journal of Abnormal and Social Psychology, 55,* 76–87.
Maciver, R. M. (1917). *Community.* London.
MacMartin, C., & Winston, A. S. (2000). The rhetoric of experimental social psychology, 1930–1960: From caution to enthusiasm. *Journal of the History of the Behavioral Sciences, 36,* 349–364.
Mahler, V. (1933). Ersatzhandlungen verschiedenen Realitatsgardes. *Psychologische Forschung, 18,* 26–89.
Malinowski, B. (1944). *A scientific theory of culture.* Chapel Hill: University of North Carolina Press.
Maller, J. B. (1929). *Cooperation and competition: An experimental study in motivation* (Teacher's College Contributions to Education).
Manicas, P. (1987). *A history and philosophy of the social sciences.* Oxford: Blackwell.
Manis, M. (1975). Comment on Gergen's "Social psychology as history." *Personality and Social Psychology Bulletin, 1,* 450–455.
Marks, G., & Miller, N. (1987). Ten years of research on the false-consensus effect: An empirical and theoretical review. *Psychological Bulletin, 102,* 72–90.
Markus, H., & Zajonc, R. B. (1985). The cognitive perspective in social psychology. In G. Lindzey & E. Aronson (Eds.), *Handbook in social psychology* (3rd ed.). New York: Random House.
Marrow, A. J. (1969). *The practical theorist: The life and work of Kurt Lewin.* New York: Basic Books.
Marsh, P., & Campbell, A. (Eds.). (1982). *Aggression and violence.* Oxford: Blackwell.
Marsh, P., Rosser, E., & Harré, R. (1978). *The rules of disorder.* London: Routledge & Kegan Paul.
Martin, E. D. (1920). *The behavior of crowds.* New York: Harper & Brothers.
May, M. A. (1950). *Toward a science of human behavior: A survey of the work of the Institute of Human Relations through two decades, 1929–1949.* New Haven, CT: Yale University Press.
Mayer, A., & Orth, J. (1901). Zur qualitativen Untersuchung der Assoziationen. *Zeitschrift für Psychologie, 8,* 1–224.
McCall, G. C., & Simmons, J. L. (1978). *Identities and interactions* (Rev. ed.). New York: The Free Press.

McClintock, C. G. (1972). *Experimental social psychology*. New York: Holt, Rinehart & Winston.
McDougall, W. (1908). *Introduction to social psychology*. New York: John W. Luce & Co.
McDougall, W. (1912). *Psychology: The science of behavior*. New York: Henry Holt.
McDougall, W. (1920). *The group mind*. New York: Putnam.
McDougall, W. (1921a). The use and abuse of instinct in social psychology. *Journal of Abnormal and Social Psychology, 16*, 285–333.
McDougall, W. (1921b). *Is America safe for democracy?* New York: Scribner's.
McDougall, W. (1928). *The group mind* (2nd ed.). New York: Putnam.
McDougall, W. (1930). William McDougall. In C. Murchison (Ed.), *History of psychology in autobiography* (Vol. 1). Worcester, MA: Clark University Press.
McGrath, J. (1978). Small group research. *American Behavioral Scientist, 21*, 651–673.
McGrath, J., & Altman, I. (1966). *Small group research: A synthesis and critique of the field*. New York: Holt.
McGuire, W. J. (1967). Some impending reorientations in social psychology: Some thoughts provoked by Kenneth Ring. *Journal of Experimental Social Psychology, 3*, 124–139.
McGuire, W. J. (1968). The nature of attitudes and attitude change. In G. Lindzey & E. Aronson (Eds.), *The handbook of social psychology* (2nd ed.). Reading, MA: Addison-Wesley.
McGuire, W. J. (1985). Toward social psychology's second century. In S. Koch & D. E. Leary (Eds.), *A century of psychology as science*. New York: McGraw-Hill.
McGuire, W. J. (1986). The vicissitudes of attitudes and similar representational constructs in twentieth century psychology. *European Journal of Social Psychology, 16*, 89–130.
McMullin, E. (Ed.). (1992). *The social dimensions of science*. South Bend, IN: Notre Dame University Press.
Mead, G. H. (1904). The relations of psychology and philology. *Psychological Bulletin, 1*, 375–391.
Mead, G. H. (1906). The imagination in Wundt's treatment of myths and religion. *Psychological Bulletin, 3*, 393–399.
Mead, G. H. (1909). Social psychology as counterpart to physiological psychology. *Psychological Bulletin, 6*, 401–408.
Mead, G. H. (1910). Social consciousness and the consciousness of meaning. *Psychological Bulletin, 7*, 397–405.
Mead, G. H. (1934). *Mind, self and society: From the standpoint of a social behaviorist* (C. W. Morris, Ed.). Chicago: University of Chicago Press.
Meade, R. D. (1986). Experimental studies of authoritarian and democratic leadership in four cultures: American, Indian, Chinese, and Chinese American. *The High School Journal, 68*, 293–295.
Meeker, B. F., & Leik, R. K. (1995). Experimentation in sociological social psychology. In K. S. Cook, G. A. Fine, & J. S. House (Eds.), *Sociological perspectives in social psychology*. Boston: Allyn & Bacon.

Menzel, H. (1957). Public and private conformity under different conditions of acceptance in the group. *Journal of Abnormal and Social Psychology, 55*, 398–402.

Merton, R. K. (1963). *Social theory and social structure.* New York: The Free Press.

Merton, R. K., & Kitt, A. (1952). Contributions to the theory of reference-group behavior. In G. E. Swanson, T. M. Newcomb, & E. L. Hartley (Eds.), *Readings in social psychology* (2nd ed.). New York: Henry Holt & Co.

Meuller, C. G. (1979). Some origins of psychology as a science. *Annual Review of Psychology, 30*, 9–20.

Meumann, E. (1907). Vorlesungen zur Einführung in die experimentelle Pädagogik und ihre psychologischen Grundlagen. (*Introductory lectures on pedagogy and its psychological basis.*) (2 vols.) Leipzig: Engelmann.

Milgram, S. (1963). Behavioral study of obedience. *Journal of Abnormal and Social Psychology, 67*, 371–378.

Milgram, S. (1974). *Obedience to authority.* New York: Harper & Row.

Mill, J. S. (1843). *A system of logic.* London: Longmans Green.

Mill, J. S. (1924). *Autobiography.* Oxford: Oxford University Press. (Original work published 1873)

Miller, A. G. (Ed.). (1972a). *The social psychology of psychological research.* New York: The Free Press.

Miller, A. G. (1972b). Role-playing: An alternative to deception? A review of the evidence. *American Psychologist, 27*, 623–636.

Miller, G. A., & Cantor, N. (1982). Review of *Human inference: Strategies and shortcomings of social judgement. Social Cognition, 1*, 83–93.

Miller, N. E., & Dollard, J. (1941). *Social learning and imitation.* New Haven, CT: Yale University Press.

Mischel, W. (1968). *Personality and assessment.* New York: Wiley.

Mixon, D. (1972). Instead of deception. *Journal for the Theory of Social Behaviour, 2*, 145–177.

Moede, W. (1914). Der Wetteifer, seine Struktur und sein Ausmass. *Zeitschrift für pädagogische Psychologie, 15*, 353–368.

Moede, W. (1920). *Experimentelle Massenpsychologie.* Leipzig: Hirzel.

Moghaddam, F. M. (1987). Psychology in three worlds as reflected by the crisis in social psychology and the move toward indigenous third-world psychology. *American Psychologist, 42*, 912–920.

Mook, D. (1983). In defence of external validity. *American Psychologist, 38*, 379–387.

Moore, W. E. (1968). Social structure and behavior. In G. Lindzey & E. Aronson (Eds.), *The handbook of social psychology* (2nd ed.). Reading, MA: Addison-Wesley.

Morawski, J. G. (1979). The structure of social psychological communities: A framework for examining the sociology of social psychology. In L. H. Strickland (Ed.), *Soviet and Western perspectives in social psychology.* New York: Pergamon.

Morris, C. (1972). *The discovery of the individual, 1050–1200.* New York: Harper & Row.

Moscovici, S. (1961). *La psychanalyse, son image et son public.* Paris: Presses Universitaires de France.

Moscovici, S. (1972). Society and theory in social psychology. In J. Israel & H. Tajfel (Eds.), *The context of social psychology: A critical assessment.* London: Academic Press.

Moscovici, S. (1981). On social representation. In J. Forgas (Ed.), *Social cognition: Perspectives on everyday understanding.* New York: Academic Press.

Moscovici, S. (1984). The phenomenon of social representations. In R. Farr & S. Moscovici (Eds.), *Social representations.* Cambridge: Cambridge University Press.

Moscovici, S. (1985). Social influence and conformity. In G. Lindzey & E. Aronson (Eds.), *The handbook of social psychology* (3rd ed.). Reading, MA: Addison-Wesley.

Moscovici, S. (1987). Answers and questions. *Journal for the Theory of Social Behavior, 17,* 513–529.

Moscovici, S. (1998a). Social consciousness and its history. *Culture and Psychology, 4,* 411–429.

Moscovici, S. (1998b, October). Why a theory of social representations? Paper presented at Conference on Social Representations: Introductions and Explorations, City University of New York Graduate School, New York. (Partially reprinted as "Why a theory of social representations?" in *Representing the social: Bridging social traditions,* K. Deaux & G. Philogene (Eds.), 2000, Oxford: Basil Blackwell)

Murchison, C. A. (Ed.) (1935). *Handbook of social psychology.* Worcester, MA: Clark University Press.

Murphy, G. (1965). The future of social psychology in historical perspective. In O. Klineberg & R. Christie (Eds.), *Perspectives in social psychology.* New York: Holt, Rinehart & Winston.

Murphy, G., & Murphy, L. B. (1931). *Experimental social psychology.* New York: Harper.

Murphy, G., Murphy, L. B., & Newcomb, T. M. (1937). *Experimental social psychology* (Rev. ed.). New York: Harper.

Newcomb, T. M. (1943). *Personality and social change: Attitude formation in a student community.* New York: Holt.

Newcomb, T. M. (1950). *Social psychology.* New York: Dryden.

Newcomb, T. M. (1951). Social psychological theory: Integrating individual and social approaches. In J. M. Rohrer & M. Sherif (Eds.), *Social psychology at the crossroads.* New York: Harper.

Newcomb, T. M. (1952). Attitude development as a function of reference groups: The Bennington study. In G. E. Swanson, T. M. Newcomb, & E. L. Hartley (Eds.), *Readings in social psychology.* (2nd ed.). New York: Henry Holt & Co.

Newcomb, T. M., & Hartley, E. L. (Eds.). (1947). *Readings in social psychology.* New York: Henry Holt & Co.

Newcomb, T. M., & Svehla, G. (1938). Intra-family relationships in attitudes. *Sociometry, 1,* 180–205.

Newcomb, T. M., Turner, R. H., & Converse, P. E. (1965). *Social psychology: The study of human interaction.* New York: Holt, Rinehart & Winston.

Nisbett, R. E., & Ross, L. (1980). *Human inference: Strategies and shortcomings of social judgment.* Englewood Cliffs, NJ: Prentice-Hall.

Nisbett, R. E., & Wilson, T. D. (1977). Telling more than we can know: Verbal reports on mental processes. *Psychological Review, 84*, 231–259.

Nye, J. L., & Brower, A. M. (Eds.). (1996). *What's social about social cognition? Research on socially shared cognition in small groups.* Thousand Oaks, CA: Sage.

Nye, R. A. (1975). *The origins of crowd psychology: Gustav Le Bon and the crisis of mass democracy in the Third Republic.* London: Sage.

Olson, T., & Christiansen, G. (1966). *The grindstone experiment: Thirty one hours.* Toronto: Canadian Friends Service Committee.

Orne, M. T. (1962). On the social psychology of the psychological experiment: With particular reference to demand characteristics and their implications. *American Psychologist, 17*, 776–783.

Orne, M. T., & Holland, C. T. (1968). On the ecological validity of laboratory deceptions. *International Journal of Psychiatry, 6*, 282–293.

Ostrom, T. (1984). The sovereignty of social cognition. In R. S. Wyer, Jr., & T. K. Srull (Eds.), *Handbook of social cognition* (Vol. 1). Hillsdale, NJ: Erlbaum.

Ovsiankina, M. (1928). Die Wiederaufnahme unterbrochenen Handlunger. *Psychologische Forschung, 1*, 302–389.

Pandora, K. (1997). *Rebels within the ranks: Psychologists' critique of scientific authority and democratic realities in New Deal America.* Cambridge: Cambridge University Press.

Park, R. E. (1902). *The crowd and the public and other essays.* Chicago: University of Chicago Press.

Park, R. E., & Burgess, E. W. (1921). *Introduction to the science of sociology.* Chicago: University of Chicago Press.

Parker, I. (1987). Social representations: Social psychology's (mis)use of sociology. *Journal for the Theory of Social Behavior, 17*, 447–469. (Special issue on social representations)

Parker, I. (1989). *The crisis in modern social psychology – and how to end it.* London: Routledge.

Parker, I., & Shotter, J. (1990). *Deconstructing social psychology.* London: Sage.

Parkovnick, S. (1998, June). Gordon Allport, civilian morale and the institutionalization of social psychology, 1940–1942. Paper presented at 1998 *Cheiron* Meeting, University of San Diego.

Parkovnick, S. (2000). Contextualizing Floyd Allport's *Social Psychology. Journal of the History of the Behavioral Sciences, 36*, 429–442.

Parsons, T. (1951). *The social system.* Glencoe, IL: The Free Press.

Parsons, T. (1968). *The structure of social action.* New York: The Free Press.

Patnoe, S. (1988). *A narrative history of experimental social psychology.* New York: Springer-Verlag.

Pepitone, A. (1976). Toward a normative and comparative social psychology. *Journal of Personality and Social Psychology, 34*, 641–653.

Pepitone, A. (1981). Lessons from the history of social psychology. *American Psychologist, 36*, 972–985.

Pepitone, A. (1999). Historical sketches and critical commentary about social psychology in the golden age. In A. Rodrigues & R. V. Levine (Eds.), *Reflections on 100 years of experimental social psychology*. New York: Basic Books.

Perrin, S., & Spencer, C. (1980). The Asch effect – a child of its time? *Bulletin of the British Psychological Society, 32*, 405–406.

Peterson, J. (1918). The functioning of ideas in social groups. *Psychological Review, 25*, 214–226.

Piaget, J. (1932). *The moral judgement of the child*. London: Paul, Trench, Truber.

Pickering, A. (1984). *Constructing quarks*. Chicago: Chicago University Press.

Planck, M. (1949). *Scientific autobiography and other papers* (F. Gaynor, Trans.). New York: Greenwood Press.

Plon, M. (1974). On the meaning of the notion of conflict and its study in social psychology. *European Journal of Social Psychology, 4*, 389–436.

Popper, K. R. (1945). *The open society and its enemies* (2 vols.). London: Routledge & Kegan Paul.

Popper, K. R. (1957). *The poverty of historicism*. London: Routledge & Kegan Paul.

Post, D. L. (1980). Floyd Allport and the launching of modern social psychology. *Journal of the History of the Behavioral Sciences, 16*, 369–376.

Potter, J., & Wetherell, M. (1987). *Discourse and social psychology: Beyond attitudes and behavior*. London: Sage.

Proshansky, H., & Seidenberg, B. (Eds.). (1965). *Basic studies in social psychology*. New York: Holt, Rinehart & Winston.

Quinn, C. O., Robinson, I. E., & Balkwell, J. W. (1980). A synthesis of two social psychologies. *Symbolic Interaction, 3*, 59–88.

Radcliffe-Brown, A. R. (1958). *Methodology in social anthropology*. Chicago: University of Chicago Press.

Radina, S. L. (1930). A pedagogical study of children entering kindergarten. *Pedologia, 4*, 468–476.

Ratzenhofer, G. (1898). *Die Sociologische Erkenntniss*. Leipzig: F. A. Brockhaus.

Reis, H. T., & Stiller, J. (1992). Publication trends in JPSP: A three-decade review. *Personality and Social Psychology Bulletin, 18*, 465–472.

Reisman, D. (1950). *The lonely crowd*. New Haven, CT: Yale University Press.

Resnick, L. B. (1991). Shared cognition: Thinking as social practice. In L. B. Resnick, J. M. Levine, & S. D. Teasley (Eds.), *Perspectives on socially shared cognition*. Washington, DC: American Philosophical Association.

Resnick, L. B., Levine, J. M. & Teasley, S. D. (Eds.). (1991). *Perspectives on socially shared cognition*. Washington, DC: American Philosophical Association.

Riley, J. W., Jr. (1947). Opinion research in liberated Normandy. *American Sociological Review, 12*, 698–703.

Ring, K. (1967). Experimental social psychology: Some sober questions about some frivolous values. *Journal of Experimental Social Psychology, 3*, 113–123.

Robinson, D. (1997). Wilhelm von Humboldt and the German university. In W. Bringmann, H. E. Lück, R. Miller, & C. E. Early (Eds.), *A pictorial history of psychology*. Chicago: Quintessence Publishing.

Rosen, B. C. (1955). Conflicting group membership: A study of parent-peer group cross pressures. *American Sociological Review, 20*, 155–161.

Rosenberg, A. (1995). *Philosophy of social science* (2nd ed.). Boulder, CO: Westview Press.

Rosenberg, M., & Turner, R. H. (Eds.). (1981). *Social psychology: Sociological perspectives*. New York: Basic Books.

Rosenberg, M. J. (1969). The conditions and consequences of evaluation apprehension. In R. Rosenthal & R. L. Rosnow (Eds.), *Artifact in behavioral research*. New York: Academic Press.

Rosenberg, S., & Jones, R. A. (1972). A method of investigating and representing a person's implicit personality theory. *Journal of Personality and Social Psychology, 22*, 372–386.

Rosenblatt, P. C., & Miller, N. (1972). Experimental methods. In C. G. McClintock (Ed.), *Experimental social psychology*. New York: Holt, Rinehart & Winston.

Rosenthal, R. (1966). *Experimenter effects on behavioral research*. New York: Appleton-Century-Crofts.

Rosenthal, R., & Rosnow, R. (1969). The volunteer subject. In R. Rosenthal & R. L. Rosnow (Eds.), *Artifact in behavioral research*. New York: Academic Press.

Rosenthal, R., & Rosnow, R. L. (1975). *The volunteer subject*. New York: Wiley.

Ross, D. (1991). *The origins of American social science*. Cambridge: Cambridge University Press.

Ross, E. A. (1906). *Social control*. New York: Macmillan.

Ross, E. A. (1908). *Social psychology: An outline and source book*. New York: Macmillan.

Ross, L. (1977). The intuitive psychologist and his short-comings: Distortions in the attribution process. In L. Berkowitz (Ed.), *Advances in experimental social psychology* (Vol. 10). New York: Academic Press.

Ross, L., Greene, D., & House, P. (1977). The "false consensus" effect: An egocentric bias in social perception and attribution processes. *Journal of Experimental Social Psychology, 13*, 279–301.

Ross, L., & Nisbett, R. E. (1991). *The person and the situation: Perspectives of social psychology*. Philadelphia: Temple University Press.

Rucci, A. J., & Tweney, R. D. (1980). Analysis of variance and the "second discipline" of scientific psychology. *Psychological Bulletin, 87*, 166–184.

Samelson, F. (1974). History, origin myth and ideology: "Discovery of social psychology." *Journal for the Theory of Social Behaviour, 4*, 217–231.

Samelson, F. (1985). Organizing for the kingdom of behavior: Academic battles and organizational policies in the twenties. *Journal of the History of the Behavioral Sciences, 21*, 33–47.

Samelson, F. (2000). Whig and anti-Whig histories – and other curiosities of social psychology. *Journal of the History of the Behavioral Sciences, 36*, 499–506.

Sampson, E. E. (1977). Psychology and the American ideal. *Journal of Personality and Social Psychology, 35*, 767–782.

Sampson, E. E. (1981). Cognitive psychology as ideology. *American Psychologist, 36*, 730–743.

Sarbin, T. R., & Allen, V. L. (1968). Role theory. In G. Lindzey & E. Aronson (Eds.), *The handbook of social psychology* (2nd ed.). Reading, MA: Addison-Wesley.

Sargent, S. S. (1965). Discussion of Gardner Murphy's paper. In O. Klineberg & R. Christie (Eds.), *Perspectives in social psychology*. New York: Holt, Rinehart & Winston.

Schachter, S. (1951). Deviation, rejection, and communication. *Journal of Abnormal and Social Psychology, 46*, 190–207.

Schachter, S. (1959). *The psychology of affiliation*. Stanford, CA: Stanford University Press.

Schachter, S. (1971). *Emotion, obesity and crime*. New York: Academic Press.

Schachter, S., & Singer, J. E. (1962). Cognitive, social and physiological determinants of emotional state. *Psychological Review, 69*, 379–399.

Schäffle, A. (1875–1878). *Bau und Leben des Socialen Körpers* (Vols. 1–4). Tübingen: H. Laupp.

Schanck, R. L. (1932). A study of a community and its groups and institutions conceived as behaviors of individuals. *Psychological Monographs, 43*, No. 195.

Schelling, T. (1978). *Micromotives and macrobehavior*. New York: Norton.

Schlenker, B. R. (1974). Social psychology and science. *Journal of Personality and Social Psychology, 29*, 1–15.

Schmitt, F. F. (1994). The justification of group beliefs. In F. F. Schmitt (Ed.), *Socializing epistemology: The social dimensions of knowledge*. Lanham, MD: Rowman & Littlefield.

Schneider, C. M. (1990). *Wilhelm Wundt's Völkerpsychologie*. Bonn: Bouvier.

Schuman, H. (1995). Attitudes, beliefs and behavior. In K. S. Cook, G. A. Fine, & J. S. House (Eds.), *Sociological perspectives in social psychology*. Boston: Allyn & Bacon.

Sears, D. O. (1968). Political behavior. In G. Lindzey & E. Aronson (Eds.), *The handbook of social psychology* (2nd ed.). Reading, MA: Addison-Wesley.

Secord, P. F. (1976). Transhistorical and transcultural theory. *Personality and Social Psychology Bulletin, 2*, 418–420.

Secord, P. F. (1982). The behavior identity problem in generalizing from experiments. *American Psychologist, 37*, 1408.

Secord, P. F. (1990). Explaining social behavior. *Theoretical and Philosophical Psychology, 10*, 25–38.

Secord, P. F., & Backman, C. W. (1964). *Social psychology*. New York: McGraw-Hill.

Secord, P. F., Backman, C. W., & Slavitt, P. R. (1976). *Understanding social life: An introduction to social psychology*. New York: McGraw-Hill.

Selltiz, C., Jahoda, M., Deutsch, M., & Cook, S. W. (1959). *Research methods in social relations* (2nd ed.). New York: Holt, Rinehart & Winston.

Selltiz, C., Wrightsman, L. R., & Cook, S. W. (1976). *Research methods in social relations* (3rd ed.). New York: Holt, Rinehart & Winston.

Shain, B. A. (1996). *The myth of American individualism: The Protestant origins of American political thought*. Princeton, NJ: Princeton University Press.

Shaw, M. (1932). A comparison of individuals and small groups in the rational solution of complex problems. *American Journal of Psychology, 19*, 491–504.

Shepard, R. N. (1987). Toward a universal law of generalization for psychological science. *Science, 237*, 1317–1323.

Shepard, R. N. (1995). Mental universals: Towards a twenty-first century science of mind. In R. L. Solso & D. W. Massaro (Eds.), *The science of the mind: 2001 and beyond*. New York: Oxford University Press.

Sherif, M. (1935). A study of some social factors in perception. *Archive of Psychology*, No. 187.

Sherif, M. (1936). *The psychology of social norms*. New York: Harper.

Sherif, M. (1948). *An outline of social psychology*. New York: Harper.

Sherif, M. (1949). The problems of inconsistency in intergroup relations. *Journal of Social Issues, 5*, 32–37.

Sherif, M. (1951). A preliminary study of intergroup relations. In J. Rohrer & M. Sherif (Eds.), *Social psychology at the crossroads*. New York: Harper.

Sherif, M. (1977). Crisis in social psychology: Some remarks towards breaking through the crisis. *Personality and Social Psychology Bulletin, 3*, 368–382.

Sherif, M., & Cantril, H. (1947). *The psychology of ego-involvements*. New York: Wiley.

Sherif, M., Harvey, O. J., White, B. J., Hood, W. R., & Sherif, C. W. (1954). *Experimental study of positive and negative intergroup attitudes between experimentally produced groups: Robbers Cave Study*. Norman: University of Oklahoma.

Sherif, M., & Sherif, C. W. (1953). *Groups in harmony and tension*. New York: Harper.

Sherman, R. C., Buddie, A. M., Dragan, K. L., End, C. M., & Finney, L. J. (1999). Twenty years of PSPB: Trends in content, design, and analysis. *Personality and Social Psychology Bulletin, 25*, 177–187.

Shibutani, T. (1955). Reference groups as perspectives. *American Journal of Sociology, 60*, 562–569.

Shook, J. R. (1995). Wilhelm Wundt's contribution to John Dewey's functional psychology. *Journal of the History of the Behavioral Sciences, 31*, 347–369.

Shotter, J. (1987). The rhetoric of theory in psychology. In W. M. Baker, M. E. Hyland, H. V. Rappard, & A. W. Staats (Eds.), *Current issues in theoretical psychology*. Amsterdam: Elsevier.

Shotter, J. (1992). Social constructionism and realism: Adequacy or accuracy? *Theory and Psychology, 2*, 175–182.

Shweder, R. A. (1990). What is cultural psychology? In J. W. Stigler, R. A. Shweder, & G. Herdt (Eds.), *Cultural psychology: Essays on comparative human development*. New York: Cambridge University Press.

Shweder, R. A., & LeVine, R. A. (Eds.). (1984). *Culture theory: Essays on mind, self and emotion*. New York: Cambridge University Press.

Siann, G. (1985). *Accounting for aggression*. London: George Allan & Unwin.

Siegel, A. E., & Siegel, S. (1957). Reference groups, membership groups, and attitude change. *Journal of Abnormal and Social Psychology, 55*, 360–364.

Sighele, S. (1892). *La foule criminelle* (translated into French by P. Vigny). Paris: F. Alcan.

Silverman, I. (1977). *The human subject in the psychological laboratory*. New York: Pergamon.

Simmel, G. (1894). Das problem der Soziologie. *Jahrbuch für Gesetzgebung, Verwaltung und Volkswistschaft, 18*, 271–277.

Simmel, G. (1959). How is society possible? In K. H. Wolff (Ed. & Trans.), *Georg Simmel, 1858–1918: A collection of essays, with translations and a bibliography*. Columbus: Ohio State University Press. (Original work published 1908)

Simons, C. W., & Pilliavin, J. A. (1972). Effect of deception on reactions to a victim. *Journal of Personality and Social Psychology, 21*, 56–60.

Sims, J. H., & Baumann, D. D. (1972). The tornado threat: Coping styles of the North and South. *Science, 17*, 1386–1392.

Singer, E. (1988). Reference groups and social evaluations. In M. Rosenberg & R. H. Turner (Eds.), *Social psychology: Sociological perspectives*. New York: Basic Books.

Singer, J. (1980). Social comparison: The process of self-evaluation. In L. Festinger (Ed.), *Retrospections on social psychology*. Oxford: Oxford University Press.

Small, A. W. (1905). *General sociology*. Chicago: Chicago University Press.

Small, A. W., & Vincent, G. E. (1894). *Introduction to the study of society*. New York: American Book Company.

Smith, M. (1945). Social situation, social behavior, social group. *Psychological Review, 52*, 224–229.

Smith, M. B. (1989). Comment on the case of William McDougall. *American Psychologist, 44*, 13–18.

Smith, R. (1997). *Norton history of the human sciences*. New York: Norton.

Solomon, M. (1994). A more social epistemology. In F. Schmitt (Ed.), *Socializing epistemology: The social dimensions of knowledge*. Lanham, MD: Rowman & Littlefield.

Spatz, C., & Underwood, B. J. (1970). A *laboratory manual for experimental psychology*. New York: Appleton-Century-Crofts.

Spence, J. T. (1987). Centrifugal and centripetal trends in psychology: Will the center hold? *American Psychologist, 42*, 1052–1054.

Spencer, H. (1855). *Principles of psychology*. London: Smith & Elder.

Spencer, H. (1870–1872). *The principles of psychology* (2nd ed., 2 vols.). London: Williams & Norgate.

Spencer, H. (1880–1896). *The principles of sociology*. New York: D. Appleton & Co.

Sprung, L., & Sprung, H. (1997). The Berlin school of Gestalt psychology. In W. G. Bringmann, H. E. Lück, R. Miller, & C. E. Early (Eds.), *A pictorial history of psychology*. Chicago: Quintessence Publishers.

Staats, A. W. (1983). *Psychology's crisis of disunity: Philosophy and method for a unified science*. New York: Praeger.

Stam, H. J., Radtke, L., & Lubek, I. (2000). Strains in experimental social psychology: A textual analysis of the development of experimentation in social psychology. *Journal of the History of the Behavioral Sciences, 36*, 365–382.

Steiner, I. D. (1974). Whatever happened to the group in social psychology? *Journal of Experimental Social Psychology, 10*, 94–108.

Steiner, I. D., & Fishbein, M. (Eds.). (1965). *Current studies in social psychology*. New York: Holt, Rinehart & Winston.

Steinthal, H. (1855). *Grammatik, Logik, und Psychology, ihre Prinzipien und ihr Verhältnis zueinander* [*Grammar, logic and psychology: Their principles and interrelationships*]. Berlin.

Stephen, C. W., & Stephen, W. G. (1991). Social psychology at the crossroads. In C. W. Stephen, W. G. Stephen, & T. F. Pettigrew (Eds.), *The future of social psychology: Defining the relationship between sociology and psychology*. New York: Springer-Verlag.

Stigler, J. W., Shweder, R. A., & Herdt, G. (Eds.). (1990). *Cultural psychology: Essays on comparative human development*. New York: Cambridge University Press.

Stocking, G. W. (1965). Editorial: On the limits of "presentism" and "historicism" in the historiography of the behavioral sciences. *Journal of the History of the Behavioral Sciences, 1*, 211–218.

Stogdill, R. M. (1950). Leadership, membership and organization. *Psychological Bulletin, 47*, 1–14.

Stouffer, S. A., Lumsdane, A. A., Lumsdaine, M. H., Williams, R. M., Smith, M. B., Janis, I. L., Star, S. A., & Cottrell, I. S. (Eds.). (1949). *The American soldier: Combat and its aftermath*. (Studies in Social Psychology in World War II No. 2). Princeton, NJ: Princeton University Press.

Stouffer, S. A., Suchman, E. A., De Vinney, L. C., Star, S. A., & Williams, R. B., Jr. (Eds.). (1949). *The American soldier: Adjustment during army life* (Studies in Social Psychology in World War II No. 1). Princeton, NJ: Princeton University Press.

Stroebe, W. (1979). The level of social psychological analysis: A plea for a more social psychology. In L. H. Strickland (Ed.), *Soviet and Western perspectives in social psychology*. New York: Pergamon.

Stroebe, W., & Kruglanski, A. W. (1989). Social psychology at epistemological cross-roads: On Gergen's choice. *European Journal of Social Psychology, 19*, 485–489.

Stryker, S. (1983). Social psychology from the standpoint of a structural symbolic interactionist: Toward an interdisciplinary social psychology. In L. Berkowitz (Ed.), *Advances in experimental social psychology* (Vol. 16). New York: Academic Press.

Suls, J. M., & Gastorf, J. (1980). Has the social psychology of the experiment influenced how research is conducted? *European Journal of Social Psychology, 10*, 291–294.

Suls, J. M., & Rosnow, R. L. (1988). Concerns about artifacts in psychological experiments. In J. G. Morawski (Ed.), *The rise of experimentation in American psychology*. New Haven, CT: Yale University Press.

Sumner, W. G. (1906). *Folkways*. Boston: Ginn & Co.

Swanson, G. E., Newcomb, T. M., & Hartley, E. L. (Eds.). (1952). *Readings in social psychology* (2nd ed.). New York: Henry Holt & Co.

Tagiuri, R. (1968). Person perception. In G. Lindzey & E. Aronson (Eds.), *The handbook of social psychology* (2nd ed.). Reading, MA: Addison-Wesley.

Tajfel, H. (1972). Experiments in a vacuum. In J. Israel & H. Tajfel (Eds.), *The context of social psychology: A critical assessment*. London: Academic Press.

Tajfel, H. (Ed.). (1978). *Differentiation between social groups: Studies in the social psychology of intergroup relations*. London: Academic Press.

Tajfel, H. (1979). Individuals and groups in social psychology. *British Journal of Social and Clinical Psychology, 18*, 173–180.

Tajfel, H. (1981). *Human groups and social categories*. Cambridge: Cambridge University Press.

Tajfel, H., & Turner, J. C. (1986). The social identity theory of intergroup behavior. In S. Worchel & W. G. Austin (Eds.), *Psychology of intergroup relations* (2nd ed.). Chicago: Nelson-Hall.

Tarde, G. (1899). *Social Laws* (H. C. Warren, Trans.). New York: Macmillan. (Original work published 1898)

Tarde, G. (1969). *L'opinion et la foule*. Paris: F. Alcan. In T. N. Clark (Ed.), *On communication and social influence*. Chicago: Chicago University Press. (Original work published 1901)

Tarde, G. (1903). *The laws of imitation* (E. C. Parsons, Trans.). New York: Holt. (Original work published 1890)

Taylor, D. M., & Brown, R. J. (1979). Towards a more social social psychology? *British Journal of Social and Clinical Psychology, 18*, 173–180.

Taylor, S. E. (1998). The social being in social psychology. In D. T. Gilbert, S. T. Fiske, & G. Lindzey (Eds.), *The handbook of social psychology*. Oxford: Oxford University Press.

Taylor, S. E., & Fiske, S. T. (1981). Getting inside the head: Methodologies for process analysis in attribution and social cognition. In J. H. Harvey, W. Ickes, & R. F. Kidd (Eds.), *New directions in attribution research*. Hillsdale, NJ: Erlbaum.

Thibaut, J. W., & Kelley, H. H. (1959). *The social psychology of groups*. New York: Wiley.

Thomas, W. I. (1904). The province of social psychology. *American Journal of Sociology, 10*, 445–455.

Thomas, W. I., & Thomas, D. S. (1928). *The child in America*. New York: Knopf.

Thomas, W. I., & Znaniecki, F. (1918). Methodological note. In *The Polish peasant in Europe and America* (Vol. 1). Chicago: University of Chicago Press.

Thorndike, E. (1913). *The original nature of man*. New York: Teachers College, Columbia University.

Thurstone, L. L. (1932). The measurement of social attitudes. *Journal of Abnormal and Social Psychology, 26*, 249–269.

Thurstone, L. L., & Chave, E. J. (1929). *The measurement of attitude*. Chicago: University of Chicago Press.

Titchener, E. B. (1910). *A textbook of psychology*. New York: Macmillan.

Titchener, E. B. (1916). On ethnological tests of sensation and perception. *Proceedings of the American Philosophical Society, 55*, 204–236.

Titchener, E. B. (1921). Wilhelm Wundt. *American Journal of Psychology, 32*, 161–178.

Tocqueville, A. de (1969). *Democracy in America* (J. P. Mayer, Ed., G. Lawrence, Trans.). New York: Doubleday. (Original work published 1830)

Tolman, E. C. (1922). Can instincts be given up in psychology? *Journal of Abnormal and Social Psychology, 17*, 139–152.

Tolman, E. C. (1923). The nature of instinct. *Psychological Bulletin, 20*, 200–218.

Tolman, E. C. (1932). *Purposive behavior in animals and men*. New York: Century.

Tolman, E. C. (1948). Cognitive maps in rats and men. *Psychological Review, 55*, 189–209.

Tomasello, M., Kruger, A. C., & Ratner, H. H. (1993). Cultural learning. *Behavioral and Brain Sciences, 16,* 495–552.

Toulmin, S., & Leary, D. E. (1992). The cult of empiricism in psychology and beyond. In S. Koch & D. E. Leary (Eds.), *A century of psychology as science.* Washington, DC: American Psychological Association.

Travis, L. E. (1925). Effect of a small audience upon eye-hand co-ordination. *Journal of Abnormal and Social Psychology, 20,* 142–146.

Triandis, H. C. (1976a). Social psychology and cultural analysis. In L. H. Strickland, F. E. Aboud, & K. J. Gergen (Eds.), *Social psychology in transition.* New York: Plenum.

Triandis, H. C. (1976b). Some universals of social behavior. *Psychological Bulletin, 4,* 1–16.

Triplett, N. (1898). The dynamogenic factors in pacemaking and competition. *American Journal of Psychology, 9,* 507–533.

Triplett, N. (1900). The psychology of conjuring deceptions. *American Journal of Psychology, 2,* 439–510.

Trotter, W. (1916). *Instincts of the herd in peace and war.* New York: Macmillan.

Turner, J. C. (1987). *Discovering the social group: A self-categorization theory.* Oxford: Basil Blackwell.

Underwood, B. J. (1949). *Experimental psychology: An introduction.* New York: Appleton-Century-Crofts.

Underwood, B. J. (1966a). *Experimental psychology.* New York: Appleton-Century-Crofts.

Underwood, B. J. (1966b). *Problems in experimental design and inference: Workbook for the first course in experimental psychology.* New York: Appleton-Century-Crofts.

Underwood, B. J. (1975). *Experimentation in psychology.* New York: Wiley.

U.S. Strategic Bombing Survey. (1946). *The effects of bombing on German morale.* Washington, DC: U.S. Government Printing Office.

van Ginneken, J. (1985). The 1898 debate on the origins of crowd psychology. *Journal of the History of the Behavioral Sciences, 21,* 375–382.

van Ginneken, J. (1992). *Crowds, psychology and politics, 1871–1899.* Cambridge: Cambridge University Press.

van Strien, P. (1997). The American "colonization" of northwest European social psychology after World War II. *Journal of the History of the Behavioral Sciences, 33,* 349–363.

Vetter, G. B., & Green, M. (1932). Personality and group factors in the making of athiests. *Journal of Abnormal and Social Psychology, 27,* 179–194.

Vico, G. B. (1984). *The new science* (Trans. T. G. Bergin & M. H. Fisch). Cornell: Cornell University Press. (Original work published 1725)

Volkart, E. H. (1971). Comments. In "Review symposium on the Handbook of Social Psychology." *American Sociological Review, 36,* 898–902.

von Eickstedt, E. (1936). *Grundlagen der Rassenpsychologie.* Stuttgart: Enke.

von Humboldt, W. (1836). *Über die Verschiedenheit des menschlichen Sprachbaus und ihren Einfluß auf die geistige Entwicklung des Menschengeschlechts.* Berlin: Acadamie der Wissenchaften.

Vygotsky, L. S. (1978). *Mind in society: The development of the higher mental processes* (M. Cole, V. John-Steiner, S. Scribner, & E. Souberman, Eds.). Cambridge, MA: Harvard University Press.

Vygotsky, L. S. (1986). *Thought and language* (A. Kozulin, Trans.). Cambridge, MA: MIT Press. (Original work published 1934)

Wallas, G. (1914). *The great society*. London: Macmillan.

Wallas, G. (1921). *Our social heritage*. New Haven, CT: Yale University Press.

Wallis, W. D. (1925). The independence of social psychology. *Journal of Abnormal and Social Psychology, 20*, 147–150.

Wallis, W. D. (1935a). The social group as an entity. *Journal of Abnormal and Social Psychology, 29*, 367–370.

Wallis, W. D. (1935b). Social history of the white man. In C. Murchison (Ed.), *Handbook of social psychology*. Worcester, MA: Clark University Press.

Waters, R. H. (1941). The J distribution as a measure of institutional strength. *Journal of Social Psychology, 13*, 413–414.

Ward, L. (1883). *Dynamic sociology*.

Watson, J. B. (1913). Psychology as a behaviorist views it. *Psychological Review, 20*, 158–177.

Watson, J. B. (1919). *Psychology from the standpoint of a behaviorist*. Philadelphia: Lippincott.

Watson, J. B., & McDougall, W. (1928). *The battle of behaviorism*. London: Trench, Trubner & Co.

Watson, G. (Ed.). (1942). *Civilian morale*. Boston: Houghton Mifflin.

Watson, W. S., & Hartmann, G. W. (1939). The rigidity of a basic attitudinal frame. *Journal of Abnormal and Social Psychology, 34*, 314–335.

Weber, M. (1978). *Economy and society* (2 vols., G. Roth & C. Wittich, Eds.). Berkeley: University of California Press. (Original work published 1922)

Wee, C. (2002). Self, other and community in Cartesian ethics. *History of Philosophy Quarterly, 19*, 255–273.

Wegner, D. M., & Vallacher, R. R. (1977). *Implicit psychology: An introduction to social cognition*. New York: Oxford University Press.

Weinreich, P. (1983). Emerging from threatened identities: Ethnicity and gender in redefinitions of ethnic identity. In G. Breakwell (Ed.), *Emerging from threatened identities*. New York: Wiley.

Wertheimer, M. (1987). *A brief history of psychology* (3rd ed.). New York: Holt, Rinehart & Winston.

West, S. G., Newsom, J. T., & Fenaughty, A. M. (1992). Publication trends in JPSP: Stability and change in topics, methods and theories across two decades. *Personality and Social Psychology Bulletin, 18*, 473–484.

Wetterstein, J. R. (1975). The historiography of scientific psychology: A critical study. *Journal of the History of the Behavioral Sciences, 11*, 157–171.

Wheeler, D., & Jordan, H. (1929). Change of individual opinion to accord with group opinion. *Journal of Abnormal and Social Psychology, 24*, 203–206.

White, M. (1973). *Pragmatism and the American mind*. New York: Oxford University Press.

Whitely, R. (1984). *The intellectual and social organization of the sciences*. Oxford: Oxford University Press.

Whiten, A. (Ed.). (1991). *Natural theories of mind.* Oxford: Basil Blackwell.

Williams, J. M. (1922). *Principles of social psychology.* New York: Knopf.

Williams, R. (1961). *The long revolution.* New York: Columbia University Press.

Williams, T. P., & Sogon, S. (1984). Group composition and conforming behavior in Japanese students. *Japanese Psychological Research, 26,* 231–234.

Willis, N. H., & Willis, Y. A. (1970). Role-playing versus deception: An experimental comparison. *Journal of Personality and Social Psychology, 16,* 472–477.

Wilson, E. O. (1975). *Sociobiology.* Cambridge, MA: Harvard University Press.

Winslow, C. N. (1937). A study of the extent of agreement between friend's opinions and their ability to estimate the opinions of each other. *Journal of Social Psychology, 8,* 433–442.

Winston, A. S. (1990). Robert Sessions Woodworth and the "Columbia Bible": How the psychological experiment was redefined. *American Journal of Psychology, 103,* 391–401.

Winston, A. S., & Blais, D. J. (1996). What counts as an experiment? A transdisciplinary analysis of textbooks, 1930–1970. *American Journal of Psychology, 109,* 599–616.

Wissler, C. (1935). Social history of the red man. In C. Murchison (Ed.), *Handbook of social psychology.* Worcester, MA: Clark University Press.

Wolff, M. (1977) Social psychology as history: Advancing the problem. *Personality and Social Psychology Bulletin, 3,* 211–212.

Wolfgang, M. E., & Ferracuti, F. (1967). *The subculture of violence: Towards an integrated theory in criminology.* London: Tavistock.

Woodward, W. R. (1982). Professionalization, rationality, and political linkages in twentieth-century psychology. In M. G. Ash & W. R. Woodward (Eds.), *Psychology in twentieth century thought and society.* New York: Cambridge University Press.

Woodworth, R. S. (1918). *Dynamic psychology.* New York: Columbia University Press.

Woodworth, R. S. (1934). *Psychology* (3rd ed.). New York: Holt.

Woodworth, R. S. (1938). *Experimental psychology.* New York: Holt.

Woodworth, R. S., & Schlosberg, H. (1954). *Experimental psychology.* New York: Holt.

Wrightsman, L. S. (1960). Effects of waiting with others on changes of level of felt anxiety. *Journal of Abnormal and Social Psychology, 61,* 216–222.

Wundt, W. (1862). *Beiträge zur Theorie der Sinneswahrnehmung* [*Contributions towards a theory of perception*]. Lepzig: Winter.

Wundt, W. (1863). *Vorlesungen über die Menschen-und Thierseele* [*Lectures on the human and animal mind*] (2 vols.). Leipzig: Voss.

Wundt, W. (1894). *Lectures on human and animal psychology* (Translation from 2nd German edition by J. E. Creighton & E. B. Titchener). New York: Macmillan.

Wundt, W. (1900–1920). *Völkerpsychologie* (Vols. 1–10). Leipzig: W. Engelmann.

Wundt, W. (1901). *Völkerpsychologie* (Vol. 1). Leipzig: Engelmann.

Wundt, W. (1902). *Outlines of psychology* (C. H. Judd, Trans.). St. Claires Shores, MI: Scholarly Press. (Original work published 1897)

Wundt, W. (1904). *Principles of physiological psychology* (E. B. Titchener, Trans.). New York: Macmillan.

Wundt, W. (1907). Über Ausfrageexperimente und über die Methoden zur Psychologie des Denkens. *Psychologische Studien, 3,* 301–360.

Wundt, W. (1908). *Logik* (Vol. 3). Stuttgart: Enke.

Wundt, W. (1911a). *Grundzüge der physiologischen Psychologie* [*Lectures on physiological psychology*] (*Vol. 3*). *Leipzig: Engelmann.*

Wundt, W. (1911b). *Völkerpsychologie* (3rd ed., vol. 1.). Leipzig: Kröner.

Wundt, W. (1916). *Elements of folk psychology: Outlines of a psychological history of the development of mankind* (E. L. Schaub, Trans.). London: George Allen & Unwin; New York: Macmillan.

Wundt, W. (1973). *The language of gestures* (A. Blumenthal, Trans.). The Hague: Mouton. (Original work published 1900)

Yardley, K. M. (1982). On engaging actors in as-if experiments. *Journal for the Theory of Social Behaviour, 12,* 291–304.

Young, K. (1925). Social psychology. In H. E. Barnes (Ed.), *The history and prospects of social sciences.* New York: Knopf.

Young, K. (1930). *Social psychology.* New York: Crofts.

Young, K. (Ed.). (1931). *Social attitudes.* New York: Henry Holt.

Zajonc, R. B. (1965). Social facilitation. *Science, 149,* 269–274.

Zajonc, R. B. (1966). *Social psychology: An experimental approach.* Belmont, CA: Brooks/Cole.

Zajonc, R. B., Heingartner, A., & Herman, E. M. (1969). Social enhancement and impairment of performance in the cockroach. *Journal of Personality and Social Psychology, 13,* 83–92.

Zavalloni, M. (1971). Cognitive processes and social identity through introspection. *European Journal of Social Psychology, 1,* 235–260.

Zeigarnik, B. (1927). Das Behalten erledigter und unerledigter Handlungen. *Psychologische Forschung, 9,* 1–85.

Zimbardo, P. G. (1999). Experimental social psychology: Behaviorism with minds and matters. In A. Rodrigues & R. V. Levine (Eds.), *Reflections on 100 years of experimental social psychology.* New York: Basic Books.

Znaniecki, F. (1925). *The laws of social psychology.* Chicago: University of Chicago Press.

Znaniecki, F. (1936). *Social actions.* New York: Farrer & Rinehardt.

# Index


Abelson, R. P., 244
Ach, N., 54
action research, 185
aggression: frustration-aggression theory of, 13–16; and leadership styles, 235; neglect of social dimensions of, 13–16; social learning theory of, 13–16, 215
Allport, F., 4, 7, 8, 13, 45, 61, 64, 99, 103, 104, 106, 107–108, 110–111, 112, 115, 116–124, 125, 126, 129, 131, 132–134, 136–137, 138, 142, 143, 145, 150, 151, 152, 155–156, 157, 164, 165, 167, 168–174, 175, 177, 178, 179, 182, 184, 187, 193, 194, 198, 199, 201, 202, 209, 210–211, 215, 221, 224, 228, 229, 239, 240, 246, 247, 249, 250, 251; account of scientific explanation, 118–120; analysis of crowd behavior, 168–171; commitment to autonomy individualism, 150; commitment to behaviorism, 132–134; critique of notion of "group mind," 116–124; definition of social psychology, 124; denial of socially engaged psychological states and behavior, 134, 168; denial that group (social) psychology distinct from individual psychology, 123, 177; experimental program of social psychology, 104–106, 122–123, 172–178; later endorsement of social causality, 125; recognition of socially engaged cognition, emotion and behavior, 155–156, 170–171; studies of social facilitation, 173–174, 175; treatment of experimental social groups as small crowds, 172–174
Allport, G., 7, 13–16, 106, 112, 115, 125–126, 131–132, 136, 137, 138, 142, 143, 145, 147, 150, 151, 153–155, 162, 168, 193, 194, 211, 218, 227, 231, 239, 244, 247, 248, 251, 252; commitment to autonomy individualism, 150–151; definition of social psychology, 7, 126; recognition of socially engaged cognition, emotion and behavior, 153–154
Altman, I., 208
*American Association for the Advancement of Atheism*, 176
*American Jewish Committee*, 187
*American Psychological Association*, 79, 104–106, 114, 252
*American Sociological Association, Section on Social Psychology*, 212, 244
*American Soldier, The*, 190–191
analysis of variance, 217
Angell, J. R., 3, 45, 92, 95, 104, 123, 133
anticipatory socialization, 219
apperception, 56
Argyris, C., 231
Aristotelean vs. Galilean modes of thought, 234, 235
*Army Information and Education Division*, 185
Aronson, E., 186, 203, 214, 229, 237, 244

Arrhenius, S. A., 141
Ash, S., 5, 28, 58, 68, 73, 77, 80, 85, 179–180, 181–182, 186, 188, 189, 192–194, 198, 199, 202, 207, 211, 212, 223, 232, 249, 250, 251; experimental studies of conformity, 179, 182; recognition of social dimensions of mind and behavior, 180, 181–183
*Asian Journal of Social Psychology*, 10
associationist psychology, 76, 77, 95
attribution theory, 216
*Authoritarian Personality, The*, 184, 244
autokinetic effect, 180
autonomy individualism, 154, 156, 159; early history in America, 149–150
Aviation Psychology Unit, 204

Back, K., 5, 186
Backman, C. W., 230
Bacon, F., 140
Bagehot, W., 96, 97, 98
Bain, A., 95
Baldwin, J. M., 57, 93, 99, 100–101, 102, 143, 152, 165, 168, 251, 254
Bales, R. F., 209
Ballachey, E. L., 159
Bandura, A., 16, 215
Banks, W. C., 237
Barber, B., 140
Barker, E., 147
Baron, R. A., 234
Bar-Tel, D., 261–263
Bartlett, F. K., 3
*Basic Studies in Social Psychology* (Proshansky & Seidenberg), 228
*Bau und Leben des Socialen Körpers* (Schäffle), 94
Bavales, A., 204
behaviorism, 104, 132–134; methodological, 122; radical, 141
*Beiträge zur Theorie der Sinneswahrnehmung* (Wundt), 48
Bellah, R. N., 148
Beloff, J., 232
Bem, D. J., 244
Benedict, R., 64
Bennington study, 190, 199, 227
Bennett, D. H., 238
Bentham, J., 95, 144, 147
Benussi, V., 202
Berkowitz, L., 16, 215
Bernard, C., 121
Bernard, L. L., 10, 18, 92, 101, 103, 104, 105–106, 111, 123, 160, 245
Blais, D. J., 224
Block, H., 5
Blondlot, R., 141
Blumenthal, A. L., 48
Blumer, H., 102, 251
Boas, F., 45, 63, 146
Bogardus, E. S., 10, 25, 39, 90, 92, 94, 101, 103, 104, 105, 123, 160, 245
Bond, M. H., 232
Boring, E. C., 220
Bosenquet, B., 109, 145
Brewer, M. B., 225, 237, 244
Bringmann, W., 44
Brock, A., 146
Brown, J. F., 12
Brücke, E., 121
Brunswik, E., 185
Brunswik, E. F., 185
Bühler, K., 54, 55
*Bureau of Applied Research* (Columbia University), 186
*Bureau of Program Surveys of the Department of Agriculture*, 185
Burrow, T., 37
bystander apathy, 210

Calhoun, J. C., 150
Campbell, A., 5, 16
Campbell, D. T., 218, 225
Cantril, H., 5
Carlsmith, J. M., 244
Carlson, H. B., 188
*Carnegie Foundation*, 14
Carr, H., 95, 104
Cartwright, D., 13–16, 185, 186, 187, 203, 205, 214
Chang, W. C., 78
Channing, W. H., 149
Chapanis, A., 236
Charters, W. W. Jr., 5, 192
Chesterton, T., 66
Chiang, L. W., 137
*Child Welfare Station* (University of Iowa), 204
Christie, R., 218, 219
Clark, K. B., 187
co-acting vs. face-to-face experimental groups, 173, 207, 209, 210
cognitive consistency theory, 216
cognitive dissonance, theory of, 206, 215

cognitive maps, 122
cognitive processes, mistaken equation of higher with social psychological, 56–57
cognitive revolution, 122, 239
Cole, M., 41, 253–258, 259
Collier, G., 94
*Commission for Community Relations*, 187, 205
*Communication and Attitude Change Program* (Yale University), 211
*Communications Research Center* (Yale University), 186
communitarian individualism, early history in America, 148–149
comparative psychology, 61
Comte, A., 68, 85–97
conceptual history, 13; and need for critical analysis of concept of social, 16–17
conformity, 159, 215, 216; cross-cultural studies of, 182; experimental studies of (Asch), 179, 181–183; interpersonal vs. social determinants of, 182; operational definition in terms of interpersonal determinants, 182
consciousness of kind, 26, 166
contexualist history, 14; vs. Whig history, 14
Converse, P., 5, 230
Cook, S., 204
Cook, T. D., 225
Cooley, C. H., 57, 93, 99, 100, 101, 102, 152, 162, 165, 168, 191, 198, 251, 254
Copernicus, N., 53
Crano, W. D., 225
crowds, 110, 116; as irrational/emotional, 156, 161, 166–167; distinguished from publics, 161, 165–166, 167
*Crowd, The* (Le Bon), 175
Crutchfield, R. S., 159, 187, 188–189, 191, 200–201, 202, 207, 209, 228, 240
cultural anthropology, 45
cultural determinism, 62
cultural learning (imitative, instructive, collaborative), 38–39
cultural psychology, 41, 67, 246, 253–259; distinguished from cross-cultural psychology, 254; distinguished from social psychology, 256–257
*Cultural Psychology: The Once and Future Discipline* (Cole), 253
*Current Studies in Social Psychology* (Steiner & Fishbein), 228
D'Andrade, R. G., 258
Danziger, K., 46, 56, 60, 202, 217, 220, 221, 222–223, 224, 226
Darley, J., 186, 203, 206, 210
Darroch, R. K., 238
Darwin, C., 95–96
Dashiell, J. F., 8, 122, 174, 177, 199, 209, 214, 215, 240, 247
Davis, K. E., 244
definition of the situation, 30
Department of Defense, 195
Department of Social Relations (Harvard University), 186
derivatively social phenomena, 88–89
Descartes, R., 135, 144
Deutsch, M., 186, 187, 203
Devine, P. G., 6
Dewey, J., 29, 93, 94, 99, 100, 102, 104, 108, 191, 254
Doctoral Program in Social Psychology (University of Michigan), 186, 187
Dollard, J., 16, 177
Doob, L. W., 16, 63
Du Bois-Reymond, E., 121
Dunlap, K., 26, 37, 39, 49, 85, 92, 99, 101, 103, 104–106, 114, 115, 117, 123, 126, 130, 160, 197, 245
Durkheim, É., 23, 41, 42, 49, 68–70, 83, 86, 87, 88, 89–90, 94, 97, 109, 110, 113, 117, 121–125, 129, 160, 162, 164, 165–166, 193, 199, 231, 245, 246, 253, 254, 259, 260–261; conception of relation between social and psychological, 74–77; concern to establish sociology as scientific discipline, 69; holistic account of social, 68–69, 70–72
*Dynamic Sociology* (Ward), 97
*Dynamic Theory of Personality, A* (Lewin), 207

Edwards, A. L., 25, 219, 220, 221
Eggleston, R. J., 234
*Elements of Folk Psychology: Outlines of a Psychological History of the Development of Mankind* (Wundt), 48, 98
*Elements of Psychology* (Wundt), 48
*Elements of Social Psychology* (Gurnee), 174
Ellison, C. E., 188

Ellwood, C. A., 10, 25, 40, 92, 97, 106–107, 132, 135, 197, 245
Elm Hollow study (Schanck), 131
Elms, A. C., 238
Emerson, W., 149–150
Espinas, A., 109
ethnic psychology, 47, 63
eugenics, 146
evolution, theory of, 95–96; contrast between accounts of Darwin and Spencer, 96
exchange theory, 216
experimental pedagogy, 104
experimentation in social psychology: atomistic assumptions of, 222–223; complaint about the artificiality of, 232; contrasted with correlational studies, 183, 220, 227; deception, 232, 233, 238, 243; demand characteristics, 232, 243; evaluation apprehension, 232; experimenter bias, 232; and external validity, 235; field experiments, 233; identity of phenomena created in, 234–237; impoverished nature of experimental social groups, 210–211; interaction effects, 232, 233, 243–244; mundane vs. experimental realism, 237; post-war commitment to not cause of neglect of social, 221–222; psychological realism, 237; randomization, 224–226; role-playing, 233, 238, 243; social dimension as source of confounding, 224; social dimensions no impediment to, 58–59; true experiments, 225; use of college students in, 218, 232
*Experimental Psychology* (Woodworth), 220, 225
*Experimental Social Psychology* (Murphy & Murphy), 63, 106, 174
*experimentelle Massenpsychologie* (Moede), 175
explanation, aggregative vs. structural, 120
explanatory dispute, between holists and individualists, 90

false consensus, 174
Faris, E., 24, 28, 40, 92, 143, 154, 191, 245, 253, 263
Farr, R., 46, 61, 143, 144, 214
Fechner, G., 214
Ferracuti, F., 5
Festinger, L., 5, 20, 186, 187, 203, 204, 205–206, 211, 212, 221, 235–236
field theory, 203, 205
Finney, R. L., 37
Fishbein, M., 228
Fiske, S. T., 214, 239, 244, 246–247, 254
Fitzhugh, G., 150
Folkman, J. R., 218
folk psychology, 48
*Food Habits Committee of the National Research Council*, 185
*Ford Foundation*, 14
Forgas, J., 46
Fouillée, A., 109
frames of reference, socially vs. individually engaged, 180
Freedman, J. L., 238
Freidrich, M., 56
French, J., 5, 186
Freud, S., 37
Functionalism, 95, 104

Gault, R. H., 112
genetic fallacy, 173
Gergen, K., 232, 238, 241–242
Gestalt psychology, 181, 200, 203, 206; Frankfurt-Berlin school contrasted with Graz school, 202
Giddings, F., 10, 25–26, 97, 165, 166, 168, 253
Gilbert, M., 82, 83
Gilliand, A. R., 188
Goffman, E., 152, 251
Goodwyn, S. A., 214, 246–247
grant-funding agencies, role of in development of American social psychology, 14
Graumann, C. F., 11, 46
Green, M., 176
Green, T. H., 109, 145, 147
Greenberg, M. S., 238
group beliefs, 261–263; definition of, 261; distinguished from personal and common beliefs, 262; relation to social and individual beliefs, 262–263
*Group Beliefs* (Bar-Tel), 245, 261
group-belongingness, 221
group cognition, 253
group-conflict, theory of, 94
group-dynamics, 185, 195, 203, 204, 215

group (social) mind, 44, 77, 91–92, 93, 98, 109–110, 132, 136, 138, 184, 200, 252; and common forms of cognition, emotion and behavior, 117, 124; and "group fallacy," 110–111, 116, 118; and group mentality, 114–115; as perceived threat to moral individuality, 145–147; association with totalitarianism, 145–146; commitment to in early American social psychology, 110; reasons for postulation of, 113–115; rejection of notion of, 112–113
*Group Mind, The* (McDougall), 98, 110, 131, 145, 158, 252
group processes, 203; decline of study in 1960s, 203
group psychology, 203; first Ph.D program in, 196
groups, experimental studies of, 195–196
group-think, 159, 212
Gumplowitz, L., 94
Gurnee, H., 122, 174

Hall, G. S., 45, 104, 114
Hamilton, D. L., 6
*Handbook of Social Psychology*, 62, 122, 174, 177, 182, 213, 214, 225, 231, 240, 244, 247, 255
Haney, C., 237
Hare, A. P., 212
Harré, R., 16
Harris, A. J., 188
Hartley, D., 95
Hartley, E. L., 187, 188, 189
Hartley, R., 219
Hartmann, G. W., 25
Hartup, W. W., 6
Haslam, A. S., 248–251
Hayes, J. R., 93
hedonistic psychology, 95
Hegel, G. W. F., 68, 77, 109, 145
Heider, F., 185, 244
Heingartner, A., 134, 228
Helmreich, R., 218
Herbart, J. F., 43, 62
Herman, E. M., 134, 228
Heelas, P., 62
Herder, J. G., 44, 62, 64
Hertzman, M., 5
Higbee, K. L., 218
Higgins, E. T., 6, 248
Hirschberg, G., 206
*History of Psychology in Autobiography*, 150
Hitler, A., 167, 185
Hobbes, T., 95, 144
Hobhouse, L. T., 54, 95–96, 98, 109
Holmes, D. S., 238
Horowitz, E. L., 28
horizontal vs vertical versions of ontological dispute between holists and individualists, 79, 81
Horowitz, I. A., 238
Horwitz, M., 186
Hovland, C. I., 186, 221, 222, 244
Howard, J. A., 244
Hull, C. L., 64, 65, 95, 177, 204, 218
*Human Inference: Strategies and Shortcomings of Social Judgement* (Nisbett & Ross), 240
*Human Relations, Institute of* (Yale), 177
Humphrey, G., 55
Hume, D., 65, 94, 95
hypothetical constructs, 121–122
Hyman, H., 78, 189, 212–213, 244

identification, interpersonal, 250
identity projects, 73
identity: self-labeling theories of, 250; social conception of, 152–153
idols of the theater, 140
imitation, 37–38, 97, 101, 104, 157, 162
impression of universality, 169–170, 174
independent variables, definition of experiments in terms of manipulation of, 217, 219–221
*Indigenous Psychologies* (Heelas & Lock), 62
indigenous psychology, 62
individual and social psychology, early development in America, 92
individual psychology, early development in America, 95
individual vs. private beliefs and attitudes, 22
individualism, 12–13, 143–144; modern concept of, 144–145; original concept of, 144, 145; social adoption of forms of, 153

individuality: communitarian conception of, 147, 148; social as developmental medium for, 251
Insko, C., 230
instinctual explanations, early American challenges to, 99
*Institute for Social Research* (University of Michigan), 186, 203
*Institute of Human Relations* (Yale University), 186, 204, 222
*Institute of Physiology* (University of Heidelberg), 48
*Institutional Psychology* (Allport, F.), 156, 171
interconditioning, 100
internal (intrinsic) vs. external (extrinsic) conceptions of the relation between the social and the psychological, 91
internal vs. external history, 13–16
interpersonal attraction, 216
*Interpersonal Relations* (Kelly & Thibaut), 208
interstimulation, 100
intrinsically social attitudes, 88, 192–193
intrinsically social phenomena, 88–89
*Introduction to Social Psychology* (Ellwood), 106, 107
*Introduction to Social Psychology* (McDougall), 97, 98, 99, 104
*Introduction to the Study of Society* (Small and Vincent), 97
*Introductory Lectures on Pedagogy and Its Psychological Basis* (Meumann), 175
introspective analysis, 57–58, 59
invariance, principle of, 63–65
*Is America Safe for Democracy?* (McDougall), 146
Israel, J., 232

Jackson, F., 141, 157
Jackson, J. M., 94
Jahoda, M., 187
James, W., 21, 29, 31, 45, 97, 99, 100, 191, 245
Janis, I., 222
Jarvie, I., 111
J-curve, 136–137, 138, 152, 157, 172, 184
Jefferson, T., 148
Jones, E. E., 13–16, 186, 203, 214, 239, 244
Joseph, J. P., 234
*Journal of Abnormal and Social Psychology*, 157, 216, 218, 255
*Journal of Experimental Social Psychology*, 215
*Journal of Personality and Social Psychology*, 216, 218, 239
*Journal of Social Psychology*, 62, 255
Judd, C. H., 40, 45, 48, 94, 245

Kane, T. R, 234
Kant, I., 44, 45
Kantor, J. R., 30, 40, 92, 130, 245
Karpf, F., 60, 95, 175
Katz, D., 12, 19, 35, 40, 92, 123, 130, 151, 152, 153, 160, 168, 174, 231
Keeler, C., 246
Kelley, H., 5, 186, 189–190, 203, 204, 205, 208, 209, 222, 244
Kendler, H., 122
Kimble, G., 65
Kitayama, S., 254
Kitt, A., 102, 192
Klotz, I. M., 141
Lee, L., 78
Koffka, K., 202
Köhler, W., 185, 202, 203, 207
Kolstad, A., 176
Kranz, D. L., 141
Krech, D., 159, 186, 187, 188–189, 191, 200–201, 202, 207, 209, 228, 240
Kruger, A. C., 18, 37, 38–39
Kruglanski, A. W., 231, 243
Kuhn, T., 140, 175
Kulp, D. H. II., 177
Külpe, O., 3, 54, 123
Kuo, Z. Y., 99
Kusch, M., 60

*Laboratory for Social Relations*, 186
laizze-faire, 95, 96, 145, 147, 151
Lamarck, J.-B. de, 96
La Piere, R. T., 22, 23, 131, 191
*La Psychoanalyse: Son Image et Son Public* (Moscovici), 260
Larson, K. S., 232
Latané, B., 210
law of effect, 95
Lazarus, M., 44

Lazerfield, P., 185
Leary, D., 46
Le Bon, G., 8, 54, 87, 94, 109, 113, 114, 115, 117, 156, 157, 160, 161, 162, 164, 165, 166, 175, 254
Lee, L., 78
Leibniz, G., 45
Leik, R. K., 244
*Les Lois de L'imitation* (Tarde), 165
*Les Lois Sociales* (Tarde), 165
level of aspiration, 204
Lévi-Bruhl, L., 109
Levine, J. M., 122, 253, 263
Lewin, K., 5, 151, 176, 181, 185, 186, 195–196, 203–205, 206–207, 210, 211, 222, 233, 234, 235, 236, 255; early cognitive and motivational studies, 204; program in social psychology, 203–210
life-space, 206–207, 209
Lindzey, G., 214
Lippitt, R., 5, 186, 187, 203, 204, 205, 207, 233, 235
Lister, J., 141
Lock, A., 62
Locke, J., 95, 135, 144
Luria, A., 44, 257, 258

Maciver, R. M., 127
MacMartin, C., 217
Marbe, K., 54
Markus, H., 239, 244, 254
Marsh, P., 16
Martin, E. D., 109
Marx, K., 68
May, M., 177
Mayer, Albert, 175
Mayer, August, 54
McDougall, W., 4, 6, 13, 25, 40, 49, 92, 97–98, 99, 101, 103, 104, 105, 106, 109, 110, 114, 115, 117, 123, 126, 127, 128, 129, 131, 145, 146–147, 150, 151, 153, 157–158, 160, 161, 193, 197, 245, 246, 251, 252, 254; communitarian conception of individuality, 147; distinction between crowd and social behavior, 157–158; eugenicist views, 146; rejection of supra-individuality and totalitarianism, 147
McGarty, C., 248–251
McGrath, J., 195, 208, 210
McGuire, W. J., 16, 239, 244, 263–264
Mead, G. H., 42, 92–94, 99, 102–103, 104, 108, 152, 229, 233, 251, 254
Mead, M., 64
Meeker, B. F., 244
Mendel, G., 141
Merton, R. K., 102, 192
Messer, A., 54
Meuller, C. G., 46
Meumann, E., 104, 175
*Middletown* (Lynd & Lynd), 149
*Middletown in Transition* (Lynd & Lynd), 149
Milgram, S., 223, 237
Mill, James, 95
Mill, J. S., 44, 62, 95, 144, 147, 149, 150
Millard, R. J., 218
Miller, A. J., 238
Miller, G., 92
Miller, N. E., 16, 177
Minton, H. L., 94
Moede, W., 175
morale, war-time studies of, 185
moral science, 97
Moreland, R. L., 122
Moscovici, S., 42, 245, 259–261
Mowrer, O. H., 16
Müller, J., 121
Münsterberg, H., 104, 175
Murchison, C., 62, 255
Murphy, G., 8, 63, 104–106, 122, 174–175, 176, 177–178, 182, 209, 215, 217, 218, 227, 234
Murphy, L. B., 8, 63, 104–106, 122, 174–175, 176, 177–178, 182, 209, 215, 217, 227, 234
Mussolini, B., 167
My Lai massacre, 237

*National Training Laboratories for Group Development*, 186
*Nature*, 141
Naturwissenschaften, causal-historical methods of contrasted with historical-cultural methods of Geistewissenschaften, 254
Neisser, U., 92–94
Newcomb, T. M., 5, 63, 68, 175, 176, 186, 187, 188, 189, 190, 192, 199–200, 217, 227, 228, 230, 249
*New School for Social Research*, 202
Newton, I., 65

Newtonian science, 65
Nisbett, R. E., 6, 240, 244, 254, 255
N-rays, 141

obedience, 159, 223
Oedipal complexes, 122
*Office of Naval Research*, 15
*Open Society and Its Enemies, The* (Popper), 146
Orth, J., 54
OSS Assessment Staff, 185
Ostrom, T. M., 6, 239, 240, 241
other-directed behavior, 159, 211
*Outline of Social Psychology, An* (Sherif), 188, 189

Pandora, K., 104–106
Park, R., 10, 97, 165, 168
Parkovnick, S., 218
Parsons, T., 186
Pasteur, L., 121
Pepitone, A., 16, 153, 186, 203, 212, 217, 227
Perrin, S., 182, 240
*Personality and Social Psychology Bulletin*, 218
*Perspectives on Socially Shared Cognition* (Resnick, Levine & Teasley), 245, 252
person perception, 216
persuasion and communication, research on, 167, 216, 221
Petras, J. R., 93
Pettit, P., 141, 157
*Philosophische Studien*, 43, 56
*Physics and Politics* (Bagehot), 98
*physiologische psychologie* (physiological psychology), 43
Piaget, J., 3, 176
Pilliavin, J. A., 238
Planck, M., 140
Plato, 68, 77, 109
pleasure principle, 95
*Polish Peasant, The* (Thomas and Znaniecki), 121–125
Popper, K. R., 146
positivism, 97
pragmatism, 12–13
prepotent responses, 168–169
Prince, M., 157
*Principia* (Newton), 65
*Principles of Behavior* (Hull), 64
*Principles of Physiological Psychology* (Wundt), 47
*Principles of Research in Social Psychology* (Crano & Brewer), 225
*Principles of Topological Psychology*, 206, 207
Profumo affair, 246
Proshansky, H., 182, 228
*Principles of Psychology* (James), 97, 100
psychological capacities, cultural constitution of, 257–258
*Psychological Info*, 213
*Psychology* (Woodworth), 220
*Psychology of Human Society* (Ellwood), 107
*Psychology: The Science of Behavior* (McDougall), 6
publish or perish, pressures to, 227

Qasselstrippe meetings, 205
quasi-experimental studies, 225

Radina, S. L., 176
Radke, M., 186, 205
Ratner, H. H., 18, 37, 38–39
Ratzenhofer, G., 94, 97
*Readings in Social Psychology* (Newcomb & Hartley), 187, 188, 189, 190, 228
reference group, 37, 78, 137, 174, 189–192, 212–213, 244
reflex arc, 104, 118
Reformation, 143, 144
Reisman, D., 159
Remmers, H. H., 188
Renaissance, 143, 144
Representations, social vs individual, 74, 76–77, 165
*Research Center for Group Dynamics*, 186, 196, 203, 204
*Research Methods in Social Relations*, 225
Resnick, L. B., 253
Reynolds, G., 94
Riecken, H. W., 5, 20, 206
Ring, K., 231
risky-shift behavior, 159
Rockefeller Foundation, 14, 177
Ross, E., 4, 10, 24, 40, 97, 98, 123, 130, 141–142, 143, 144, 145, 160, 165, 168, 245
Ross, L., 6, 186, 203, 240, 244, 255
Rosser, E., 16
Rothschild, B. H., 238

Rousseau, J.-J., 145
Ruble, D. N., 6
*Russell Sage Foundation*, 14

Samelson, F., 9, 97, 125, 210, 227
Sampson, E. E., 153
Sargent, S., 218
Schachter, S., 5, 20, 186, 203, 206, 213
Schachter-Singer experiment, 163–164, 206
Schäffle, A., 94, 97, 109
Schanck, R. L., 12, 19, 22–23, 35, 40, 92, 123, 130, 131, 137, 151, 152, 153, 160, 245
Schaub, E. L., 48
Schopler, J., 230
Schuman, H., 244
scientific revolution, 140
Scripture, E., 56
Sears, R. R., 16
Second World War, as catalyst for intellectual cooperation among social scientists, 185
Secord, P. F., 230
Seidenberg, B., 182, 228
self-observation, 55
Shain, B. A., 148
Shepard, R. N., 65
Sherif, M., 5, 12, 57, 58, 66, 85, 151, 153, 176, 179–181, 188, 189, 191, 207, 208, 212, 231, 241, 249, 251; experimental study of group norms, 179–181; recognition of social dimensions of human psychology and behavior, 180
Shibutani, T., 213
Shook, J. R., 46
Shweder, R. A., 255, 258
Siann, G., 16
Siegel, A. E., 5
Siegel, S., 5
Sighele, S., 109, 157, 160, 166
Simmel, G., 19, 41, 62, 84, 85, 87, 94, 97, 123, 155, 156, 160, 194, 233, 245, 256
Simons, C. W., 238
Singer, J, 203, 213
Singer, E., 212–213
Skinner, B. F., 95, 141
Slavitt, P. R., 230
Small, A., 94, 95, 97
small group research, 208, 209, 210, 212, 215
Smith, M., 209
Smith, P. B., 232
social: as aspect of the biological, 178; holist vs individualist accounts of the, 68–69, 87; relation to cultural, 255–256
social act; and circular social behavior, 103; Mead's definition of, 94, 102–104
social action, 81–83; and interpersonal behavior, 81–82, 83–84; and imitation, 82–83; and generality, 83
*Social Animal, The* (Aronson), 229
social atmospheres, 207, 235
social behavior: altruistic and selfish, 33–34; and collective goals, 32–33; as analogous to crowd behavior, 156–157, 160–161; changing conceptions of, 7; conflict and competitive, 34; contrasted with individual behavior (as socially vs individually engaged), 18, 30–31; cooperative, 18, 34; definition of, 20; distinguished from crowd behavior, 161–164; interpersonal conception of, 7, 124; not equivalent to common/plural behavior, 31–32; not equivalent to interpersonal behavior, 31–32, 103; rational choice theories of, 33, 89; sociobiological theories of, 33, 89; traditional vs. rational, 166
social beliefs/attitudes, 137; as common beliefs/attitudes, 125–126, 138, 153–155; conditionality of, 24–25, 193; conservative nature of, 21; contrasted with individual beliefs/attitudes (as socially vs. individually engaged), 20–23; definition of, 20; development of instruments for measuring, 184; as directed to both social and non-social objects, 27–28; early American studies of, 5; not equivalent to common/general beliefs/attitudes, 23–24, 193; of Methodists and Baptists towards forms of baptism, 22–23; occupational, 25; orientation to misrepresentation of beliefs/attitudes of members of social group, 36; political, 25; and public opinion, 167, 216; religious, 25; and scientific theories, 138–141; and small group research, 246–248; social anchoring of, 211; as social dimensions of personality, 28; and social prejudice/stereotyping, 28, 72–74, 126; types of group to which oriented, 25

social cognition, 215, 244; asocial conception of, 239–241; contemporary definition of, 5, 6; explained by principles of individual cognitive psychology, 6, 246; paradigm seen as resolution to 1970s crisis, 239; popularity of paradigm in 1980s, 239; and small group research, 246–248; sovereignty of, 239
*Social Cognition* (Fiske & Taylor), 239
*Social Cognition*, 239, 240
social ("collective" or "group") cognition, emotion and behavior: abandonment of empirical/experimental study in post-war period, 221; and animal psychology, 18, 135; association with notion of supraindividual group/social mind, 128–130; as casualty of American individualism, 265; claim that cannot be investigated experimentally, 46–47, 222, 227; and conformity, 151; and cultural learning, 38–39; definition of, 18–19; examples of illustrative only, 35–36; experimental analysis of, 179, 184, 192, 218–219, 224; and experimental methods, 123; and fashion, 20; implications for scientific and moral judgements, 138–141; incapable of being explained/predicted via principles of individual psychology, 127–128; intrinsic relation to social groups, 198, 202; as involuntary/regimented, 151–152; as justification of distinctive social/group psychology, 127–128; link to sense of identity, 189; no essential connection with notion of supraindividual group/social mind, 130–131; no in principle impediments to the objective and experimental study of, 264; no intrinsic connection to socialism, 151; not necessarily conscious/reflective, 37; as objects of explanation in social psychology, 40; oriented to cognition, emotion and behavior of non-membership groups, 36–37; oriented to variety of different social groups, 191–192; as pathological, 152, 171, 172, 211; and social comparison, 211; and social learning, 37–38; socially shared, 19; as source of individuality, 145; theoretical accounts in post-war period, 187–196; as threat to moral individualism/autonomy, 138–143, 157, 159; as threat to rationality, 140–142, 171; treated as independent variables (subject variables) in experiments in, 220–221, 225
social consciousness, 26, 85, 132, 239; as rationale for distinctly social psychology, 26–27; as interpersonal consciousness, 124
social control, 97
*Social Control* (Ross), 142
social Darwinism, 96, 151
social dimensions of science, 139
social emotion: contrasted with individual emotion (as socially vs. individually engaged), 34–35; definition of, 34; not equivalent to common/plural emotions, 34; social and non-social objects of, 34–35; and universality, 35
social entities, as analogous to biological entities, 113, 117, 118
social evolution, 96–97
social facilitation, 92, 215
social facts, 72–74; demarcation of, 69–70; and generality, 69, 73; and imitation, 72–73; as statistical facts about social groups, 70
social groups, 19–20; assimilated with crowds/mobs, 164, 165; contrasted with aggregate groups, 78–80, 199–200, 249, 250; distinguished from crowds/mobs, 162–163; identity over time, 113–114; in-groups, 130; interactive definition of, 201, 209; primary and secondary, 101; psychological nature of, 84–85; relation to social forms of cognition, emotion and behavior, 194–195; as supraindividuals, 109–110, 117
social identity, theory of, 246, 248
social individuals, 194, 202
social influence, contrasted with interpersonal (crowd) influence, 115
social interaction, 230
social interactionism, 37, 99–104, 152, 244; analogous to Allport's individualistic form of social psychology, 102–104; distinguished from distinctively social form of American social psychology, 101–108

social interests, theory of, 94
social learning: not equivalent to cultural learning, 38–39; not equivalent to interpersonal learning, 37–38, 100, 102; treated as interpersonal learning, 177–178, 215
social norms: experimental studies of (Sherif), 179–181; and the social mind, 180–181
social projection, 170
social psychological: appropriateness of experimental investigation of the, 46–47; distinctive early American conception of the, 39–40; early conception of the, 2–4; early European conception as equivalent to early American conception of the, 3; "individualization" of the, 10–11, 144; joint constitution (singularity) of the, 198; neglect of the early conception of the, by American social psychologists, 5–7, 40, 245, 264; personification of the, 109, 121, 138, 145; as proper intersection of the social and the psychological, 90; quantitative vs qualitative dimensions of the, 222–223; reasons for neglect of the early conception of the, 7–8; reification of the, 109, 112, 121–122, 138, 146; social vs. asocial/individualistic conceptions of the, 8–9; supraindividual conception of the, 11–12, 40, 46, 47, 49–53, 113
social psychological explanation: cultural/historical variance of, 65–67, 233; as form of psychological explanation, 76–77
social psychological states and behavior: denial that distinct from individual psychological states and behavior, 178–179; as form and object of explanation in social psychology, 40; oriented to represented cognition, emotion and behavior of non-membership groups, 36–37; oriented to a variety of social groups, 28–29, 195–196; pre- and post-war recognition of, 183; as psychological properties of individual persons, 36; as socially vs. individually engaged, 1, 20, 193–194, 195; and social mind, 193
social psychology: American dominance of post-war, 186; artificiality of experiments in, 104–106; asocial theoretical and experimental paradigm of, 179, 183, 184, 214–215, 221; as behavioral science, 158, 187; call for more integrated/interdisciplinary, 187; of the cockroach, 228; cognition of contrasted with social psychology of cognition, 248; complaints about asocial nature of, 231; complaints about fragmentation in, 231; contrasted with individual psychology, 2, 3, 75–76, 101; crisis in, 94, 231–233, 239, 241; denials of crisis in, 238; early American textbooks in, 97–98; early development in America, 94–108; experimentation in, 12, 46–47, 167, 176, 214–215, 216; experimentation in sociological, 215; Gestalt-psychological approach to, 201–202; and indigenous psychology, 62–65; influence of emigration of European academic refugees on development of American, 185–186; Lewinian program in, 203–210; master-problem of, 40, 198; neglect of the social in American, 10–13; post-war commitment to experimentation in American, 216–222, 227; post-war development of American, 186–187, 196, 214–216; primary subject matter of, 90; psychological vs. sociological forms of, 10, 40, 90–92, 187, 197, 212, 244; Research Advisory Office in, 218; revival of interest in social in, 245; social constructionist movement in, 239, 241–243; subject-matter of, 197, 200; theoretical and empirical achievements of twentieth century American, 9; treated as form of individual psychology, 4, 108, 177–179, 199, 200, 228–230, 239
*Social Psychology* (Allport, F.), 45, 111, 132–134, 168, 171, 174
*Social Psychology* (Asch), 188, 192, 198
*Social Psychology* (Dunlap), 105
*Social Psychology* (Ross), 4, 97
*Social Psychology at the Crossroads* (Rohrer and M. Sherif), 199
*Social Psychology of Groups, The* (Thibaut & Kelley), 208
*Social Psychology, Sociological Perspectives*, 212, 244

social representations, 24, 42, 245, 259–261; and cultural psychology, 261; holistic rhetoric of theory of, 260; six senses of "social" in, 263–264; as socially oriented/grounded, 259–260, 261
*Social Science Research Council*, 14
social self, 29
sociality and individuality, relation between, 251
socially shared cognition, 252–253
society: definition of, 20; as social group, 78
*Society for the Study of Social Issues*, 188
*Sociological Perspectives in Social Psychology*, 244
sociology: early development in America, 91–92, 93; subject-matter of, 89–90
Spencer, C., 182, 232
Spencer, H., 68, 95, 96, 97, 147, 150, 151
Springfield College, 204
Stanford-Binet intelligence test, 258
Stanley, J. C., 225
Steiner, I. D., 228, 238
Steinthal, H., 44, 45
stereotypes, 216
Stogdill, R. M., 209
Stouffer, S. A., 5, 180, 190–191
Stroebe, W., 243
structural dynamics, 137
structural explanation, 89–90
structuralism, 95
*Suicide* (Durkheim), 70
Summer, W. G., 96
*Survey Research Center*, 186, 203
Svehla, G., 188
symbolic interactionism, 93–94, 122

Tajfel, H., 231, 232, 245, 248, 250
Tarde, G., 1, 8, 54, 82, 83, 87, 94, 97, 100, 109, 156, 157, 160, 161, 162, 164, 165, 166, 175, 177, 254
Taylor, S. E., 239, 244, 246
Teasley, S. D., 253
Tedeschi, J. T., 234
*Theory and Problems of Social Psychology* (Krech & Crutchfield), 187, 188–189
theory of mind, 38
Thibaut, J., 186, 203, 208, 209
Thomas, D. S., 196
Thomas, W. I., 10, 12, 26–27, 30, 40, 42, 92, 97, 101, 103, 104, 123, 125, 160, 196, 241, 245, 259
Thorndike, E. L., 99
Titchener, E. B., 3, 45, 63, 64, 92, 95, 110, 123, 218, 254; views on *Völkerpsychologie*, 4
Tocqueville, A. H. de, 148, 149
Tolman, E. C., 95, 99
Tomasello, M., 18, 37, 38–39
Torres-Straits Expedition, 4
Triandis, H. C., 232
Triplett, N., 174
Trotter, W., 96
Turner, J. C., 246, 248, 250, 251
Turner, R. H., 230
Tweney, R., 44

Underwood, B., 225
universality, principle of, 63–65
utilitarianism, 95, 147

Vallacher, R. R, 6
verbal reports (of beliefs and attitudes), 22–23
Vetter, G. B., 176
Vico, G., 44, 62, 64
Vincent, G. E., 94, 95, 97
vitalism, 121
Volkart, E. H., 5, 231
*Völkerpsychologie*, 2, 3, 4, 41, 42, 43–45, 47–53, 55, 56, 59–61, 62, 63, 64, 94, 104, 109, 146, 254, 256, 257; best translation of, 43, 48–49, 94; and causal-explanatory inference, 50–53; comparative-historical methods of, 49–53; and indigenous psychology, 62–65; individualistic assumption of Wundt's, 60–61; laws of development in Wundt's, 60; as supplement to experimental psychology, 43, 50–53; synchronic dynamical vs. diachronic historical analysis in Wundt's, 59–60; Wundt's early interest in, 47–48
*Völksgeist*, 44
von Eickstedt, E., 146
von Helmholtz, H., 47
von Humboldt, W., 43–44
*Vorlesungen über die Menschen-und Thierseele* (Wundt), 48
Vygotsky, L., 44, 254, 257

Wallas, G., 96
Wallis, W. D., 40, 92, 101, 103, 104, 109, 113, 119, 123, 126, 128, 129, 160, 193, 245
Ward, L., 94, 97
*War Department, Information and Education Section*, 221
Warren, J., 149
Watson, G., 187
Watson, J. B., 95, 99, 104–106, 132, 133, 218
Watson, W. S., 25
Watt, H. J., 54
Weber, M., 32, 33, 41, 68, 69, 80–84, 85, 87, 94, 111, 112, 117, 121, 123, 124, 160, 162, 175, 193, 199, 201, 245, 248; essential agreement with Durkheim on nature of social, 81–84; individualistic account of social contrasted with Durkheim's holistic account, 80–81
Wee, C., 144
Wegner, D. M., 6
Wells, M. G., 218
Wertheimer, M., 185, 202, 203, 207
*What's Social About Social Cognition?* (Nye & Brower), 245, 246, 248
White, R. K., 5, 207, 233, 235
Williams, R., 145
Willis, N. H., 238
Willis, Y. A., 238
Wilson, T. D., 237, 244
Winslow, C. N., 188
Winston, A. S., 217, 220, 224
Winthrop, J., 148
Witasek, S., 202
Wolfe, H., 56
Wolfgang, M. E., 16
Woodruff, C. L., 5
Woodworth, R. S., 55, 99, 220, 225, 227
Wundt, W., 2–3, 4, 41, 42, 43–53, 54, 55, 57, 58, 62, 63, 64, 70, 74, 94, 95, 98, 104, 109, 110, 123, 128, 146, 218, 222, 227, 245, 246, 254, 255, 256, 257; as founder of scientific psychology, 43; objections to Würzburg experiments on higher cognitive processes, 54–59; second psychology contrasted with first psychology, 255; social vs. individual psychology, 45
Würzburg School, 54–58

Young, K., 10, 40, 92, 245

Zajonc, R. B., 16, 134, 175, 215, 228, 229, 239, 244
Zander, A., 186
*Zeitschrift für Völkerpsychologie und Sprachwissenschaft*, 44
Zimbardo, P., 186, 203, 237
Znaniecki, F., 26–27, 125, 241, 259; definition of social psychology, 125